Springer Series in Optical Sciences

Volume 195

Springer Series in Optical Sciences

The Springer Series in Optical Sciences, under the leadership of Editor-in-Chief William T. Rhodes, Georgia Institute of Technology, USA, provides an expanding selection of research monographs in all major areas of optics: lasers and quantum optics, ultrafast phenomena, optical spectroscopy techniques, optoelectronics, quantum information, information optics, applied laser technology, industrial applications, and other topics of contemporary interest.
With this broad coverage of topics, the series is of use to all research scientists and engineers who need up-to-date reference books.
The editors encourage prospective authors to correspond with them in advance of submitting a manuscript. Submission of manuscripts should be made to the Editor-in-Chief or one of the Editors. See also www.springer.com/series/624

More information about this series at http://www.springer.com/series/624

Stefan Nolte · Frank Schrempel
Friedrich Dausinger
Editors

Ultrashort Pulse Laser Technology

Laser Sources and Applications

Springer

Editors
Stefan Nolte
Institute of Applied Physics
Friedrich-Schiller-University Jena
Jena
Germany

Friedrich Dausinger
Dausinger + Giesen GmbH
Stuttgart
Germany

Frank Schrempel
Institute of Applied Physics
Friedrich-Schiller-University Jena
Jena
Germany

ISSN 0342-4111 ISSN 1556-1534 (electronic)
Springer Series in Optical Sciences
ISBN 978-3-319-17658-1 ISBN 978-3-319-17659-8 (eBook)
DOI 10.1007/978-3-319-17659-8

Library of Congress Control Number: 2015940734

Springer Cham Heidelberg New York Dordrecht London

Printed on acid-free paper

Springer International Publishing AG Switzerland is part of Springer Science+Business Media (www.springer.com)

Preface

Today, ultrafast lasers providing pulses with a duration of a several femtoseconds (1 fs = 10^{-15} s) up to a few picoseconds (1 ps = 10^{-12} s) are commercially available and are gaining more and more attention in the area of precise microstructuring. This is due to the fact that they enable completely new machining approaches impossible by applying conventional tools. The short interaction times result in strong non-equilibrium processes. Consequently, thermal and mechanical damage can be minimized and processing with practically no melt even in metal processing becomes possible resulting in sub-μm precision. The high intensities achievable when focusing ultrashort laser pulses even of moderate pulse energies allow for nonlinear interaction processes, which are the basis for processing transparent materials. Application examples are, e.g. the cutting of fragile glass structures for displays, the structuring of thin-films in photovoltaics to increase the solar cell efficiency, and the fabrication of LEDs or computer chips. Fantastic opportunities open up in the field of medicine. Here, for example highly precise cuts with minimal damage inside the human eye enable absolutely new therapy opportunities for cataract or presbyopia treatment. Thus, ultrashort laser pulse systems are universal tools with a plethora of applications in precise micro-manufacturing and medicine.

However, all these opportunities come along with a multitude of various challenges. On the one hand the system technology is rather complex and on the other hand the nonlinear effects during beam propagation and during the interaction process require fundamental understanding and precise control. Thus, although this technology is promising significant advantages, it is often termed not robust, not reliable, expensive, slow and inefficient and its implementation in real production processes is hindered. To overcome these obstacles and to explore the full potential of ultrashort laser pulses, the German Federal Ministry of Education and Research (BMBF) has launched a comprehensive research initiative on ultrashort pulse laser technology. This includes the development of innovative, inexpensive and powerful new ultrashort pulse laser systems as well as the appropriate system technology for beam control and steering. Apart from the research on laser sources a strong focus was on the development of the interaction processes itself, closely linked to specific application examples. This research effort has been pursued by 23 companies and 13 research institutions, which closely worked together within 9 joint research projects. The research was supported with more than 24 million Euros by the BMBF. A similar amount of money was spent in addition by the industrial partners involved. The major results of this research initiative are summarized in this book. In order to provide a coherent picture, the individual contributions are grouped thematically rather than by the original projects. Laser source development is covered in the first section, followed by a section describing fundamental interaction aspects, processing strategies as well as beam shaping, scanning and other relevant system technology. Finally, specific applications from surgery, THz generation to electron source generation for X-ray sources complete the overview.

The remarkable progress in ultrashort pulse technology and its applications as described here would not have been possible without the dedication of the companies and institutes, especially the researchers involved, but also the generous support by the German Federal Ministry of Education and Research. The editors and authors of this book express their sincere thanks to all of these.

Stefan Nolte
Jena — Frank Schrempel
Stuttgart — Friedrich Dausinger

Contents

Part II Materials Processing

Contributors

Ilya Alexeev Institute of Photonic Technologies, Friedrich-Alexander-Universität Erlangen-Nürnberg, Erlangen, Germany; Erlangen Graduate School in Advanced Optical Technologies (SAOT), Friedrich-Alexander-Universität Erlangen-Nürnberg, Erlangen, Germany

Antonio Ancona Institute for Photonics and Nanotechnologies, (IFN)-CNR U.O. S. Bari, Physics Department "M. Merlin", Bari, Italy

Dominik Bauer TRUMPF Laser GmbH + Co. KG, Schramberg, Germany

M. Baumgartl Institut Für Angewandte Physik, Friedrich-Schiller-Universität Jena, Jena, Germany

Peter Bechtold Institute of Photonic Technologies, Friedrich-Alexander-Universität Erlangen-Nürnberg, Erlangen, Germany; Erlangen Graduate School in Advanced Optical Technologies (SAOT), Friedrich-Alexander-Universität Erlangen-Nürnberg, Erlangen, Germany

Yunpei Deng Max-Planck Institute of Quantum Optics, Garching, Germany

Felix Dreisow Institute of Applied Physics, Abbe Center of Photonics, Friedrich Schiller University of Jena, Jena, Germany

Sven Döring Institute of Applied Physics, Abbe Center of Photonics, Friedrich-Schiller-Universität Jena, Jena, Germany

Jan Düsing Laser Zentrum Hannover e.V, Hannover, Germany

T. Eidam Friedrich-Schiller-Universität Jena, Institut Für Angewandte Physik, Jena, Germany

Moritz Emons Institut Für Quantenoptik, Leibniz Universität Hannover, Hannover, Germany

B. Ewers Deutsches Zentrum für Luft-und Raumfahrt e.V., Institut für Aerodynamik und Strömungstechnik, Göttingen, Germany

S. Fritzler Imaging & IT Division, Components and Vacuum Technology, Siemens AG Healthcare Sector, Erlangen, Germany

Uwe Griebner Max-Born-Institut für Nichtlineare Optik und Kurzzeitspektroskopie, Berlin, Germany

Xun Gu Max-Planck Institute of Quantum Optics, Garching, Germany

Florian Harth Photonik-Zentrum Kaiserslautern e.V., Kaiserslautern, Germany

Oliver H. Heckl TRUMPF Laser- und Systemtechnik GmbH, Ditzingen, Germany

Dieter Hoffmann Fraunhofer-Institut für Lasertechnik ILT, Aachen, Germany

Alexander Horn Laser Zentrum Hannover e.V, Hannover, Germany

C. Jansen Philipps-Universität Marburg, Experimentelle Halbleiterphysik, Marburg, Ehemals, Germany

C. Jördens Metrology Department, Volkswagen AG, Wolfsburg, Germany

Reinhard Kienberger Max-Planck Institute of Quantum Optics, Garching, Germany

Jochen Kleinbauer TRUMPF Laser GmbH + Co. KG, Schramberg, Germany

Peter Klopp Max-Born-Institut für Nichtlineare Optik und Kurzzeitspektroskopie, Berlin, Germany

Ulrich Klug Laser Zentrum Hannover e.V, Hannover, Germany

M. Koch Phillips-Universität Marburg, Experimentelle Halbleiterphysik, Marburg, Germany

Dietmar Kracht Laser Zentrum Hannover e.V, Hannover, Germany

Martin Kraus Robert Bosch GmbH, Stuttgart, Germany

Ferenc Krausz Max-Planck Institute of Quantum Optics, Garching, Germany

N. Krumbholz Technische Entwicklung der Volkswagen AG, Wolfsburg, Germany

Jens König Robert Bosch GmbH, Corporate Sector Research and Advance Engineering Laser Material Processing and Joining Technology (CR/APJ2), Stuttgart, Germany

Andreas Lenk Continental Automotive GmbH, Limbach-Oberfrohna, Germany

J. Limpert Institut Für Angewandte Physik, Friedrich-Schiller-Universität Jena, Jena, Germany

Johannes L'huillier Photonik-Zentrum Kaiserslautern e.V., Kaiserslautern, Germany

Torsten Mans Fraunhofer-Institut für Lasertechnik ILT, Aachen, Germany

Gilad Marcus Max-Planck Institute of Quantum Optics, Garching, Germany

Thomas Metzger Max-Planck Institute of Quantum Optics, Garching, Germany

Andreas Michalowski Robert Bosch GmbH, Stuttgart, Germany

Javier Gonzalez Moreno Laser Zentrum Hannover e.V, Hannover, Germany

Uwe Morgner Institut Für Quantenoptik, Leibniz Universität Hannover, Hannover, Germany

Stefan Nolte Institute of Applied Physics, Abbe Center of Photonics, Friedrich Schiller University of Jena, Jena, Germany; Fraunhofer Institute for Applied Optics and Precision Engineering IOF, Jena, Germany

Uwe Oberheide Augenklinik Am Neumarkt, Cologne, Germany

B. Ortaç Bilkent Univ, UNAM Inst Mat Sci & Nanotechnol, Ankara, Turkey

Ludger Overmeyer Laser Zentrum Hannover e.V, Hannover, Germany; ITA, Leibniz Universität Hannover, Garbsen, Germany

Guido Palmer Institut Für Quantenoptik, Leibniz Universität Hannover, Hannover, Germany

O. Peters Menlo Systems GmbH, Martinsried, Germany

M. Plötner Fraunhofer IOF, Jena, Germany

Stephan Roth Bayerisches Laserzentrum GmbH, Erlangen, Germany; Erlangen Graduate School in Advanced Optical Technologies (SAOT), Friedrich-Alexander-Universität Erlangen-Nürnberg, Erlangen, Germany

Peter Russbueldt Fraunhofer-Institut für Lasertechnik ILT, Aachen, Germany

F. Röser Helmholtz-Zentrum Dresden-Rossendorf, Dresden, Germany

Hakan Sayinc Laser Zentrum Hannover e.V, Hannover, Germany

M. Scheller Optical Sciences, University of Arizona, Tucson, Arizona, USA

B. Scherger Active Safety, Delphi Deutschland GmbH, Wuppertal, Germany

Stefan Schippel Layertec GmbH, Mellingen, Germany

Michael Schmidt Institute of Photonic Technologies, Friedrich-Alexander-Universität Erlangen-Nürnberg, Erlangen, Germany; Bayerisches Laserzentrum GmbH, Erlangen, Germany; Erlangen Graduate School in Advanced Optical Technologies (SAOT), Friedrich-Alexander-Universität Erlangen-Nürnberg, Erlangen, Germany

Marcel Schultze Institut Für Quantenoptik, Leibniz Universität Hannover, Hannover, Germany

Silvia Schumacher Laserzentrum Hannover e.V., Hannover, Germany

Viktor Schütz Laser Zentrum Hannover e.V, Hannover, Germany

Steffen Sommer Dausinger + Giesen GmbH, Stuttgart, Germany

M. Stecher Carl Zeiss Laser Optics GmbH, Productmanagement, Oberkochen, Germany

M. Strotkamp Fraunhofer-Institut für Lasertechnik ILT, Aachen, Germany

Dirk Sutter TRUMPF Laser GmbH + Co. KG, Schramberg, Germany

Oliver Suttmann Laser Zentrum Hannover e.V, Hannover, Germany

A. Tünnermann Friedrich-Schiller-Universität Jena, Institut Für Angewandte Physik, Jena, Germany

Thorsten Ulm Photonik-Zentrum Kaiserslautern e.V., Kaiserslautern, Germany

Dmitrij Walter Manz AG, Reutlingen, Germany

Dieter Wandt Laser Zentrum Hannover e.V, Hannover, Germany

Sascha Weiler TRUMPF Inc., Farmington, CT, USA

Markus Weyers Ferdinand-Braun-Institut, Leibniz-Institut für Höchstfre quenztechnik, Berlin, Germany

R. Wilk Menlo Systems GmbH, Martinsried, Germany

Bodo Wojakowski ALLTEC GmbH, Selmsdorf, Germany

J. Wueppen Fraunhofer-Institut für Lasertechnik ILT, Aachen, Germany

Maik Zimmermann Bayerisches Laserzentrum GmbH, Erlangen, Germany; Erlangen Graduate School in Advanced Optical Technologies (SAOT), Friedrich-Alexander-Universität Erlangen-Nürnberg, Erlangen, Germany

Martin Zorn JENOPTIK Diode Lab GmbH, Berlin, Germany

Part I
Ultra Short Pulse Laser Sources

Chapter 1
High Energy/High Repetition Rate Laser Pulses from Yb Based Solid State Oscillators with Cavity-Dumping and Regenerative Amplifiers

Moritz Emons, Guido Palmer, Marcel Schultze, Uwe Morgner, Hakan Sayinc, Dieter Wandt and Dietmar Kracht

Abstract Various scientific and industrial laser applications profit from the latest progress of directly diode pumped high-power femtosecond oscillators, in particular applications such as laser micromachining, nonlinear spectroscopy, or laser surgery. These oscillators provide pulse energies in the microjoule regime at 100 kHz to megahertz repetition rates, replacing complex and expensive chirped-pulse amplification (CPA) systems. Regarding power and energy scaling, Yb-doped materials are of great interest, offering the opportunity of high-power diode-pumping plus the generation of femtosecond pulses. In this article, we report on recent progress in the generation of femtosecond pulses with high energies either from oscillators or from regenerative amplifiers. In the first part of this chapter we discuss solid state oscillators with cavity-dumping and solitary pulse shaping followed by the concept of oscillators in the positive dispersion regime. The principle of thin-disk oscillator technology and its suitability for cavity-dumping is discussed, and the latest results from our two-crystal oscillator are presented, followed by some selected applications. In the second part of this chapter the progress of regenerative thin-disk amplifier technology will be discussed including latest results from thin-disks of potassium yttrium and lutetium monoclinic double tungstate oxide.

M. Emons (✉) · G. Palmer · M. Schultze · U. Morgner
Institut Für Quantenoptik, Leibniz Universität Hannover, Welfengarten 1, 30167 Hannover, Germany
e-mail: emons@iqo.uni-hannover.de

H. Sayinc · D. Wandt · D. Kracht
Laser Zentrum Hannover e.V, Hollerithallee 8, 30419 Hannover, Germany
e-mail: h.sayinc@lzh.de

S. Nolte et al. (eds.), *Ultrashort Pulse Laser Technology*,
Springer Series in Optical Sciences 195, DOI 10.1007/978-3-319-17659-8_1

1.1 Solid State Oscillators with Cavity-Dumping

In this section we report on passively mode-locked Ytterbium-based oscillators with MHz repetition rate cavity-dumping in different dispersion regimes allowing for the generation of microjoule pulses directly from a single oscillator. When the laser is operated in the solitary regime the current limit in energy scalability is given by the exceeding nonlinearity from several microjoules of intracavity pulse energy [1]. For even higher pulse energies it is necessary to either significantly reduce the nonlinear propagation length or to operate the oscillator with less peak power. The latter approach can be realized with the positive dispersion concept. The spectral bandwidth of the resulting positively chirped pulses is explicit broader than in the solitary regime [2], which allows for the generation of femtosecond pulses after external compression.

When starting to scale the pulse energy in femtosecond oscillators with cavity-dumping the simplest approach for the generation of pulse energies exceeding 1 μJ is based on a single bulk-crystal (see Fig. 1.1). The $KY(WO_4)_2$-crystal can either be directly pumped by a free emitting laser diode or a fiber-coupled laser diode with up to 30 W at 980 nm. Typical resonator lengths which have been realized yield repetition rates around 16–19 MHz.

This repetition rate already offers reasonably high pulse energies and makes electronic switching unproblematic. Electro-optical cavity-dumping is achieved by the combination of a Pockels-cell (electro-optical modulator with BBO-crystals) and a thin-film polarizer (TFP). This method is the same for all following cavity-

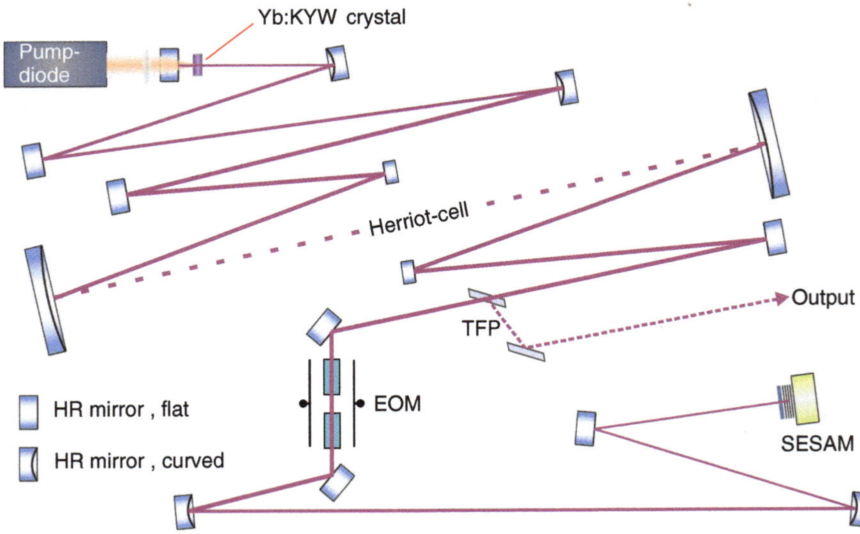

Fig. 1.1 Schematic of the laser setup, *EOM* electro-optical-modulator (BBO-Pockels-cell), *TFP* thin-film polarizer, *SESAM* semiconductor saturable absorber mirror, *Herriott-cell* number of reflections per mirror depends on cavity length (not shown)

dumped systems. The saturable absorber mirror stabilizes the solitary mode-locking against cw- and Q-switch-operation. Beside the N_m-optical axis (1030 nm) the bulk-laser can also be operated with polarization parallel to the N_p-optical axis at a wavelength of 1040 nm. The crystal length is between 1 and 2 mm with a typical Yb-doping concentration is of 5 at.%.

1.1.1 Solitary Regime in Bulk-Oscillators

For maximum internal pulse energies around 2.3 μJ a negative group-delay-dispersion of $-9200\ \mathrm{fs}^2$ is required to compensate for the nonlinearities (self-phase-modulation) in the solitary setup similar to that shown in Fig. 1.1 [2]. With cavity-dumping efficiencies of around 60 % the laser emits pulse energies of 1.35 μJ with pulse durations of 340 fs (time-bandwidth-product, TBP: 0.34). The attributed power spectrum is depicted in Fig. 1.2. The prominent sidebands are identified as Kelly sidebands from the perturbation of the soliton with the periodicity of the oscillator round trip frequency [3, 4].

Numerical simulations based on the soliton theory have led to a deep understanding of the pulse evolution and the dynamic behavior in the system [5]. They revealed that three dumping frequency regimes can be defined: the relaxation, the intermediate and the transient frequency regime for low (<20 kHz), medium and high (>80 kHz) dumping rates, respectively. For low repetition rates the time between the dumping is long compared to the period of the relaxation oscillations which allows the laser to relax to its steady state between two dumping events. In the ‘intermediate’ regime with a dumping rate close to the relaxation oscillation frequency (around 40 kHz), instability and subharmonic behavior is observed. At dumping frequencies higher than 80 kHz the laser has no time to relax before the

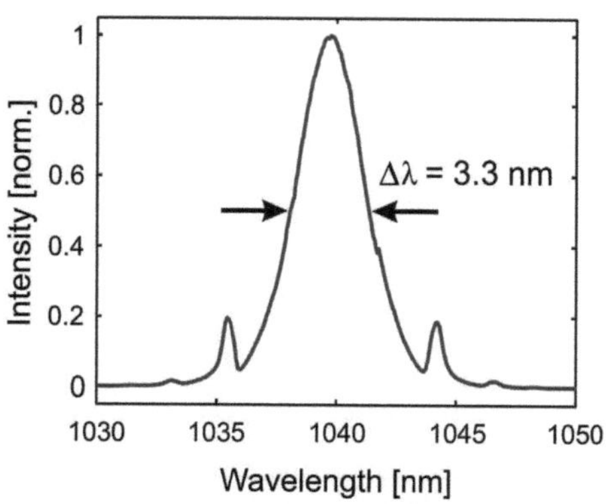

Fig. 1.2 Power spectrum of the solitary bulk laser clearly revealing Kelly sidebands which de-rive from the disturbance of the soliton-like pulses caused by the dumping. The laser was operated with polarization parallel to the N_p-optical axis resulting in a central wavelength of 1040 nm

next dumping occurs. The latter allows for dumping-frequencies from roughly 300 kHz to more than 1 MHz closing the gap between oscillators and solid-state amplifier systems [5].

1.1.2 The Chirped Pulse Oscillator (CPO)

For pulsed operation and a compact configuration the resonator was stretched to a total length of 8.64 m (repetition rate of 17.35 MHz) with a Herriott-type multipass cell (see Fig. 1.1). In contrast to the solitary regime a total material dispersion of +4250 fs^2 per round-trip for all participating optics was estimated. By changing type and number of the negative dispersive mirrors the magnitude of the net intracavity dispersion could be easily changed [6].

By using a SESAM stable self-starting cw-mode-locking operation of the laser has been achieved even though there is no pulse shaping mechanism from solitary pulse propagation. To sustain stability it was necessary to raise the pump power simultaneously while increasing the dumping ratio to compensate for the dumping losses. Stable mode-locking could be observed with dumping ratios up to 45 %. At that point the maximum output energy was beyond 2 μJ which is a factor of 1.4 higher compared to the solitary version from Sect. 1.1.1. We operated the oscillator in four different dispersion configurations, namely 250, 750, 1250 and 2250 fs^2 for the total GDD per round trip. The resulting power spectra are compared in Fig. 1.3 on a logarithmic scale. The distinctive rectangular shape is typical for chirped-pulse oscillators.

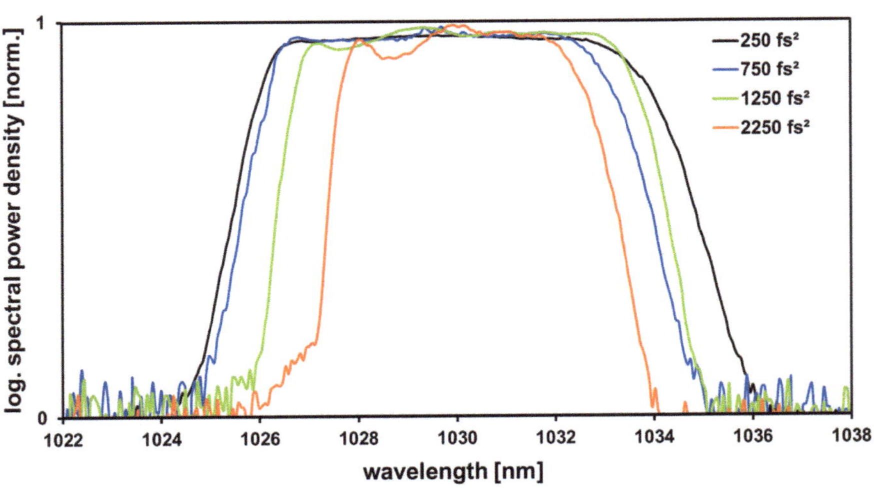

Fig. 1.3 Optical power spectrum of the output pulses for the different dispersion regimes

Beginning with a large dispersion of 2250 fs^2, the spectral bandwidth becomes significantly broader with decreasing net-dispersion, while the available output power remained constant for all four different dispersion values. The broadest bandwidth (FWHM) of 8 nm was observed at 250 fs^2. A further decrease of the GDD results in unstable pulsing. With a dispersion of 250 fs^2 the duration of the chirped output pulses was measured to be in the range of 3 ps (4.4 ps autocorrelation width). The measured rms-noise was below 1 % with a beam parameter M^2 of less than 1.1. The contrast ratio between the dumped pulses and the background exceeded 500:1.

Laser operation at 250 fs^2 is optimal when subsequent external pulse compression is used, since it provides the broadest spectrum enabling the shortest available pulse duration. In this case two fused silica transmission gratings with 1250 lines per millimeter were employed. The compressed pulses were analyzed with an intensity autocorrelator. By calculating the pulse duration directly from the power spectrum thus obtains a Fourier-limit of 360 fs (490 fs autocorrelation width). The measured autocorrelation reveals a FWHM of 570 fs resulting in a pulse duration of approximately 420 fs. The residual chirp led back to misalignments of the rather complex compression setup.

1.1.3 Thin-Disk Oscillator in the Solitary Regime

Using the previously discussed solitary mode-locked bulk oscillators (see Sect. 1.1.1) in combination with the cavity-dumping technique, a further scaling of the energy is limited by thermal effects and nonlinear contributions of the relatively long laser crystals. For this reason, the thin-disk concept is used instead to investigate the possibility for the generation of even higher pulse energies [1]. The thin-disk laser consists in principle of a very thin laser medium in combination with an optical multipass pump scheme. By using fibers for transport and homogenization of the pump light, low cost diode lasers can easily be coupled into the multipass pump system. Therefore the demands on brightness of the pumping diodes are comparably low. Due to the thin crystal, the homogeneous pump profile and the axial cooling geometry, such lasers are very resistant to thermal lensing [7, 8]. The schematic setup is similar to Fig. 1.1. The thin-disk acts as a folding mirror enabling four bounces per roundtrip. To maximize the absorbed pump power, a commercial thin-disk pump module is used which allows for 24 passes through the gain medium [9].

The 110 μm thin Yb:KYW disk is pumped with up to 50 W at 976 nm. Output coupling of the pulses is realized as before with a BBO-Pockels-cell in combination with a TFP. Mode-locking is induced by a SESAM with a modulation depth of approximately 1 % and a saturation fluence of about 90 $\mu J/cm^2$. Due to the high nonlinearity contributions from the Pockels-cell, the thin-disk and the ambient air, a total negative dispersion of 40,000 fs^2 is inserted into the cavity by means of dispersive mirrors (GTI) to fulfill the soliton condition. The maximum output power

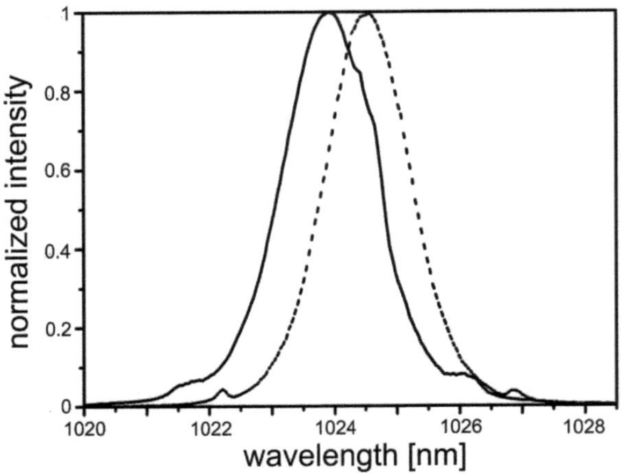

Fig. 1.4 Oscillator output spectrum with air (*black solid*) and helium (*black dotted*)

equals 2.34 W at 1 MHz repetition rate leading to pulse energies of 2.34 μJ in combination with nearly Fourier limited pulse durations of about 700 fs. Figure 1.4 shows the corresponding spectrum (solid line). To minimize the nonlinearities from the ambient air, the resonator laserbox is purged with helium in a second step. Here, stable operation with more than 3 μJ of pulse energy and durations of 680 fs have been obtained (spectrum in Fig. 1.4, dashed line).

Compared to the bulk concept, two main drawbacks of the thin-disk concept with cavity-dumping can be pointed out. On the one hand side, the dumping ratios are relatively low (~25 % in contrast to more than 50 % in the bulk concept). As a reason for this, the low single pass gain is found to be the main limitation (see Sect. 1.1.4). Supplementary the pulse duration is relatively long (~700 fs) compared to 400 fs in the bulk setup. Here, etalon and hole-burning effects lead to unstable operation for bandwidths larger than 1.8 nm [10]. Thus, the available gain bandwidth has to be restricted by means of a birefringent filter (BRF) inside the cavity.

1.1.4 Limits of Cavity-Dumping in Thin-Disk Oscillators

The maximum achievable peak power from the oscillator is limited by the B-Integral in the cavity. By using Yb:YAG thin-disk modules this limitation can be overcome and pulse energies up to 11 μJ at a repetition rate of 4 MHz have been demonstrated in SESAM mode-locked oscillators with pulse durations of 790 fs [11]. As elaborated in Sect. 1.1.3 the thin-disk oscillator with cavity-dumping based on Yb:KYW resulted in pulse energies up to 3 μJ at a repetition rate of 1 MHz with pulse durations of 680 fs [1]. Currently, the limiting factor for the pulse energy in this setup is the dumping ratio which, at 24 %, is far lower than in the previous experiments with bulk material (see Sect. 1.1.1) where the maximum dumping ratio easily exceeds 50 %.

To understand the reasons for this limitation and to find ways of improving the performance of our laser we started to investigate the theoretical properties of such laser systems. In this section we present a theoretical study of cavity-dumped systems with special regard to how the implementation of a thin-disk scheme affects the dynamic properties of the cavity-dumped laser. The comparison between the numerical model and experiments allows for determining inherent limits of the current setups and finding possible ways for improvement.

It is possible to accurately describe the dynamics of a SESAM mode-locked laser by using three differential equations which describe the temporal dynamics of the pulse envelope, the laser gain and the absorber loss. The detailed equations and parameters are given in [12]. The excellent agreement between numerical simulations and experimental data forms the basis for the theoretical investigation of the limiting factors for cavity-dumping. From this treatment we expect a deeper understanding of the pulse shaping mechanisms and some guidelines how to extract some higher fraction of the intracavity pulse energy. Not surprisingly, the numerical model revealed the same limitation for the maximum dumping ratio for the thin-disk setup between 25 and 30 % with the parameter sets close to the experimental values. Higher dumping ratios result in Q-switching instabilities, and no stable cw mode-locked laser operation was possible. Nevertheless, the question is, what is the reason for the big difference in the maximum dumping ratio between the bulk and thin-disk setup? Looking at the two laser setups, the main difference in terms of the gain dynamics is the mode radius in the gain medium. While the mode radius in an Ytterbium bulk laser is typically around 100 μm, the same parameter in a thin-disk setup is typically more than 5 times larger. Indeed Fig. 1.5 reveals that the key factor for the stability of a cavity-dumped mode-locked laser is the mode size in the gain medium.

In the simulations we varied the mode radius from 540 μm down to 200 μm. Starting from a plateau at around 18 % for large mode radii the stability border grows linearly to above 80 % as the mode radius is reduced. It is also important to note that the small signal gain was kept constant while the mode size was scanned. In the experiment this is only possible up to a certain point because the small signal gain is directly linked to the pump power and the pump power density is limited by saturation effects in the gain medium and the damage threshold of the disk. With our used disk the maximum pump power density is limited to 4 kW/cm^2 which would be reached at a mode radius of 350 μm. So for even smaller pump radii the pump power would have to be decreased which would still allow for an increased dumping ratio but at the expense of a reduction in pulse energy. In the more realistic scenario, starting with the parameters of the thin-disk laser and continuously reducing both pump power and laser mode area on the disk one eventually ends up with a system very similar to the bulk laser described in Sect. 1.1.1. At this end the maximum dumping ratio could easily exceed 50 % but the pulse energy would be decreased to about 1 μJ. In a qualitative sense the importance of the mode size can be understood simply by analyzing its influence on the small signal gain and the

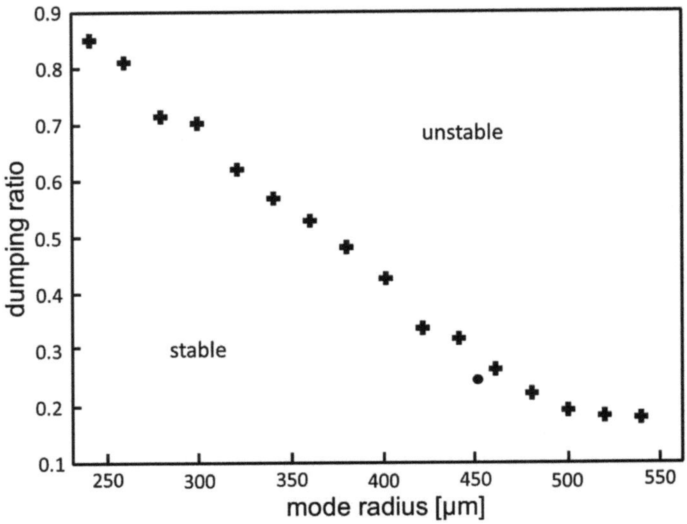

Fig. 1.5 Calculated maximum dumping ratio for stable cw mode-locking operation in the thin-disk laser as a function of the beam radius on the laser disk. The *black dot* marks the parameters of the experiment from Sect. 1.1.3

saturation energy of the laser. A larger mode radius on the disk leads to an increase in saturation energy at the expense of a smaller gain. As described by Hönninger et al. [13] the pulse energy E_p necessary to avoid Q-switching in a SESAM mode-locked laser without cavity-dumping can be calculated using the following equation

$$E_p^2 > E_{sat,A} E_{sat,L} \Delta R, \tag{1.1}$$

where ΔR is the SESAM reflectivity modulation depth, $E_{sat,A}$ is the saturation energy of the absorber, and $E_{sat,L}$ that of the laser gain medium. Here, an increase in the saturation energy of the laser will result in an increased threshold for stable cw mode-locked operation. For a thin-disk laser like the one described in [11] this is of little consequence since the low output coupling ratio ensures that the intracavity pulse energy is usually high enough to avoid Q-switching instabilities. In the case of a cavity-dumped laser however the situation is different, since the small-signal stability analysis behind (1.1) is not valid anymore. From our model the numerical value for the threshold was found to be up to an order of magnitude higher in the case of a cavity-dumped laser.

Still (1.1) provides a good starting point for a systematic investigation of the cavity-dumped thin-disk oscillator. We started by varying two key parameters of the model, the mode area on the gain medium and the modulation depth of the SESAM. The left part of Fig. 1.6 shows the maximum dumping ratio for different modulation depths ΔR and different mode radii r_{gain} color coded between blue and red. It can be seen that, in accordance with the discussion of (1.1), higher

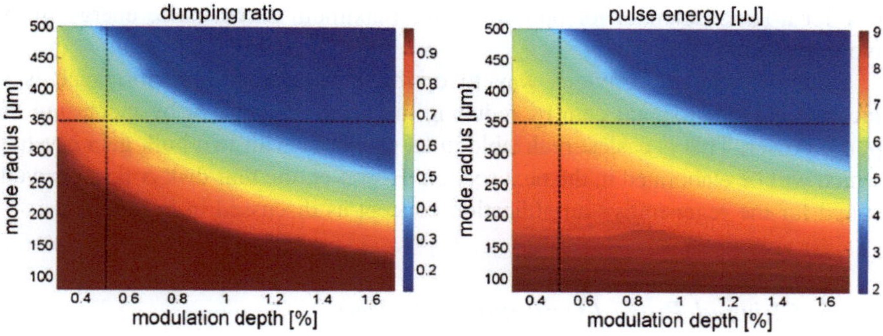

Fig. 1.6 *Left* Maximum dumping ratio as a function of the mode radius on the disk and the modulation depth of the SESAM. *Red color* denotes a high ratio, *blue color* a low ratio. *Right* Resulting out-coupled pulse energies in μJ for the same set of parameters. *Red color* denotes high energy, *blue color* low energy. The *dashed lines* indicate experimental limits. The maximum power density, i.e. the minimal mode radius on the disk (*horizontal*) and the minimal modulation depth for stable mode-locking (*vertical*) leave only the *upper right* quadrant experimentally accessible

modulation depths lead to lower possible dumping ratios. But nevertheless even for small modulation depths the mode radius is still the dominant parameter. For modulation depths exceeding 1.5 % the maximum dumping ratio depends almost exclusively on the mode radius.

The right part of Fig. 1.6 shows the maximum stable pulse energy in units of μJ that would be coupled out with the dumping ratios shown in the left. The dashed lines indicate some empirical experimental limitations: The vertical line stems from the fact that for modulation depths lower than 0.5 % only cw operation of the laser was possible as the loss modulation was not high enough to achieve stable mode-locking. The horizontal line represents the power density damage limit of the Yb: KYW disk of 4 kW/cm^2. So the accessible regime is limited to the upper right quadrant in the two plots. By choosing optimum parameters pulse energies around 6 μJ should be achievable. It also needs to be pointed out that even if the limitation of the power density on the disk could be overcome, any substantial increase in pump power would lead to other problems. To fulfill the soliton condition in our cavity-dumped oscillator (which is currently done by using home designed chirped mirrors) we need to accumulate a second order dispersion (GDD) of −40,000 fs^2. Since any increase in intracavity power results in an increase in the nonlinearities in the cavity, even larger values of the GDD would be required for compensation, which might be hard to achieve by mirrors. One solution could be laser operation in the positive dispersion regime (see Sect. 1.1.2).

The third variable given in (1.1) is the saturation energy of the absorber which forms a third accessible degree of freedom. Unfortunately, in the experiment it has been found that the mode size on the SESAM is limited to a very narrow range roughly between 270 and 450 μm. For smaller mode radii the SESAM is prone to damage because of the high intensities, while for larger radii no stable cw-mode locking was possible. For this reason the mode area on the SESAM does not seem to offer a large potential for optimizing the maximum output pulse energy of the mode-locked thin-disk laser.

In conclusion, the obvious way to optimize the dumping performance is to increase the small-signal gain by reducing the beam diameter or by increasing the pump power. This is easiest realizable in an oscillator with two bulk gain crystals, which will be elaborated in the next section. However, the thin-disk concept is well-suited for the regenerative amplification concept as discussed in Sect. 1.2.

1.1.5 Chirped-Pulse Oscillator with Two Yb:KYW Bulk Crystals

As discussed in Sect. 1.1.4 energy scaling concerning cavity-dumped systems is restricted to bulk-based laser technology since the thin-disk scheme does not allow for sufficient dumping efficiencies. For higher energies it is necessary to increase the pump power. However a single bulk-crystal laser is limited in terms of both applicable pump power and thermal load. For instance when applying pump powers of more than 30 W to pump diameters of 300 μm the laser clearly exhibits anisotropic beam distortion and a strong tendency to transversal multi-mode operation. Especially the anisotropic properties of the Yb:KYW crystal lead to asym-metric effects such as elliptical beam profiles and astigmatism, respectively. In the worst case the crystal will break under the thermal load. In order to prevent the gain medium from such events the pump power can be divided onto two or even more crystals. In that way higher pump power levels become possible. Moreover this approach allows for a higher net-gain which is ideal to reach high cavity-dumping efficiencies.

For such an oscillator the laser setup equals the previous bulk-laser configuration (see Fig. 1.1) except that there are two separate Yb:KYW-crystals each pumped by a fiber coupled laser diode at 981 nm (Fig. 1.7) [14]. With a total absorbed pump power of almost 20 W this system reaches pulse energies of 7 μJ (average power: 7 W) at 1 MHz dumping frequency. At this point the maximum dumping efficiency is around 67 % which is—as expected—even better than for the single-bulk-laser.

With respect to the power values the oscillator works with an optical-to-optical efficiency of almost 20 % which is outstanding for mode-locked oscillators with cavity-dumping. The net GDD is at +3500 fs^2 where mode-locking is the most stable over the widest pulse energy range. The GDD can be altered by varying the amount of negative dispersive mirrors (GTI). The Gaussian shaped chirped pulses inside the resonator feature FWHM-durations of 14 ps.

The power spectrum is displayed in Fig. 1.8. Calculating the Fourier-limit from the spectrum yields a pulse duration of 261 fs. Low-loss pulse compression into the femtosecond regime can be realized by means of the compressor setup from Fig. 1.7.

This compact configuration uses one transmission grating (here: 1250 lines/mm) and gives rise to losses below 13 %. This compressor allowed for measured pulse durations as short as 416 fs at the best compressor alignment. With respect to the

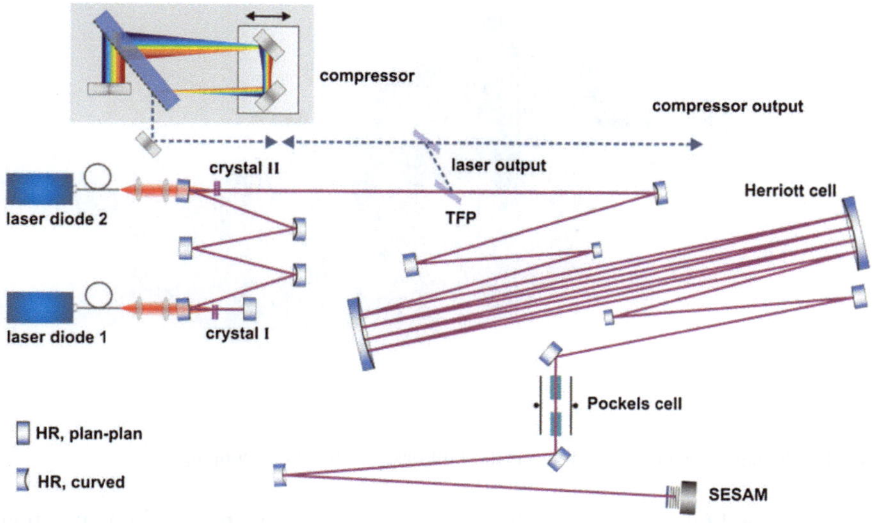

Fig. 1.7 Schematic of the two-crystal laser and the external compressor

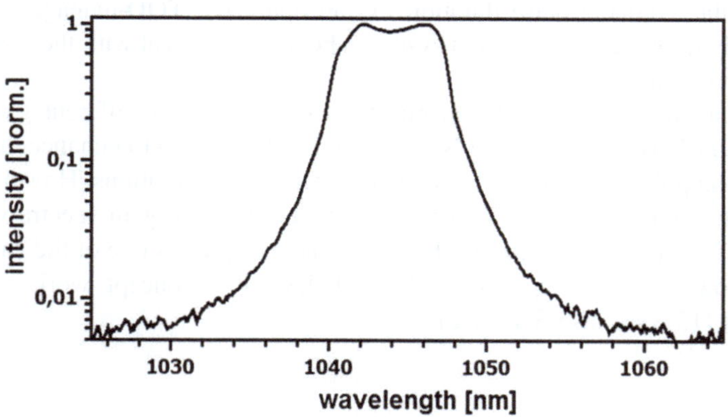

Fig. 1.8 Power spectrum of the two-crystal bulk laser

investigated system the Fourier-limited pulse duration could not be reached due to higher order dispersion. The residual third order dispersion (TOD) from the grating was calculated to be around +0.007 ps^3. By means of calculations and dispersion measurements one resonator mirror could be identified as a further high TOD-source where a TOD of +0.017 ps^3 was accumulated. Nonetheless pulse energies of 6.1 μJ were measured behind the compressor yielding a total peak power of 12 MW. Apart from the pulse properties very good power noise properties (rms < 1 %) and beam characteristics (M^2: 1.1) have been observed. This is due to the compact setup (0.9 × 0.5 × 0.3 m) and the sealed laser box.

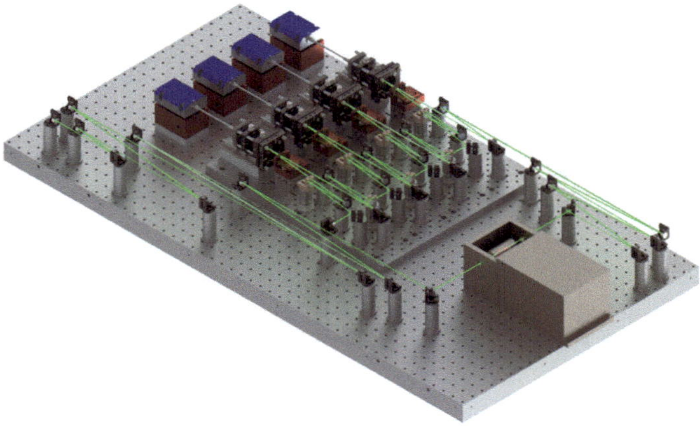

Fig. 1.9 Schematic picture of a four-crystal bulk-laser with cavity dumping

Current and future research will account for even higher energies implying more than two crystals at higher pump power levels. As part of this work no particular energy scaling limitation could be found and multi-crystal approaches seem feasible (see principle setup illustrated in Fig. 1.9). There are also intentions to further decrease the available pulse duration. As an optimized TOD-management would most likely improve the presented system other concepts deal with the combination of different gain spectra.

The fundamental idea is to operate the oscillator under different gain spectra simultaneously (i.e. the N_p-, and N_m-spectrum in Yb:KYW) to enhance the spectral width of the pulses and in turn to generate shorter pulse durations. However, recent experiments indicate that the combination of the different gain spectra within the resonator is cannot be stabilized. The laser dynamics prefers one of the gain spectra and the other one is suppressed. Nevertheless, this concept works well with amplifiers [15] (see also Sect. 1.2).

1.1.6 *Applications*

The laser systems described before have been employed in some selected applications such as waveguide writing, two-photon polymerization or as seed sources for different amplification procedures.

1.1.6.1 Waveguide Writing

With the pulses of the contemplated lasers a straightforward fabrication of optical waveguides in transparent materials is possible. When these pulses are tightly focussed inside a bulk material, the intensity in the focal volume becomes high

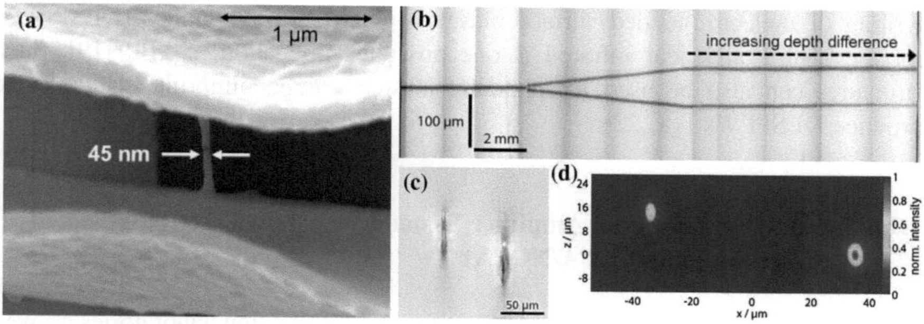

Fig. 1.10 **a** SEM images of a free hanging 2PP line structure [18]. **b** Simultaneously written 3D Y-Splitter with linearly increasing depth difference on the right half. **c** Cross-sectional microscope views of the end faces of the 3D coupler, the depth difference at the end is 13.8 μm. **d** Near field profiles of the output beams [19]

enough to cause absorption through nonlinear processes. This leads to a localized, permanent refractive index change, allowing direct fabrication of guiding structures. Unlike systems with lower repetition rates, femtosecond laser writing with 1 MHz repetition rate in general gives rise to thermal diffusion and heat accumulation effects which lead to an increase in the waveguide diameter and an increase in the refractive index contrast, while reducing the waveguide losses.

Fused silica is a common material for waveguide writing; Although we did not observe heat accumulation here with the laser parameters available, a significant refractive index increase has been achieved, which is high enough to produce multiple parallel waveguides, complex couplers [16], and 3D splitters in a single sweep (see Fig. 1.10b–d) [17]. These structures provide single mode guiding and variable coupling/splitting ratios for light at 976 and 1550 nm. Using special beam shaping methods various photonic devices are easily achievable just by software control for complex but flexible 3D photonic networks, e.g. for optical sensing in lab-on-a-chip devices.

1.1.6.2 Two-Photon Polymerization (2PP)

The two-photon polymerization (2PP) process involves temporal as well as spatial overlap of photons, resulting in a nonlinear absorption in a highly localized volume. The absorbed photons induce chemical reactions between starter molecules and monomers within the medium, inducing the polymerization. Using liquid materials polymerization results in localized solidification of the fluid.

Since the region for the light/matter interaction is limited within the focal volume one can fabricate any computer-generated 3D structure, by producing a trace of polymerized material, if the laser focus moves within such photosensitive material. Because of a quadratic intensity dependence of the two-photon-absorption probability and the well-defined polymerization threshold resolutions of less than 50 nm in the fabricated structures can be reached [20, 21] (see Fig. 1.10a). Direct laser-writing of dielectric-loaded surface plasmon–polariton wave-guides (DLSPPWs) for the visible and near infrared is one possible application of 2PP [19]. Such structures can also be used for realizing masters for nanoimprint techniques to produce DLSPPWs [18, 22].

1.1.6.3 Seed Oscillators for Amplifier Systems (Fiber Amplifier, OPA/NOPA)

The transfer from new laser technology into the application laboratories is well established. Laser systems like the above mentioned are ideal seed sources for miscellaneous amplifier applications. With the oscillator discussed in Sect. 1.1.1 it was possible to pump an optical parametric amplifier (OPA) directly by a femtosecond oscillator first time [23]. Wavelength-tunable pulses in the signal—(0.65–0.85 μm) and idler range (1.4–2.5 μm) are generated at a repetition frequency of 1 MHz. Pulses with 30 nJ of energy and a duration of 16 fs are achieved from a super-continuum seed generated in a sapphire plate.

Many different applications in physics, chemistry, biology, and medicine profit from ultrafast lasers with high pulse energies, from lasers with short pulses, and from lasers with high pulse repetition rates. The literature is full of reports on lasers which can fulfill one or two of the desired requirements. But some applications such as nonlinear bio-imaging or spectroscopy need them simultaneously, and here one laser system which is capable of meeting all three goals simultaneously was demonstrated: a noncollinear optical parametric amplifier pumped by a MHz/μJ-oscillator with cavity-dumping (MHz-NOPA) [24]. This system can produce few-cycle pulses with MHz repetition rates and pulse energies/peak powers well beyond the typical oscillator level. Finally, it could be easily scaled with the occurred progress in pump laser technology. In particular, the single-stage amplification of a cavity-dumped Yb:KYW laser oscillator with a Yb-doped rod-type fiber allows the generation of 420 fs pulses with a pulse energy of 9 μJ at a1 MHz repetition rate [25]. Driven by these pulses a noncollinear optical parametric amplifier (NOPA) delivering sub 10 fs pulses with 420 nJ of pulse energy was realized [26]. The ultra-broadband seed used for this system is based on stable white-light generation from the 420 fs long pulses in a Yttrium aluminium garnet (YAG) plate.

1.2 Regenerative Amplification with Yb:KYW and Yb:KLuW Thin-Disks

In Sect. 1.1.4 it has been elaborated that the thin-disk concept is limited for use in cavity-dumped oscillators. However, it is well-suited for the amplification of short pulses. For gain media with low single pass gain, a multi pass configuration is necessary to achieve high amplification factors. The most flexible multi pass amplifier setup is a regenerative amplifier. In principle regenerative amplification is divided in two periods in the temporal domain. In a first period the amplifier is pumped without seeding. This regeneration period is followed by an amplification period, where a single seed pulse is switched in and out of the amplifier.

Placing the amplifying medium in an optical resonator offers the possibility for multiple amplifier cavity roundtrips. In this way, the gain factor can easily be controlled by the number of cavity roundtrips and very high amplification factors can be achieved. The switching unit can be realized by an electro-optic modulator (EOM) in combination with a polarizer. As seed pulses usually ps and fs pulses are used. In general laser crystals for such setups are cuboids and thin-disks [27].

The thin-disk geometry offers the well-known possibility to decrease the intensity by scaling the spot size on the disc. Furthermore, the efficient thermal management of thin-disks makes them the most promising candidates for high power amplifiers. Especially in ultra-short pulse operation, thin-disk regenerative amplifiers (RA) emitting at 1 μm are convenient tools for a variety of applications like micromachining and waveguide writing in transparent media (see also Sect. 1.1.6).

However, less work has been done in the development of RA emitting dechirped pulse durations below 200 fs. The main challenge to achieve such short pulse durations is to overcome the gain narrowing effect, i.e. the spectral narrowing during amplification due to the spectral gain profile [28]. Several techniques are used to balance the spectral narrowing, such as spatially dispersive amplification [29], regenerative spatial shaping by intracavity elements [30] and seed pulse shaping in phase and amplitude [31, 32].

Another simple method to balance gain narrowing is to use different gain media to achieve a broad effective gain bandwidth. In particular for biaxial crystals the effective gain bandwidth can easily be increased by amplifying the pulses with polarization parallel to the different optical axes. In the wavelength region of 1 μm potassium yttrium and lutetium monoclinic double tungstate oxide (Yb:$KY(WO_4)_2$, Yb:$KLu(WO_4)_2$) are promising candidates for ultra-short pulsed regenerative amplifiers with and without gain combining [15, 33].

A chirped pulse amplifier was set up consisting of an ytterbium fiber based ultra-short pulse oscillator, a fiber based stretcher and pre-amplifier, a RA based on a Yb:$KY(WO_4)_2$ and a Yb:$KLu(WO_4)_2$ thin-disk and a GRISM (Grating-Prism) compressor. For the experiments where the Yb:KYW thin-disk was used as gain medium, the output pulses of the pre-amplifier had a central wavelength of 1034 nm, a pulse duration of 31 ps and a pulse energy of 9 nJ. The pulses of the seed oscillator could be dechirped to the Fourier limited pulse duration of 152 fs by using the GRISM compressor.

The designed RA setup, shown in Fig. 1.11, allowed 12 passes through the thin-disk per cavity roundtrip. Gain combining was done in every single roundtrip by rotating the polarization through 90°. Thus, the amplifying optical axis was changed between N_p and N_m. The thin-disk had an Yb concentration of 10 at.%, a diameter of 7 mm and a thickness of 102 μm. As pump source a laser diode was used, which had a wavelength of 981 nm and a maximum pump power of 78 W.

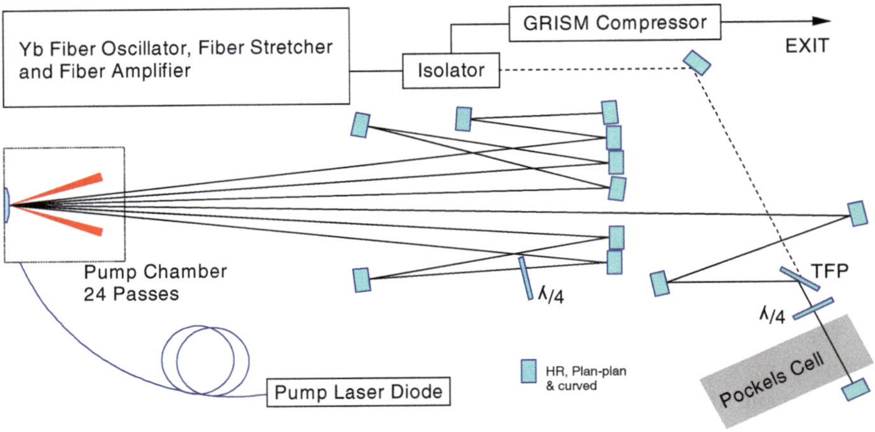

Fig. 1.11 Setup of the thin-disk based regenerative amplifier system

A Pockels-cell consisting of two β-barium borate crystals (BBO) together with a thin-film polarizer served as electro optical switch, which allowed repetition rates up to 200 kHz. The thin-disk was placed in a pump chamber consisting of a collimator for the pump diode fiber and a parabolic mirror, which enabled 24 passes of the pump beam through the thin-disk. The outcoupled pulses were dechirped with a GRISM compressor, which consisted of a grating and a prism and enabled a simultaneous compensation of second- and third order dispersion [34].

The RA was operated at a repetition rate of 20 kHz. The output pulse energy of the amplifier was easily scaled by increasing the number of cavity roundtrips. A variation between 38 and 53 roundtrips resulted in a pulse energy change from 114 to 500 μJ, measured behind the GRISM compressor. The optical spectrum of the output pulses at pulse energies of 500 μJ is shown at the left side in Fig. 1.12. A bandwidth of 9.3 nm and a central wavelength of 1031 nm were calculated by the second moment normalized to the FWHM of a Gaussian shape. The spectral modulation indicates a satellite pulse, which had its origin in an etalon between the surfaces of the half wave plate in the resonator. From the optical spectrum we calculated a Fourier limited pulse duration of 185 fs. The corresponding autocorrelation trace is shown on the right side of Fig. 1.12. It has a FWHM of 268 fs corresponding to a pulse duration of 185 fs, which is the Fourier limited pulse duration.

In a second experiment an Yb:KLuW thin-disk was used in a similar setup without polarization rotation [35]. For this experiment, the seed pulses had a pulse energy of 4 nJ, a pulse duration of 43 ps and a dechirped pulse duration of 170 fs. The central wavelength of the seed pulses was 1026 nm and was matched to the emission maximum of the thin-disk.

The output pulse energy of the amplifier was again scaled by increasing the roundtrip number. A variation from 28 to 44 cavity roundtrips resulted in a pulse energy change from 50 μJ to a maximum of 400 μJ. The pulse durations were

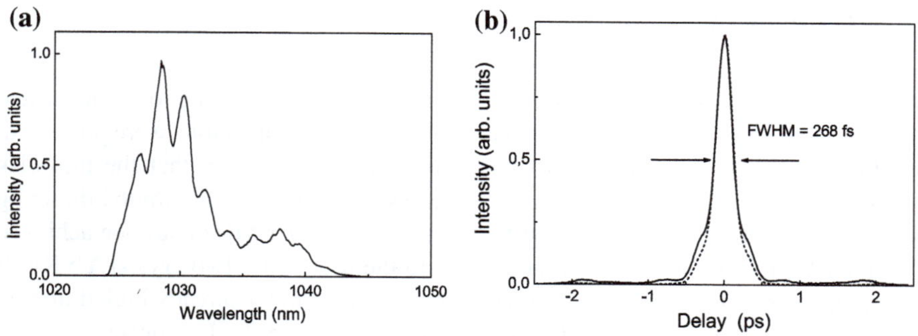

Fig. 1.12 Optical spectrum of the output pulses (*left*) and corresponding autocorrelation (*right*) at a pulse energy of 500 μJ

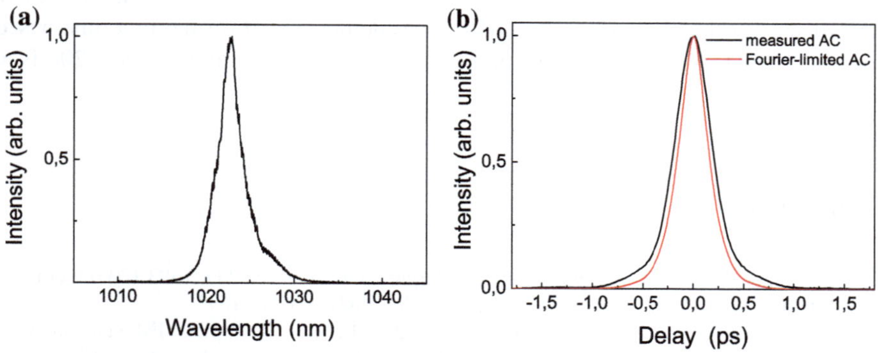

Fig. 1.13 Power spectrum (*left*) and the autocorrelation function together with Fourier limited autocorrelation (*right*) at the maximum pulse energy

nearly independent from the roundtrip number and remained between 190 and 214 fs. The power spectrum and the autocorrelation function at the maximum pulse energy are depicted in Fig. 1.13. The spectral FWHM was 2.9 nm and the central wavelength 1022.7 nm while the autocorrelation trace had a FWHM of 426 fs. A deconvolution factor of 2.16 was calculated, resulting in a pulse duration of 197 fs, which is 21 % above the Fourier-limit. Additionally, the measurement of pulse durations at different repetition rates between 20 and 125 kHz at constant pump power level yielded nearly similar values around 210 fs.

These results revealed the ability of operating the developed RA at different round trip numbers as well as at different repetition rates without a significant change of the dechirped pulse durations.

In conclusion ultrafast thin-disk regenerative amplifiers were demonstrated with the active media Yb:KYW and Yb:KLuW. In the case of Yb:KYW, a pulse duration below 200 fs was achieved at a pulse energy of 500 μJ. A regenerative thin-disk amplifier based on Yb:KLuW was demonstrated for the first time, delivering a pulse energy of 400 μJ at a pulse duration of 197 fs.

1.3 Conclusion

In this chapter results on Yb-based cavity-dumped solid state oscillators and regenerative amplifiers supplemented by some selected applications were presented.

In the case of cavity-dumped oscillators it has been shown that the thin-disk concept is not well suited compared to bulk lasers because of the limited dumping ratio. Nevertheless 790 fs pulses with energies of 3 μJ at 1 MHz could be achieved from an Yb:KYW thin-disk oscillator. Due to the usage of a two crystal Yb:KYW setup in combination with the CPO concept an increase in the already high dumping ratio of the bulk material could be obtained ending up in 7 μJ of pulse energy at 1 MHz repetition rate and a compressed pulse duration of 416 fs.

For the presented regenerative amplifiers Yb:KYW and Yb:KLuW thin-disks were used as gain media. By gain combining of the axes N_m and N_p a dechirped pulse duration of 185 fs was achieved at a pulse energy of 500 μJ for the Yb:KYW thin-disk. Another regenerative amplifier was demonstrated for the first time based on a an Yb:KLuW thin-disk, resulting in dechirped pulse durations below 200 fs at a pulse energy of 400 μJ.

References

1. G. Palmer, M. Siegel, A. Steinmann, U. Morgner, Opt. Lett. **32**(11), 1593 (2007). doi: 10.1364/OL.32.001593. http://ol.osa.org/abstract.cfm?URI=ol-32-11-1593
2. A. Killi, A. Steinmann, J. Dörring, U. Morgner, M.J. Lederer, D. Kopf, C. Fallnich, Opt. Lett. **30**(14), 1891 (2005). doi:10.1364/OL.30.001891. http://ol.osa.org/abstract.cfm?URI=ol-30-14-1891
3. S.M.J. Kelly, **28**, 806 (1992). doi: 10.1049/el:19920508. http://ieeexplore.ieee.org/stamp/stamp.jsp?tp=&arnumber=133147&isnumber=3651
4. N.J. Smith, K.J. Blow, I. Andonovic, **10**, 1329 (1992). doi:10.1109/50.166771.ieeexplore.ieee.org/xpls/abs_all.jsp?arnumber=166771&tag=1
5. A. Killi, U. Morgner, **12**(15), 3397 (2004). http://www.opticsexpress.org/abstract.cfm?URI=oe-12-15-3397
6. G. Palmer, M. Emons, M. Siegel, A. Steinmann, M. Schultze, M. Lederer, U. Morgner, Opt. Express **15**(24), 16017 (2007). doi:10.1364/OE.15.016017. http://www.opticsinfobase.org/abstract.cfm?URI=oe-15-24-16017
7. A. Giesen, in *Proceedings of SPIE*, vol. 5332 (2004), pp. 212–227. doi:10.1117/12.547973. http://spie.org/x648.html?product_id=547973
8. A. Giesen, in *European Symposium on Optics and Photonics for Defence and Security* (International Society for Optics and Photonics, 2004), pp. 112–127. doi:10.1117/12.578272. http://proceedings.spiedigitallibrary.org/proceeding.aspx?articleid=853172
9. TGSW Stuttgart, http://tgsw-photonics.de/scheibenlasertechnologie. http://tgsw-photonics.de/scheibenlasertechnologie
10. R. Paschotta, J.A. der Au, G.J. Spühler, S. Erhard, A. Giesen, U. Keller, **B 72**, 267 (2001). doi:10.1007/s003400100486. http://link.springer.com/article/10.1007%2Fs003400100486?LI=true
11. S.V. Marchese, C.R. Baer, A.G. Engqvist, S. Hashimoto, D.J. Maas, M. Golling, T. Südmeyer, U. Keller, **16**(9), 6397 (2008). http://www.opticsexpress.org/abstract.cfm?URI=oe-16-9-6397

12. M. Siegel, G. Palmer, A. Steinmann, M. Pospiech, U. Morgner, **15**(25), 16860 (2007). doi:10.1364/OE.15.016860. http://www.opticsexpress.org/abstract.cfm?URI=oe-15-25-16860
13. C. Hönninger, R. Paschotta, F. Morier-Genoud, M. Moser, U. Keller, **B 16**, 46 (1999). doi:10.1364/JOSAB.16.000046. http://www.opticsinfobase.org/josab/abstract.cfm?id=35627
14. G. Palmer, M. Schultze, M. Emons, A.L. Lindemann, M. Pospiech, D. Steingrube, M. Lederer, U. Morgner, **18**(18), 19095 (2010). doi:10.1364/OE.18.019095. http://www.opticsexpress.org/abstract.cfm?URI=oe-18-18-19095
15. U. Buenting, H. Sayinc, D. Wandt, U. Morgner, D. Kracht, **17**, 8046 (2009). doi:10.1364/OE.17.008046. http://www.opticsinfobase.org/oe/abstract.cfm?uri=oe-17-10-8046
16. M. Pospiech, M. Emons, A. Steinmann, G. Palmer, R. Osellame, N. Bellini, G. Cerullo, U. Morgner, **17**(5), 3555 (2009). http://www.opticsexpress.org/abstract.cfm?URI=oe-17-5-3555
17. M. Pospiech, M. Emons, B. Väckenstedt, G. Palmer, U. Morgner, **18**(7), 6994 (2010). doi:10.1364/OE.18.006994. http://www.opticsexpress.org/abstract.cfm?URI=oe-18-7-6994
18. A. Seidel, C. Reinhardt, T. Holmgaard, W. Cheng, T. Rosenzveig, K. Leosson, S. Bozhevolnyi, B. Chichkov, IEEE Photonics J. **2**(4), 652 (2010). doi:10.1109/JPHOT.2010.2056490. http://ieeexplore.ieee.org/xpls/abs_all.jsp?arnumber=5512549&tag=1
19. C. Reinhardt, A. Seidel, A. Evlyukhin, W. Cheng, R. Kiyan, B. Chichkov, Appl. Phys. A: Mater. Sci. Process. **100**(2), 347 (2010). doi:10.1007/s00339-010-5872-0. http://link.springer.com/article/10.1007%2Fs00339-010-5872-0?LI=true#page-1
20. M. Emons, K. Obata, T. Binhammer, A. Ovsianikov, B.N. Chichkov, U. Morgner, Opt. Mater. Express **2**(7), 942 (2012). doi:10.1364/OME.2.000942. http://www.opticsinfobase.org/ome/abstract.cfm?URI=ome-2-7-942
21. V. Paz, M. Emons, K. Obata, A. Ovsianikov, S. Peterhänsel, K. Frenner, C. Reinhardt, B. Chichkov, U. Morgner, W. Osten, J. Laser Appl. **24**(4), 042004 (2012). doi:10.2351/1.4712151. http://jla.aip.org/resource/1/jlapen/v24/i4/p042004_s1
22. A. Seidel, C. Ohrt, S. Passinger, C. Reinhardt, R. Kiyan, B.N. Chichkov, J. Opt. Soc. Am. B **26**(4), 810 (2009). doi:10.1364/JOSAB.26.000810. http://josab.osa.org/abstract.cfm?URI=josab-26-4-810
23. A. Killi, A. Steinmann, G. Palmer, U. Morgner, H. Bartelt, J. Kobelke, Opt. Lett. **31**(1), 125 (2006). doi:10.1364/OL.31.000125. http://ol.osa.org/abstract.cfm?URI=ol-31-1-125
24. A. Steinmann, A. Killi, G. Palmer, T. Binhammer, U. Morgner, Opt. Express **14**(22), 10627 (2006). doi:10.1364/OE.14.010627. http://www.opticsexpress.org/abstract.cfm?URI=oe-14-22-10627
25. A. Steinmann, G. Palmer, M. Emons, M. Siegel, U. Morgner, Laser Phys. **18**, 527 (2008). doi:10.1134/S1054660X08050010. http://dx.doi.org/10.1134/S1054660X08050010
26. M. Emons, A. Steinmann, T. Binhammer, G. Palmer, M. Schultze, U. Morgner, Opt. Express **18**(2), 1191 (2010). doi:10.1364/OE.18.001191. http://www.opticsexpress.org/abstract.cfm?URI=oe-18-2-1191
27. C. Hönninger, I. Johannsen, M. Moser, G. Zhang, A. Giesen, U. Keller, Appl. Phys. B **65**, 423 (1997). doi:10.1007/s003400050291
28. P. Raybaut, F. Balembois, F. Druon, P. Georges, IEEE J. Quantum Electron. **41**(3), 415 (2005). doi:10.1109/JQE.2004.841930. http://ieeexplore.ieee.org/xpls/abs_all.jsp?arnumber=1397888
29. N.B. Chichkov, U. Bünting, D. Wandt, U. Morgner, J. Neumann, D. Kracht, Opt. Express **17**(26), 24075 (2009). doi:10.1364/OE.17.024075. http://www.opticsexpress.org/abstract.cfm?URI=oe-17-26-24075
30. C.P.J. Barty, G. Korn, F. Raksi, C. Rose-Petruck, J. Squier, A.C. Tien, K.R. Wilson, V.V. Yakovlev, K. Yamakawa, Opt. Lett. **21**(3), 219 (1996). doi:10.1364/OL.21.000219. http://ol.osa.org/abstract.cfm?URI=ol-21-3-219
31. C.W. Hillegas, J.X. Tull, D. Goswami, D. Strickland, W.S. Warren, Opt. Lett. **19**(10), 737 (1994). doi:10.1364/OL.19.000737. http://ol.osa.org/abstract.cfm?URI=ol-19-10-737
32. A. Monmayrant, B. Chatel, Rev. Sci. Instrum. **75**(8), 2668 (2004). doi:10.1063/1.1771492. http://ieeexplore.ieee.org/xpl/freeabs_all.jsp?arnumber=5001085&abstractAccess=no&userType=inst

33. A. Buettner, U. Buenting, D. Wandt, J. Neumann, D. Kracht, Opt. Express **18**(21), 21973 (2010). doi:10.1364/OE.18.021973. http://www.opticsexpress.org/abstract.cfm?URI=oe-18-21-21973
34. S. Kane, J. Squier, J. Opt. Soc. Am. B **14**(3), 661 (1997). doi:10.1364/JOSAB.14.000661. http://josab.osa.org/abstract.cfm?URI=josab-14-3-661
35. H. Sayinc, U. Buenting, D. Wandt, J. Neumann, D. Kracht, Opt. Express **17**(17), 15068 (2009). doi:10.1364/OE.17.015068. http://www.opticsexpress.org/abstract.cfm?URI=oe-17-17-15068

Chapter 2
High Power Femtosecond Diode Lasers

Thorsten Ulm, Florian Harth and Johannes L'huillier

Abstract We present a concept for a femtosecond laser based on a passively mode-locked semiconductor diode laser and a tapered amplifier. The absorption in the monolithically integrated quantum well absorber is controlled by a reverse voltage bias. Excellent mode-locking stability was observed without any rf modulation applied to gain current or absorber voltage. To avoid fast gain saturation and strong nonlinear pulse distortions within the tapered amplifier the technique of chirped pulse amplification is applied. In contrast to common chirped pulse amplification setups the oscillator emits pre-chirped pulses and a stretcher stage can be omitted. A pulse duration of 267 fs was achieved after compression in the colliding pulse mode-locking regime. For the first time we adapted this technique to the generation of tailored chirped pulses and investigated the influence of the collision point in an asymmetric two-section oscillator. This diode laser system is suited as a ultrafast pulse source for high power bulk or fiber amplifiers and paves a road to highly integrated laser systems.

List of Abbreviations and Symbols

ACF	Autocorrelation function
ASE	Amplified spontaneous emission
COD	Catastrophic optical damage
CPA	Chirped pulse amplification
CPML	Colliding pulse mode-locking
DBR	Distributed bragg reflector

T. Ulm (✉) · F. Harth · J. L'huillier
Photonik-Zentrum Kaiserslautern e.V., Kohlenhofstrasse 10, 67663 Kaiserslautern, Germany
e-mail: thorsten.ulm@pzkl.de

F. Harth
e-mail: florian.harth@pzkl.de

J. L'huillier
e-mail: johannes.lhuillier@pzkl.de

S. Nolte et al. (eds.), *Ultrashort Pulse Laser Technology*,
Springer Series in Optical Sciences 195, DOI 10.1007/978-3-319-17659-8_2

DFB	Distributed feedback
GD	Group-delay
GDD	Group-delay dispersion
MOPA	Master-oscillator power-amplifier
QW	Quantum well
SCPML	Self-colliding pulse mode-locking
SPM	Self-phase modulation
TA	Tapered amplifier
$A_M(\tau)$	Measured autocorrelation function
$A_S(\tau)$	Autocorrelation function of $I_S(t)$
E_{pulse}	Pulse energy
$f(t)$	Pulse shape function
F^*	Pulse shape factor
$I_S(t)$	Temporal intensity of a bandwidth-limited pulse with the spectrum S
λ	Wavelength
$\Delta\lambda$	Spectral bandwidth
M^2	Beam-quality factor
n	Refractive index
ν_{rep}	Repetition rate
ω	Angular frequency
ω_{inst}	Instantaneous angular frequency
ω_0	Carrier frequency
P	Power
$P^{ase}_{\max}$	Maximum ASE power at zero input
P_{av}	Average power of a pulse train
P_{in}	Optical input power
P_{out}	Optical output power
P_{peak}	Pulse peak power
P_{sat}	Saturation power of a tapered amplifier
$P^{sig}_{\max}$	Maximum signal output power at gain saturation
$\varphi(\omega)$	Spectral phase
$\frac{\partial\varphi}{\partial\omega}$	Group-delay (GD)
$\frac{\partial^2\varphi}{\partial\omega^2}$	Group-delay dispersion (GDD)
R	Reflectivity
Δs	Optical path length
$S(\omega)$	Optical pulse spectrum
Δt	Delay
τ	Temporal delay
τ_p	Pulse duration (FWHM)
U_{abs}	DC absorber voltage bias

2.1 Introduction

Among the various types of lasers operating in scientific or industrial applications diode lasers are outstanding by the fact, that these devices allow the direct transformation of electrical energy into coherent radiation. The use of diode lasers as pump sources opened up a variety of applications for mode-locked solid-state lasers [1–4] in science [5–8] and production [9–11]. Femtosecond lasers using gain materials like Cr:LiSAF, Yb:YAG, Yb:KYW [2] or Nd:YLF (see Chap. 1) have been successfully demonstrated in the past and various concepts like bulk (Chaps. 1 and 3), disk (Chaps. 4 and 7) and fiber lasers (Chap. 5) have been studied intensively. Diode-pumped solid-state lasers are also important as pump source for the titanium-sapphire laser. Frequency-doubled Nd-doped solid-state-lasers have mostly replaced larger and less energy-conserving argon ion lasers and made the titanium-sapphire laser to a dominating device in the field of ultrashort laser pulses.

Most applications of femtosecond lasers—e.g. material processing [9–11] (see also Part 2), nonlinear optics [12] (see also Chap. 6), fluorescence imaging [13], medical applications [14] or THz generation (see Chap. 15)—are based on the concentration of light at a specific point in space and time. Therefore, a suitable beam source has to deliver short pulses with a high pulse energy in a nearly diffraction limited beam.

Today, commercially available titanium-sapphire and diode pumped fiber lasers provide short pulses with high pulse energy and an excellent spatial beam quality. Beside the great importance of diode lasers for pumping solid-state lasers, the diode laser itself can be used as a source of femtosecond pulses. For two decades [15–24] femtosecond lasers based on electrically pumped diode lasers have been investigated. The main drawback of existing ultrafast diode lasers is their limited energy per pulse, which is far below the pulse energy necessary for industrial applications like material processing. Indeed, bulk solid-state or fiber amplifiers stages can be used to boost the pulse energy. Due to their fabrication process, diode lasers paves a road to highly integrated beam sources easily controllable via electronic interfaces [25].

This chapter focusses on our progress in edge-emitting femtosecond diode lasers achieved within the *FEMTO-DIODE* project (2004–2008). Further information on the recent progress achieved with femtosecond semiconductor disk lasers can be found in Chap. 7 in this book. The basic principles of semiconductor lasers are treated in various textbooks, for example [26–30].

If a diode laser does not contain bandwidth-limiting structures like a distributed Bragg reflector (DBR) or distributed feedback (DFB) grating [31], its amplification bandwidth can exceed 40 nm. A material of such high bandwidth should be well suited for the amplification of femtosecond pulses. However, the generation of mode-locked pulses in diode lasers is complicated by the fact that the shape of high-intensity ultrashort pulses is modified in time and space after travelling trough a semiconductor medium. These changes are caused by the strong coupling of the charge carrier density, the microscopic charge carrier dynamic [32] and the refractive index [33]. A changing refractive index causes a phase shift of the propagating pulse. As the pulse itself modifies gain, loss and index of the medium these effects are referred to as *self-amplitude* and self-phase modulation (SPM). The largest changes in the refractive index occurs, if the inversion of a gain material is completely depleted or if the absorption is saturated.

SPM effects can be damped or even prevented, if the population of the quantum well (QW) states is restored within the pulse duration. The gain recovery inside the QW levels proceeds on two different time scales (Fig. 2.1): The carrier injection from the doped regions into the active area is driven by the external electric field applied to the diode. Between two consecutive pulses no photons are present inside the gain section, and according to the classical rate-equation approach [29], the inversion build-up time is equal to the carrier life time. Common values given in literature are in the range of several 100 ps [35, 36]. This dynamic causes the *slow self-phase modulation*. The re-population of single QW states from nearby states is driven by carrier-carrier and carrier-phonon scattering and occurs on a timescale of a few 100 fs [37, 38]. This dynamics leads to the *fast self-phase modulation*. For pulse durations above 1 ps—as in Fig. 2.6a, b—the fast SPM can be neglected. For pulse durations below 1 ps the gain drastically decreases, because depleted states cannot be filled fast enough and only few carriers contribute to the amplification. Earlier investigations [34] revealed, that semiconductor lasers are inapplicable to amplify femtosecond pulses to high energies due to fast gain saturation effects.

In order to overcome this inherent physical limitation of diode laser amplifiers, we used the technique of chirped pulse amplification (CPA). CPA was first invented for high power solid-state amplifiers, but has also been applied to fiber lasers or optical parametric amplifiers (OPA) [39–43]. In principle, CPA protects high power amplifiers from high peak powers by elongating the pulse and thus reducing its peak power. Long pulses can be amplified to high pulse energies without reaching the damage threshold of the amplifying crystal. The pulses are compressed to their

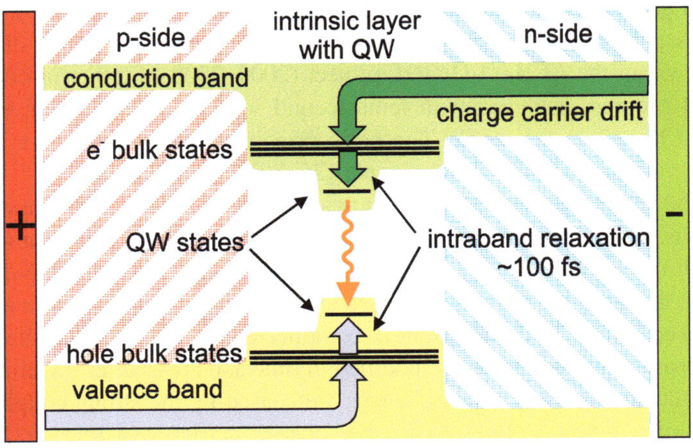

Fig. 2.1 Gain recovery by carrier injection and carrier relaxation in a QW diode laser [34]

original duration afterwards. Most stretcher and compressor setups exploit the fact, that changes of the spectral phase of the pulse are linked to changes of the temporal pulse profile by fourier transform. These types for setups are therefore called *Fourier domain pulse shaper* [44–46].

Within the second section we discuss a high power femtosecond laser system based on diode lasers and present the complete experimental setup. The third section deals with the oscillator. Here, passive mode-locking by a waveguide saturable absorbers in an external cavity. Different cavity designs and mode-locking schemes are discussed. The subject of the fourth section is the amplification of ultrashort pulses in tapered amplifiers. After amplification, a grating compressor forms femtosecond pulses, which is explained in the fifth section. The sixth section comprises a short discussion of the spatial beam quality of the compressed pulses.

2.2 Experimental Setup

For the generation of femtosecond pulses we choose a three stage setup (Fig. 2.2) consisting of a master oscillator for pulse generation, a high-power tapered amplifier and a pulse compressor. The combination of a master oscillator (MO) and a power amplifier (PA) is referred to as MOPA system [47]. A mayor advantage of a MOPA system compared to a broad area laser is the separation of the pulse formation and the generation of the high output power. This allows an excellent

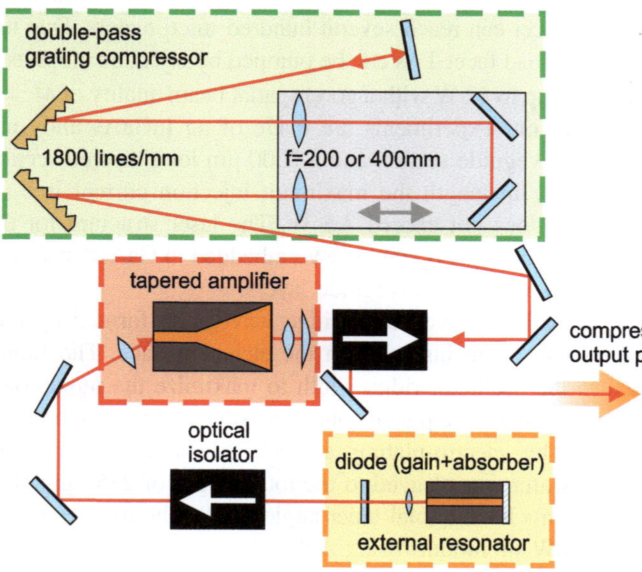

Fig. 2.2 Experimental setup containing picosecond oscillator (*yellow box*), tapered amplifier (*red box*) and grating compressor for femtosecond pulse compression (*green box*)

control of both processes and an optimization of oscillator and amplifier for their specific tasks. Please note that—compared to traditional CPA—we can omit a stretcher stage because our pulses leave the oscillator pre-chirped.

The total length of the oscillator chip is 1300 μm. The waveguide is divided into a gain and an absorber section, which are connected to separated contact wires. The chemical composition and epitaxial structure of both sections are identical. A reverse DC voltage bias applied to the absorber section controls the absorption and saturation characteristics [48, 49]. All diode lasers used in our setup were manufactured by the "*Ferdinand-Braun-Institut*" in Berlin, Germany. We investigated oscillators prototypes with absorber lengths of 80, 100 and 200 μm. To guarantee a high spectral bandwidth the laser contains neither a DBR nor a DFB grating. The layer structure was grown using metal-organic vapour-phase epitaxy (MOVPE) and soldered epi-side up on a c-mount. The active region consists of an InGaAs double quantum well (DQW) embedded in GaAsP spacer layers. The p- and n-doped 1800 nm thick $Al_{0.45}Ga_{0.55}As$ waveguide layers are sandwiched between 450 nm $Al_{0.70}$ $Ga_{0.30}As$ cladding layers.

Due to the super large optical cavity (SLOC) structure of the 3600 nm thick waveguide layers, the vertical far field angle is reduced to 20° (FWHM). The lasers have a ridge-waveguide for the lateral mode confinement with an effective index step of $\Delta n_{eff} = 5 \times 10^{-3}$. The width of the ridge is 3 μm. Aspherical lenses with 4.5 mm focal length are used to couple the waveguide to the external resonator . The facet anti-reflection coatings have a residual reflectivity of $R \leq 5 \times 10^{-4}$ [23].

For amplification a tapered amplifier (TA) was used. A TA consists of a short single mode input waveguide acting as a mode-filter and a tapered section for high power generation [50]. The width of the input facet is a few micrometers, while the width of the output facet can reach several hundred micrometers. Due to the large contact area and the broad facet TAs can be pumped by injection currents of several Ampere and can emit up to 12 W with a good spatial beam quality of $M^2 = 1.2$ [51].

The TAs used for our experiments are made of an InGaAs alloy and have a 750 μm long ridge waveguide section and a 2000 μm long tapered section. Due to their relatively short gain length the maximum injection current is 6 A and the maximum cw output does not exceed 2.5 W. The laser structure for the tapered amplifier is formed by an InGaAs single QW embedded in GaAsP spacer layers and 800 nm thick AlGaAs n- and p-waveguide layers.

The index guiding is achieved by a ridge waveguide formed by reactive ion etching and depositing of an insulator on the etched surface. The ridge width is 3 μm and equal to the oscillator ridge width to maximize the input coupling efficiency. The metallization on the p-side contact was formed by evaporating a Ti–Pt–Au multilayer and by electro-plating a thick Au layer. After thinning and n-metallization the wafer was cleaved to the total length of 2750 μm. The devices used in the experiments have a total taper angle of 6°. The devices were mounted epi-side down on CuW submounts using AuSn. This subassembly was soldered on C-mounts using PbSn. The n-side was contacted by wire bonding.

Anti-reflective coatings on the facets prevent lasing. The residual reflectivity is $R < 10^{-4}$. To compensate for the astigmatism caused by the different diffraction angles along the slow (horizontal) and fast axis (vertical) a combination of a 4.5 mm collimator lens and a cylindrical lens of 80 mm focal length is used.

An optical isolator was inserted between oscillator and TA. Without this isolator amplified spontaneous emission ASE generated in the TA could reach the oscillator and lead to unstable mode-locking.

After leaving the amplifier the pulses pass the compressor. For our work we used a grating compressor, because diffraction gratings provide a large angular dispersion and generated a large *group-delay dispersion* (GDD) along a relatively short beam path. Additionally the GDD can be easily adapted by changing the grating distance. A detailed explanation of grating stretchers and compressors can be found in [39, 52–54]. We used gratings with 1800 grooves/mm blazed for a wavelength of 1 μm. The gratings are optimized for an incident angle near (±10°) the Littrow angle. The diffraction efficiency is 94 % for the first diffraction order. The compressor contains a 1:1 telescope assembled of two 200 or 400 mm plan-convex lenses mounted on a translation stage together with two folding mirrors. Depending on the position of this stage the GDD of the compressor has a positive or negative sign. Therefore, the sign and amount of the linear chirp can be obtained from the compressor geometry. An end-mirror sends the pulse backwards through the compressor on the same path to eliminate the spatial chirp introduced by the gratings.

A second optical isolator protects the TA output facet from reflections and is used to separate the counter propagating compressed pulses from the amplified pulses. The output beam is guided to the beam diagnostics afterwards. We used an intensity autocorrelator, a double-grating spectrometer with 0.02 nm resolution and a 26 GHz rf spectrum analyser for pulse characterization.

In the following sections the individual components of the system are described and characterized in detail.

2.3 Mode-Locked Oscillators

The mode-locked oscillator is based on a single-stripe laser. The laser radiation is confined within a single-mode waveguide typically several micrometers in diameter [26]. Single-stripe lasers can deliver optical powers up to one watt with a good spatial beam quality [55]. The maximum output power is limited by the area of the optical facets. Electronic surface states lead to absorption and to a significant thermal load at the facet, which can cause a degradation or even the destruction of the facet. These phenomena are often referred to as *catastrophic optical damage* (COD) [56].

For pulse generation the saturable absorber concept has drawn a lot of interest since the early 1980s [15] and has been studied intensively [57–60]. The main advantage of this approach is, that the absorber can be integrated on the chip [18, 61–65]. The main challenge is to control the saturation fluence of the absorber. In mode-locked solid state lasers this is easily achieved by adapting the spot size on the absorber. However, this technique is not feasible for absorbers within a waveguide, since their mode diameter is given by the fabrication process and cannot be adapted during operation. A simple and sophisticated solution of this problem is to apply a reverse voltage bias to a quantum well absorber, as explained in Sect. 2.3.1.

We investigated two different cavity designs, that lead to two different types of mode-locking: The first oscillator type was anti-reflection coated at the gain side and left uncoated at the absorber side. The resonator is formed by an external mirror and the uncoated facet. The pulse overlaps with itself after being reflected at the absorber facet. This mode-locking technique is therefore called Self-colliding Pulse Mode-locking (SCPML). The second oscillator type was anti-reflection coated on both facets and placed in a linear resonator formed by two external mirrors. Since the absorber is placed near the center point of the cavity, two counter-propagating pulses are formed and collide at the absorber. This type of operation is called Colliding Pulse Mode-locking (CPML). Both mode-locking setups are discussed in Sect. 2.3.3, while the details of the saturable absorber are described in Sect. 2.3.1 and the fundamentals of the pulse evolution in Sect. 2.3.2.

2.3.1 Waveguide Saturable Absorbers

A saturable absorber can be easily integrated in a diode laser by using separate contact pads for gain and absorber sections or simply by leaving the absorber section unpumped. The modulation depth of an absorber has to be adapted to the laser system, which is usually done by changing the spot size on the absorber. However, if the absorber is integrated in a waveguide the mode diameter is fixed. In this case the absorption properties can be changed by applying a reverse bias voltage. The main effect of a reverse voltage is to shorten the carrier lifetime within the QW (see Fig. 2.3a) [48, 66]. Electrons and holes are created by photon absorption. The density of electron-hole-pairs decays due to recombination (1), thermal excitation (2) over the QW barrier, or tunneling the barrier. A reverse voltage reduces the rate of recombinations since electrons and holes are drawn to different sides of the QW. It increases the rate of thermal emission and tunneling, because the electric field reduces the width and height of the barrier. As a result the carrier lifetime within the QW and the absorption recovery time is shortened.

This can be easily observed by comparing the absorption spectrum of a 100 μm long absorber for different reverse voltage biases shown in Fig. 2.3b [49]. For voltages below 2 V electrons and holes form a bounded pair (*exciton*). If p- and n-side of the absorber are shortcut (0 V), the curve shows a clear and sharp exciton resonance, which is first broadened and finally disappears with increasing voltage. This behaviour is caused by the reduced carrier lifetime in the QW and the break-up

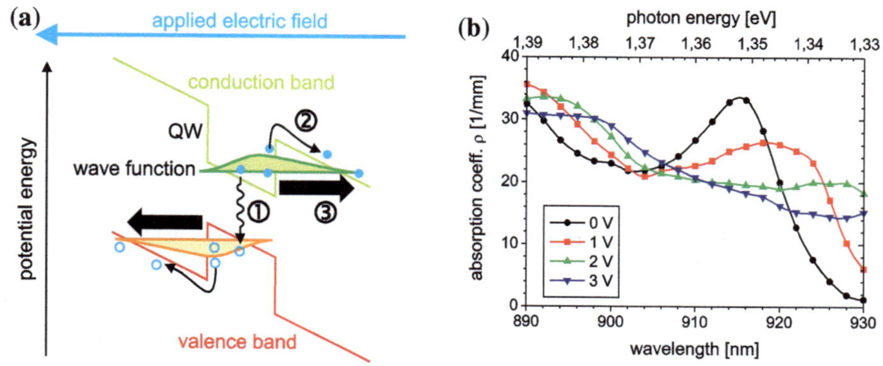

Fig. 2.3 **a** Sweep-out of charge carriers off a quantum well by *1* recombination, *2* thermal emission and *3* tunneling. **b** Absorption of a quantum well under DC reverse voltage bias [48, 66]. The diagram shows the exciton resonance and its disappearing due to high reverse bias voltages [49]

of the exciton binding [49]. From the experimental data we found that stable mode-locking is achieved for voltages above 3 V in 80 μm long absorbers and above 1 V in 200 μm long absorbers.

2.3.2 *Pre-chirped Pulses*

To generate a femtosecond pulse typically a careful dispersion compensation has to be applied when the laser cavity is designed. Without dispersion compensation (e.g. by a prism compressor) a femtosecond pulse circulating inside the cavity would be more and more stretched. Finally spectral components at the low-intensity head and tail of the pulse will be clipped by the saturable absorption or gain saturation. As the consequence, the pulse degenerates to a picosecond pulse with low spectral bandwidth.

In contrast to solid-state laser materials (like Nd or Yb doped crystals) semiconductor lasers provide a much higher gain per length and are operated at output couplings of 80 % or more. This drastically reduces the photon lifetime inside the cavity and also reduces spectral narrowing caused by cavity filtering effects. Furthermore, the spectral bandwidth of a pulse can be strongly increased by SPM. This allows the creation of broadband pulses without dispersion compensation inside the cavity. The pulses have typically durations of several picoseconds and a strong wavelength chirp [17, 67].

A qualitative explanation of this phenomenon is given in Fig. 2.4. The figure shows a two-section diode laser containing a gain and an absorber section as used in our experiments. The waveguide does not contain any dispersion compensation. The delayed re-filling of depleted QW states by carrier scattering and carrier injection leads to a smaller gain for shorter pulses, while the absorption decreases

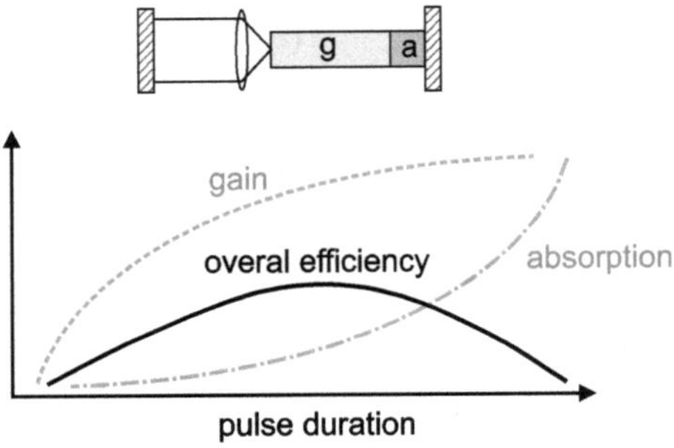

Fig. 2.4 Principle variation of the overall efficiency of a single stripe laser containing a gain and an absorber section as a function of the pulse width

for shorter pulses due to saturation. This results in a maximum net gain for pulses with several picoseconds duration.

2.3.3 Cavity Designs

The oscillators used for the two mode-locking techniques are two-section devices with a gain and an absorber section. Both oscillator types have an identical epitaxial composition but different facet coatings. Details of the semiconductor components are summarized in Sect. 2.2. Here, we describe the details of the cavity designs.

2.3.3.1 Self-colliding Pulse Mode-Locking (SCPML)

The cavity design used for SCPML is depicted in Figs. 2.2 and 2.4. The saturable absorber section is located at the end of the cavity. The pulse is reflected at the uncoated absorber facet and collides with itself (compare to the "gain-at-the-end" design used in many mode-locked solid-state lasers). The Fresnel reflectivity of the uncoated facet is 32 %, due to the refractive indices of GaAs ($n = 3.6$) and air ($n = 1$).

As mentioned in Sect. 2.1, the inversion is restored by carrier injection into the active zone within several 100 ps, after a pulse has left the gain section. To assure a complete gain recovery between consecutive pulses a pulse repetition time of 250 ps was chosen. This corresponds to a repetition rate of 4 GHz. Because the oscillator chip is only 1300 μm long, the cavity must be extended, in order to achieve a repetition rate of ≈4 GHz. The gain section facet is anti-reflection coated (see Sect. 2.2) in order to couple the light to the external resonator. The external resonator mirror has a reflectivity of 30 % and was used as output coupling mirror. It should be mentioned that reflectivities of less than 20 % are sufficient to operate the laser at gain saturation. The spectral and temporal broadening of the pulse is therefore only weakly dependent on the output coupling.

2.3.3.2 Colliding Pulse Mode-Locking (CPML)

Most investigations of colliding pulse mode-locking (CPML) [68–70] were aimed at the generation of ultrashort bandwidth-limited pulses. In contrast to these previous investigations, we now demonstrate the benefit of CPML for the formation of strongly chirped picosecond pulses and the tailoring of the chirp for amplification and compression [23]. A sketch of the oscillator geometry is shown in Fig. 2.5a. We used single stripe lasers with asymmetric sectioning embedded in a linear resonator formed by two plane mirrors M_a and M_g at the gain and the absorber side and the collimating lenses (see Fig. 2.5a).

The great advantage of this cavity design is the possibility to shift the collision point with respect to the optical chip simply by moving the mirrors. This can be used to optimize the chirp. Depending on the location of pulse collision the strong saturation of gain or absorption has different impact on the pulse shaping within the oscillator. Usually the sections in CPML diode lasers are aligned that way, that the pulses collide within the absorber [71]. This situation is depicted in the upper part of Fig. 2.5a, where the absorber section is placed exactly at the cavity center C_0. Since the absorber length (80 μm) is much smaller than the spatial pulse length (≈500 μm) the absorber length can be neglected and its position is denoted by A. The collision occurs in the center C_0 of the optical path L between the mirrors M_a and M_g.

For a proper colliding pulse mode-locking the collision point (C) has to be at—or at least close to—the absorber section (A). If the pulses collide inside the absorber section the rapid generation of new charge carriers leads to a decrease of the refractive index. If the collision occurs inside the gain section, the depletion of the inversion results in an increasing refractive index. Neither amplitude nor phase changes arise from a pulse collision in air, i.e. outside the waveguide. The dynamics of the refractive index changes is closely connected to the frequency chirp of the

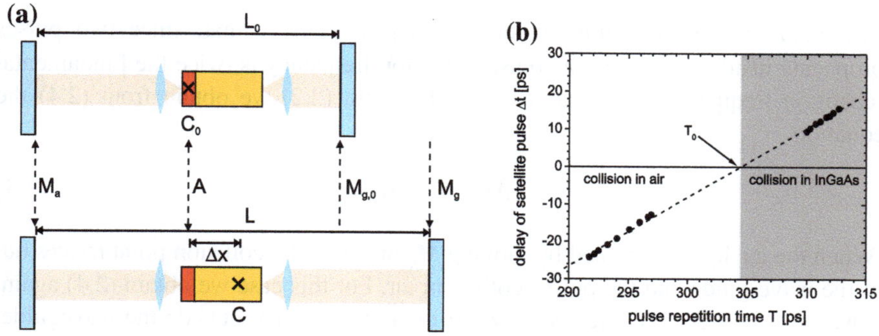

Fig. 2.5 **a** The pulse collision point can be shifted by moving one of the mirrors of the external resonator. For the definition of labels see text. **b** Delay Δt between satellite pulse and main pulse for different positions of the cavity mirror next to the gain section. The shift of this mirror causes a shift of the collision point and changes the cavity round-trip time T

pulse [72]. We observed a significant influence of the collision point on the chirp of the pulse. As the absolute value of the repetition rate is not critical for our experiments, the cavity length does not need to be constant. Therefore we shifted the collision point by moving mirror M_g and kept the position of M_a fixed.

To quantify the location of pulse collision we choose the optical path length $\Delta s = n\Delta x$ between collision point and absorber (see Fig. 2.5a). When the absorber section is located at the collision point C_0, the relation

$$\overline{M_a A} = \frac{1}{2} L_0 \tag{2.1}$$

holds. If the right mirror is moved outwards (as depicted in the lower part of Fig. 2.5a), the collision point C shifts to the right to a distance of Δx from the absorber. The absorber is now bleached twice: once by the clockwise and once by the counter-clockwise propagation pulse. If the temporal separation of both pulses is longer than the recovery time ($\approx$10 ps) [48, 49, 66, 73], this causes a formation of satellite pulses that can be clearly observed in the autocorrelation trace. The delay Δt of satellite and main pulse is given by

$$2n\Delta x = 2\Delta s = c\Delta t, \tag{2.2}$$

with n denoting the refractive index of the waveguide ($n \approx 3.6$). The center point of the cavity with length L is now located at C and therefore the relation

$$\overline{M_a A} + n\Delta x = \frac{1}{2} L \tag{2.3}$$

holds. Solving this relation for Δx we obtain together with (2.1)

$$2n\Delta x = 2\Delta s = L - L_0 = c\left(\frac{1}{\nu_{rep}} - \frac{1}{\nu_{rep,0}}\right). \tag{2.4}$$

$\nu_{rep} = c/L$ denotes the pulse repetition frequency. Note that, since two pulses propagate inside the cavity, the pulse repetition frequency is twice the fundamental repetition frequency of the cavity. Together with (2.2) we obtain from (2.4) the equation

$$\Delta t = T - T_0. \tag{2.5}$$

When the cavity is shortened by moving M_g inwards, the collision point moves out of the wave-guide and the pulses collide in air. For this case we obtain (2.4) again, if we set $n = 1$. Δx becomes negative, if the pulses collide outside the waveguide.

With (2.4) the separation Δx of collision point and absorber section can be calculated from the repetition frequency ν_{rep}, if $\nu_{rep,0}$ is known and as long as we keep the mirror M_a fixed. For the repetition rate $\nu_{rep,0}$ the pulses meet at the absorber section and Δx becomes zero. If Δx is larger than the pulse length, its value can be determined directly from the autocorrelation trace by measuring the delay Δt of main and satellite pulse and using (2.2). $\nu_{rep,0}$ can therefore be obtained from Fig. 2.5b which shows the satellite pulse delay Δt for different repetition times T. At a repetition rate of $\nu_{rep,0} = 3.285$ GHz the pulses collide exactly within the absorber section.

Figure 2.5b shows the linear dependency of Δt and T. The inclination of the line is not equal to one, as predicted by (2.5), but has a value of 1.8. In our opinion this is caused by the fact that the absorber remains transparent up to 10 ps after a pulse has passed [48, 49, 66, 73]. Therefore a larger delay of the main and the satellite pulse occur. Further, a pulse pair with larger delay experiences more gain, because the larger delay allows more gain to be build up after the passage of the first pulse.

2.4 Amplification of Mode-Locked Pulses

The composition and the geometry of the tapered amplifiers (TA) used for our experiments are summarized in Sect. 2.2. In the following we discuss the impact of gain saturation effects on the pulse shape. We although present a phenomenological model for the amplifier performance.

2.4.1 Gain Saturation

To achieve a good signal-to-ASE ratio and a high output power laser amplifiers are usually operated at gain saturation. Gain saturation occurs, if the inversion in the gain medium is completely depleted and the population of the upper and lower laser level is equal. Diode amplifiers are high gain devices and thus already small input powers of several milliwatts are sufficient to operate an amplifier in the regime of gain saturation. Unfortunately, gain saturation causes nonlinear phase changes that affect the temporal and spatial pulse shape [35, 74, 75].

According to the Kramers-Kronig relations [76] changes of the imaginary part of the susceptibility (gain/absorption) are coupled to changes of the real part. The temporal change of the refractive index $n(t)$ leads to a momentary frequency [36, 77]

$$\omega_{inst} = \omega_0 \left(1 - \frac{z}{c} \frac{dn(t)}{dt} \right). \tag{2.6}$$

ω_0 is called the *mean* or *carrier frequency* of the pulse. The modulation of the temporal phase by index changes caused by the pulse itself is called self-phase modulation (SPM) [37, 72, 78, 79]. If the SPM is strong enough to create phase terms of quadratic or higher order, this leads (in general) to a pulse broadening in the time and spectral domain [52]. Therefore strong SPM effects are usually unwanted. However, we saw in Sect. 2.3.2 that we can take advantage of these effects to generate pre-stretched pulses.

If the SPM is strong enough to create cubic terms in the temporal phase, two waves with the same frequency are emitted at two different times. According to the relative phase of both waves, constructive or destructive interference may occur. As a result, a cubic temporal phase may generate minima within the pulse spectrum [75]. For pulse compression the generation of cubic phase terms is highly unwanted, because spectral gaps limit the spectral bandwidth and rises the minimum pulse duration after compression.

2.4.2 Amplifier Performance

Figure 2.6a shows the average output power of the 2750 µm long tapered amplifier pumped by an injection current of 4 A and thermoelectrically cooled to 20 °C. For comparison cw radiation and 4 ps long mode-locked pulses with a repetition rate of 4 GHz were injected into the TA. The center wavelength is 920 nm. It can be clearly seen in Fig. 2.6a that the average output power is lower for mode-locked than for cw radiation.

To describe the experimental data we use a phenomenological model [80], that takes into account the limitation of the output power and the suppression of the amplified spontaneous emission (ASE) by gain saturation at high input powers.

$$P_{out} = \underbrace{P^{ase}_{\max} \cdot \exp\left(-\frac{P_{in}}{P_{sat}}\right)}_{\text{ASE noise}} + \underbrace{P^{sig}_{\max} \cdot \left[1 - \exp\left(-\frac{P_{in}}{P_{sat}}\right)\right]}_{\text{signal}} \tag{2.7}$$

$P^{ase}_{\max}$ is the ASE power at zero input power and $P^{sig}_{\max}$ is the maximum output power. If the input power P_{in} equals the saturation input power P_{sat} the output power reaches $(1 - e^{-1}) = 63\,\%$ of its maximum value. Please note that (2.7) is only

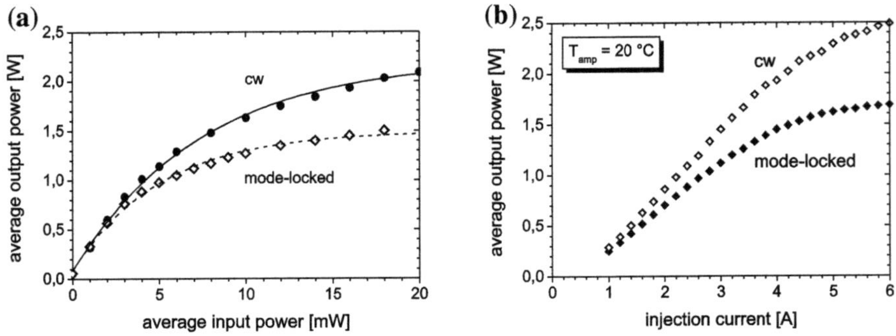

Fig. 2.6 Average output power of the 2750 µm long tapered amplifier **a** pumped by 4 A DC current, **b** seeded with 15 mW of average power. The duration of the mode-locked pulses is 4 ps

strictly valid for cw radiation. In the mode-locked regime most of the ASE is emitted between two consecutive pulses and the ASE noise is then independent of the input power. However (2.7) turns out to be valid for repetition rates in the GHz range, too, as can be seen in Fig. 2.6a. Fitting (2.7) to the data, we obtain that for mode-locked pulses $P^{sig}_{\max}$ is reduced by a factor of 33 % compared to cw radiation. The saturation input power decays from 7.4 mW (cw) to 5.7 mW (mode-locked).

The decay of the maximum output power for ultrashort pulses cannot be compensated by a higher injection current, as Fig. 2.6b reveals, and is caused by the delayed re-population of the QW states inside the active zone. As discussed in Sect. 2.1, the re-population of depleted QW states from nearby states is driven by carrier-carrier and carrier-phonon scattering and occurs at a time scale of several 100 fs [37]. However, the QW levels can only store a limited amount of charge carriers, i.e. they store a limited amount of pump energy. Re-filling of depleted QW states occurs by carrier injection from the doped regions into the active area. This involves a spatial carrier transport and happens on a time scale of several ten picoseconds [49]. Thus the inversion cannot be restored within a single mode-locked pulse and only few carriers contribute to pulse amplification. As can be clearly seen in Fig. 2.6a, b, compared to cw radiation the gain is significantly reduced for mode-locked pulses.

2.5 Pulse Compression

After amplification the pulses are compressed using a grating compressor (see Fig. 2.2). The temporal shape of the amplitude is changed by manipulating the phase of the pulse in the spectral domain [44, 52, 81, 82]. The beam path length between the grating determines the effect of the compressor stage on the spectral phase of the pulse. Usually the temporal elongation of a pulse is mainly caused by a non-vanishing second derivative of the spectral phase $\varphi(\omega)$. In such a case the pulse is said to have a *group-delay dispersion* (GDD). This means that the instantaneous frequency $\omega_{inst}(t)$ becomes a function of time [36, 52]. If $\omega_{inst}(t)$ rises with time, the pulse is *up-chirped*, if it falls the pulse is *down-chirped*. Pulses in the visible and NIR range are up-chirped after travelling through an optical medium (e.g. glass, sapphire,...), because "red" components are less delayed than "blue" components. A GDD unequal to zero causes a temporal elongation

$$\Delta t = \frac{2\pi c}{\lambda^2}\frac{\partial^2 \varphi}{\partial \omega^2}\Delta\lambda \tag{2.8}$$

of the pulse, where $\Delta\lambda$ denotes the spectral width.

In the following we present the important operation parameters used for optimization and discuss the shortest pulse compression results we achieved with both mode-locking techniques.

2.5.1 Self-colliding Pulse Mode-Locking (SCPML)

To achieve short pulse durations after compression the operation parameters have to be adapted to the pulse compressor. During the SCPML experiments the absorber voltage (U_{abs}) turned out to be the most important parameter for pulse formation, because its value determines the value of absorption and the recovery time of the absorber. Low reverse voltages lead to a long absorption recovery time, incomplete mode-locking and low-bandwidth pulses. Large reverse voltages cause fast sweep-out of charge carriers from the QWs [66], which creates a non-quadratic chirp. Thus, we optimized U_{abs} to minimize the pulse duration after compression. The shortest pulses were achieved for $U_{abs} \approx 4$ V.

Furthermore, we observed that shorter absorber sections deliver shorter pulses after compression. This is a result of two different effects: First, oscillators with short absorbers have higher gain and therefore a higher intracavity power, which leads to a stronger spectral bandwidth due to SPM. Second, short absorbers introduce less high-order phase distortions and the resulting "smooth" chirp can be compensated almost completely by the grating compressor.

Figure 2.7 shows the autocorrelation function (ACF), optical and rf spectra of our shortest pulses generated by SCPML. The oscillator consists of a 1220 μm long gain section and a 80 μm long absorber section. The gain injection current is 120 mA and the absorber current is 17 mA. The average output power of the oscillator is 19.8 mW, its repetition rate is 4 GHz and the pulse duration is 6.1 ps (Gaussian shape).

The tapered amplifier was pumped by a 4 A injection current. Oscillator and TA were kept to 20 °C by a peltier cooler. The TA is operated at gain saturation ($P_{in} = 16.6\,\text{mW}$) and delivers 1.56 W of average output power.

As can be seen from Fig. 2.7c the ACF of the compressed pulse is neither equal to a Gaussian nor a sech^2 shape. To retrieve the true pulse duration from the ACF an estimation of the pulse shape is made using the spectrum $S(\omega)$ of the compressed pulse. Note that the measured spectrum $S(\omega)$ represents the spectral intensity, not the electrical field. In a first approximation, we assumed a constant spectral phase and calculate the shortest possible pulse by a Fourier transform of $\sqrt{S(\omega)}$. This bandwidth-limited pulse $I_S(t)$ would be 255 fs long.

In a second step the intensity autocorrelation $A_S(\tau)$ is calculated from $I_S(t)$ as function of the autocorrelator delay time τ. The resulting FWHM of $A_S(\tau)$ is 372 fs. The relation of pulse duration and autocorrelation width is 0.69. This value is in between the values for Gaussian (0.71) and sech^2 pulses (0.65).

Indeed, the measured ACF of the compressed pulse is broader than the calculated one because of uncompensated high-order spectral phase terms. We made the assumption that the real and the calculated pulse differ in length but are similar in shape. In a third step the calculated ACF $A_S(\tau)$ is fitted to the measured ACF $A_M(\tau)$ by stretching it along the τ-axis. The fitted ACF is in good agreement with the measured ACF near the center of the pulse. From the fitted ACF a pulse duration (FWHM) of 365 fs can be obtained.

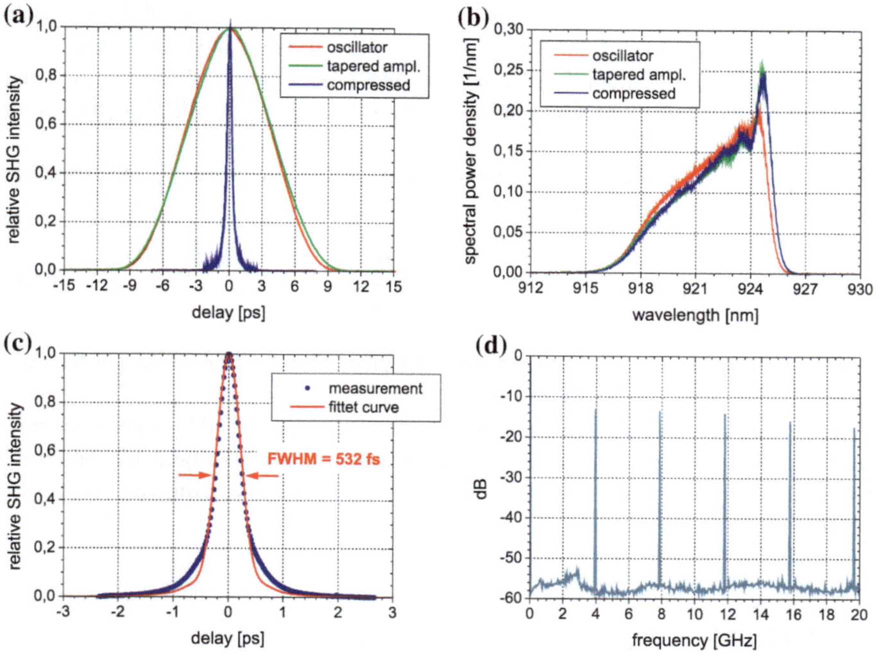

Fig. 2.7 **a** Autocorrelation functions and **b** spectra of the pulses produced by the oscillator, tapered amplifier and grating compressor. **c** Measured and fitted autocorrelation function of the compressed pulse. **d** Rf spectrum of the pulse train after pulse compression. For further details see text

The peak power after compression depends on the pulse shape, which is described by the *pulse shape factor* F^*. The *pulse shape function* $f(t)$ is then given by the normalized temporal intensity $I_S(t)$ or instantaneous optical power $P(t)$

$$f(t) = \frac{I_S(t)}{I_{peak}} = \frac{P(t)}{P_{peak}}. \tag{2.9}$$

Using $f(t)$ we obtain the pulse energy

$$E_{pulse} = \int_{\text{one pulse}} P(t)dt = P_{peak} \int_{\text{one pulse}} f(t)dt \stackrel{!}{=} \frac{P_{peak} \cdot \tau_p}{F^*} \tag{2.10}$$

with the pulse shape factor

$$F^* = \left[\frac{1}{\tau_p} \int_{\text{one pulse}} f(t)dt \right]^{-1}. \tag{2.11}$$

From $I_S(t)$ we obtained a value of $F^* = 0.8854$. The same F^* is valid for a sech^2-pulse as predicted by the theory of passive mode-locking [83]. However, please note that the exact temporal pulse shape cannot be identified based on a certain value of F^*. For this more advanced techniques like FROG or SPIDER [84, 85] are required.

The pulse peak power can be obtained from the average power P_{av}, the repetition rate ν_{rep} and the pulse duration τ_p by

$$P_{peak} = F^* \frac{P_{av}}{\nu_{rep}\,\tau_p}. \tag{2.12}$$

After the compression an average power of 851 mW and a peak power of 516 W is reached. The pulse energy is 213 pJ.

From the compressor geometry we obtained that the MOPA system generates up-chirped pulses [80]. Therefore the compression could also be achieved simply by a pair of diffraction gratings. This allows a much more compact compressor set-up.

2.5.2 Colliding Pulse Mode-Locking (CPML)

When the laser is operated in the CPML regime we changed Δs between −1200 and +800 μm. Positive values denote a collision within the chip ($n = 3.6$), negative values a collision in air ($n = 1$). The reflectivities of the cavity mirrors are $R_a = 30\,\%$ and $R_g = 99.3\,\%$, the gain injection current is 100 mA and the absorber voltage bias is changed from 2 to 6 V. The repetition rate changes from 3.16 to 3.42 GHz depending on the position of mirror M_g. The current applied to the tapered amplifier is 4 A and the amplifier is operated at gain saturation. For each set of parameters the pulse compressor is optimized to the shortest pulse duration after compression.

The minimum pulse width is reached for a collision in air, approximately 400 μm away from the absorber. If the collision point is located near the absorber but inside the gain section, both gain and absorption saturation influences the temporal phase of the pulse. This leads to a rather complicated, non-quadratic spectral phase, which cannot be compensated by the grating compressor. If the pulses collide in air near the absorber, the absorber gives the pulse a strong up-chirp. Since the pulses have little overlap in the gain section the phase changes are mainly influenced by absorption saturation. This leads to a nearly quadratic spectral phase and a nearly linear frequency chirp, which can be efficiently removed in the compressor.

Figure 2.8 reveals that the value of the absorber voltage is much less important than the choice of the right collision point to generate short pulses. There is not much difference between the minimal pulse duration achieved with 4 and 6 V. We thus choose a reverse voltage of 4 V for further experiments to limit the absorber

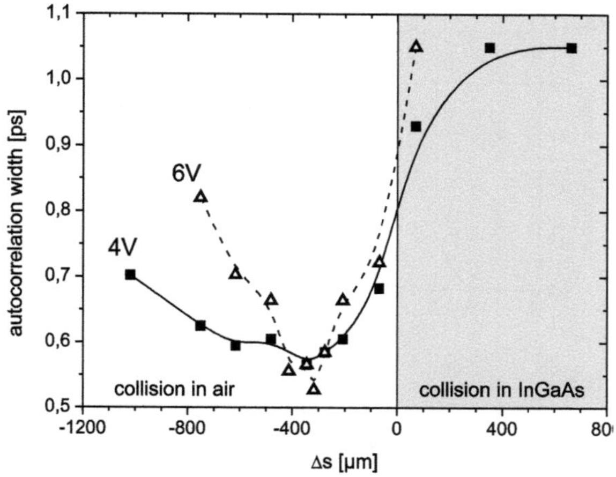

Fig. 2.8 Autocorrelation width (FWHM) of the compressed pulse in dependence of the reverse bias voltage and the optical path length Δs between collision point and absorber section. Spline curves are added as a guide to the eye

leakage current. In the next step we optimized the reflectivity of the cavity mirrors R_g and R_a in order to further decrease the duration of the compressed pulse. The values $R_g = R_a = 40\ \%$ yield the shortest pulses after compression.

Figure 2.9 shows the autocorrelation functions (ACF), the optical spectra and the rf spectra of the shortest pulse achieved in our CPML experiments. The oscillator gain current was 120 mA, the reverse absorber voltage was 4 V. The oscillator emits a pulse train with a repetition rate of 3.326 GHz and an average output power of 20 mW. Assuming a Gaussian shaped pulse a pulse duration of 6.2 ps can be obtained from the FWHM of the ACF. The peak power and pulse energy is 0.9 W and 6 pJ, respectively. The pulses collide in air, 563 µm behind the absorber section.

The tapered amplifier increases the average power to 1.45 W. The pulse is slightly elongated to 7.6 ps, but is still Gaussian shaped. The peak power reaches 54 W and the pulse energy rises to 436 pJ. Due to diffraction losses at the compressor gratings and transmission losses inside the optical isolator the average power after compression is 708 mW.

As can be seen from Fig. 2.9c the ACF of the compressed pulse is not of Gaussian or sech^2 shape. In order to retrieve the true pulse duration from the ACF we used the method described in Sect. 2.5.1. A bandwidth-limited pulse $I_S(t)$ would be 148 fs long. The relation of pulse duration and autocorrelation width is 0.6604, which is close to the value for sech^2 pulses (0.6482). The fitted ACF is in good agreement with the measured ACF near the center of the pulse. From the FWHM of the fitted ACF (408 fs) a pulse duration of 267 fs can be obtained. The pulse duration is 1.8 times above the Fourier limit (148 fs). This indicates, that the chirp is not completely removed by the compressor, because quadratic and higher-order dispersion of a grating compressor cannot be adjusted independently [39, 52]. Thus higher-order phase terms remain after compression leading to a non-fourier-limited pulse.

The measured ACF indicates a small satellite pulse at a delay of approximately 1 ps. This delay agrees well with the spectral maxima with a separation of 2.74 nm,

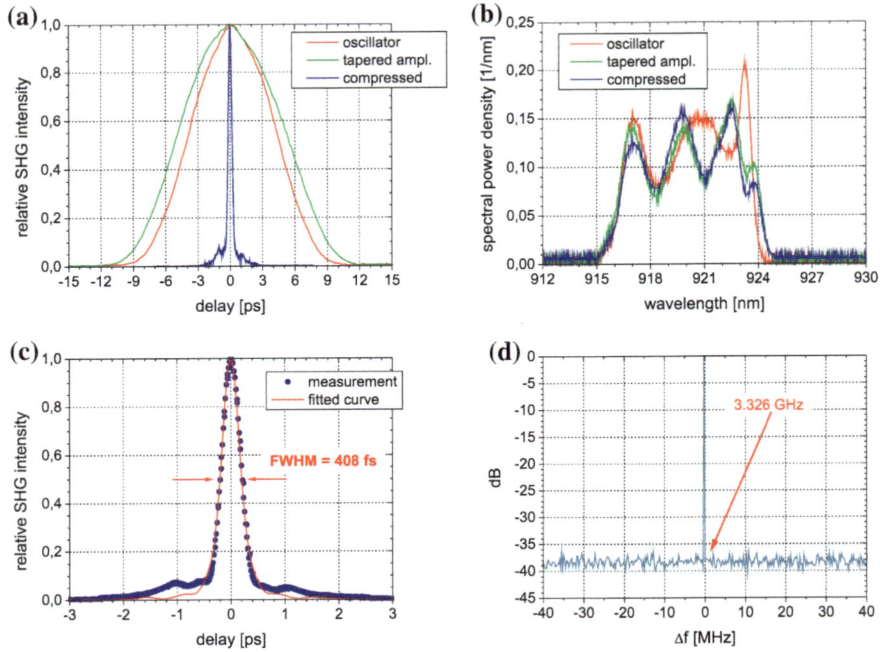

Fig. 2.9 **a** Autocorrelation functions and **b** spectra of the pulses produced by the oscillator, tapered amplifier and grating compressor. **c** Measured and fitted autocorrelation function of the compressed pulse. **d** Rf spectrum of the pulse train after pulse compression. For further details see text

that can be seen in the spectra of the amplified and compressed pulse. Hence the satellite pulse is created in the MOPA system and is not an artefact of pulse amplification or compression.

From $I_S(t)$ we obtained a value of $F^* = 0.83$. This shows the similarity of the compressed pulse to a sech^2-pulse ($F^* = 0.88$), which is predicted by the theory of passive mode-locking [83]. We calculated a peak power of 661 W for the compressed pulse. After pulse compression a pulse energy of 213 pJ was achieved. In Fig. 2.9d the RF spectrum of the compressed pulse train is shown. From the rf signal we conclude that the amplitude modulation is less than 36 dB.

2.6 Beam Quality

Usually TAs offer a better spatial beam quality along their fast axis, because of the strong optical confinement between the guiding layers. As the current injection cannot be made perfectly homogeneous, the beam quality along the slow axis is strongly affected by filamentation, a distortion of the beam by regions of larger and

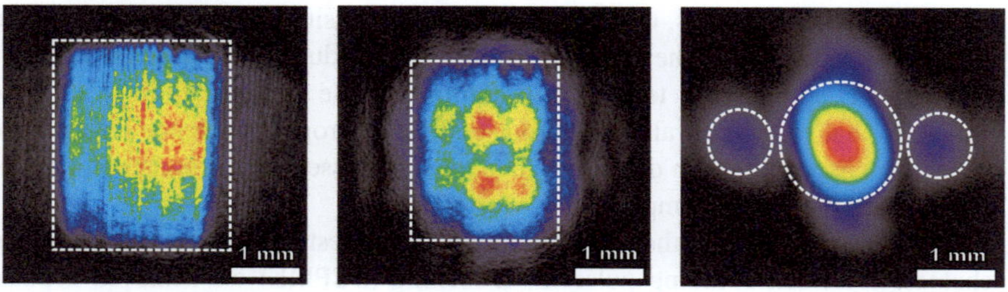

Fig. 2.10 Beam cross section after the tapered amplifier (*left*), the compressor (*center*) and at the focus behind a lens system of 500 mm focal length (*right*). The fast axis is aligned vertically, the slow axis horizontally. The shape of the beam cross section is indicated by a white *dashed line*

smaller refractive index. An example is shown in Fig. 2.10a. The mean M^2 values were measured to be

$$\langle M_X^2 \rangle = 1.33 \pm 0.15$$
$$\langle M_Y^2 \rangle = 1.71 \pm 0.12.$$

The beam profile was recorded by a moving 5 μm wide slit through the beam. The measurement was repeated with four lenses with focal lengths of 40–80 mm to eliminate systematic errors by lens imperfections.

Most important for laser applications is the beam shape in the focal plane, depicted in Fig. 2.10c. The focus cross section is the 2D fourier transform of the collimated beam. The rectangular shape of the facet causes side lobes around the fundamental mode at the center. To increase its quality we spatially filtered the beam using a slit of variable width located in the focal plane of a 200 mm lens. The slit was mounted perpendicular to the slow axis. The minimum M_Y^2 is 1.36 along the slow axis after the compressor. This value is slightly worse than for cw radiation ($M_Y^2 = 1.22$). 15 % of power is lost due to the filtering. 85 % of the radiation is emitted as a diffraction limited beam.

2.7 Conclusion

This chapter presented an overview on our recent progress on the way to a compact femtosecond diode laser source with high output power and good spatial beam quality. Passive mode-locking was enforced by a monolithically integrated saturable absorber section on the chip. The behaviour of the QW absorber was controlled by a reverse DC voltage bias.

A mayor challenge for amplifying ultrashort pulses in diode lasers is the delayed gain recovery within the quantum wells due to microscopic carrier dynamics. This leads to a strong gain saturation and weak output performance of the amplifier for pulse durations below 1 ps. Changes in the carrier density also causes spectral and temporal broadening of the pulse by self-phase modulation. To overcome this problem, we applied the technique of chirped pulse amplification. We chose oscillators that emit spectrally broad pulses with a strong wavelength chirp. The pulses were approx. 6 ps in duration and were compressed by a grating compressor after leaving the tapered amplifier.

In order to minimize the pulse duration we investigated two mode-locking schemes. With self-colliding pulse mode-locking (SCPML) we achieved a pulse duration of 365 fs and a peak power of 516 W at a repetition rate of 4 GHz. Using colliding pulse mode-locking (CPML) the pulse duration could be lowered to 276 fs by optimizing the location of pulse collision. The peak power was 661 W at a repetition rate of 3.3 GHz. The optimum collision point is located in air near the absorber facet.

We achieved a M^2 value of 1.33 along the fast axis of the tapered amplifier. The M^2 value along the slow axis can be improved to 1.36 by spatial filtering. 85 % of the radiation is emitted in a diffraction limited beam.

Acknowledgments The authors would like to thank G. Erbert, A. Klehr and H. Wenzel (Ferdinand-Braun-Institut, Berlin) for providing the diode laser components and detailed informations on the structure of these devices as well as for many helpful discussions. We also thank R. Wallenstein for his continuous support and interest in our work. This work was funded by the German Ministry of Education and Research (project number 13 N 8568).

References

1. E. Sorokin, I. Sorokina, E. Wintner, Appl. Phys. B **72**(1), 3 (2001)
2. C.T.A. Brown, M.A. Cataluna, A.A. Lagatsky, E.U. Rafailov, M.B. Agate, C.G. Leburn, W. Sibbett, New J. Phys. **6** (2004)
3. C. Hönninger, R. Paschotta, M. Graf, F. Morier-Genoud, G. Zhang, M. Moser, S. Biswal, J. Nees, A. Braun, G. Mourou, I. Johannsen, A. Giesen, W. Seeber, U. Keller, Appl. Phys. B **69**(1), 3 (1999)
4. J. Kleinbauer, R. Knappe, R. Wallenstein, Appl. Phys. B **80**(3), 315 (2005)
5. R.J. Levis, G.M. Menkir, H. Rabitz, Science **292**, 709 (2001)
6. T.W. Hänsch, Rev. Mod. Phys. **78**, 1297 (2006)
7. M. Aeschlimann, M. Bauer, D. Bayer, T. Brixner, F.J.G. de Abajo, W. Pfeiffe, M. Rohmer, C. Spindler, F. Steeb, Nature **446**, 301 (2007)
8. P. Eckle, A.N. Pfeiffer, C. Cirelli, A. Staudte, R. Dörner, H.G. Muller, M. Büttiker, U. Keller, Science **322**, 1525 (2008)
9. F. Dausinger, F. Lichtner, H. Lubatschowski, *Femtosecond Technology for Technical and Medical Applications* (Springer, 2004)
10. F. Korte, J. Serbin, J. Koch, A. Egbert, C. Fallnich, A. Ostendorf, B. Chichkov, Appl. Phys. A **77**(2), 229 (2003)
11. J. König, S. Nolte, A. Tünnermann, Opt. Express **13**(26), 10597 (2005)
12. F. Ruebel, P. Haag, J.A. L'huillier, Appl. Phys. Lett. **92**(1), 011122 (2008)
13. W.R. Zipfel, R.M. Williams, W.W. Webb, Nat. Biotechnol. **21**(11), 1369 (2003)
14. A. Vogel, J. Noack, G. Hüttman, G. Paltauf, Appl. Phys. B **81**(8), 1015 (2005)

15. J.P. van der Ziel, W.T. Tsang, R.A. Logan, R.M. Mikulyak, W.M. Augustyniak, Appl. Phys. Lett. **39**(7), 525 (1981)
16. P.J. Delfyett, L.T. Florez, N. Stoffel, T. Gmitter, N.C. Andreadakis, Y. Silberberg, J. P. Heritage, G.A. Alphonse, IEEE J. Quantum Electron. **28**(10), 2203 (1992)
17. P.J. Delfyett, A. Dienes, J.P. Heritage, M.Y. Hong, Y.H. Chang, Appl. Phys. B **58**(3), 183 (1994)
18. J. Yu, M. Schell, M. Schulze, D. Bimberg, Appl. Phys. Lett. **65**(19), 2395 (1994)
19. K.A. Williams, M.G. Thompson, I.H. White, New J. Phys. **6,** 179 (2004)
20. K. Kim, S. Lee, P.J. Delfyett, Opt. Express **13**(12), 4600 (2005)
21. T. Schlauch, M. Li, M. Hofmann, A. Klehr, G. Erbert, G. Tränkle, Electron. Lett. **44**(11) (2008)
22. P. Klopp, U. Griebner, M. Zorn, A. Klehr, A. Liero, M. Weyers, G. Erbert, Opt. Express **17** (13), 10820 (2009)
23. T. Ulm, A. Klehr, G. Erbert, F. Harth, J.A. L'huiller, Appl. Phys. B **99**(3), 409 (2010)
24. T. Ulm, F. Harth, A. Klehr, G. Erbert, J. L'huillier, Proc. SPIE, Semicond. Lasers Laser Dyn. V **8432**, 84320Y (2012)
25. F. Harth, T. Ulm, M. Lührmann, R. Knappe, A. Klehr, T. Hoffmann, G. Erbert, J.A. L'huillier, Opt. Exp. **20**(7), 7002 (2012)
26. R. Diehl, *High-Power Diode Lasers. Topics in Applied Physics* (Springer, Berlin, 2000)
27. P. Vasil'ev, *Ultrafast Diode Lasers* (Artech House Inc, Norwood, 1995)
28. L.A. Coldren, S.W. Corzine, *Diode Lasers and Photonic Integrated Circuits* (Wiley, New York, 1995)
29. O. Svelto, *Principles of Lasers*, 4th edn. (Plenum Press, New York, 1998)
30. T. Suhara, *Semiconductor Laser Fundamentals* (Marcel Dekker, New York, 2004)
31. A. Yariv, M. Nakamura, IEEE J. Quantum Electron. **13**(4), 233 (1977)
32. A. Othonos, J. Appl. Phys. **83**(4), 1789 (1998)
33. C.H. Henry, R.A. Logan, K.A. Bertness, J. Appl. Phys. **52**(7), 4457 (1981)
34. T. Ulm, H. Fuchs, J.A. L'huillier, A. Klehr, B. Sumpf, E. Gehrig, Opt. Commun. **281**(8), 2160 (2007)
35. A. Mecozzi, J. Mørk, J. Opt. Soc. Am. B **14**(4), 761 (1997)
36. G.P. Agrawal, N.A. Olsson, IEEE J. Quantum Electron. **25**(11), 2297 (1989)
37. E. Gehrig, O. Hess, A. Volland, G. Jennemann, I. Fischer, W. Elsäßer, J. Opt. Soc. Am. B **21**(9), 1638 (2004)
38. W.W. Chow, H.C. Schneider, W. Koch, C.H. Chang, L. Chrostowski, C.J. Chang-Hasnain, IEEE J. Quantum Electron. **38**(4), 402 (2002)
39. S. Backus, C.G. Durfee, M.M. Murnane, H.C. Kapteyn, Rev. Sci. Instrum. **69**(3), 1207 (1998)
40. G. Cerullo, S.D. Silvestri, Rev. Sci. Instrum. **74**(1), 1 (2003)
41. A. Galvanauskas, P.A. Krug, D. Harter, Opt. Lett. **21**(14), 1049 (1996)
42. A. Galvanauskas, A. Hariharan, D. Harter, M.A. Arbore, M.M. Fejer, Opt. Lett. **23**(3), 210 (1998)
43. R. Butkus, R. Danielius, A. Dubietis, A. Piskarskas, A. Stabinis, Appl. Phys. B **79**(6), 693 (2004)
44. A.M. Weiner, Prog. Quantum Electron. **19**(3), 161 (1995)
45. A.M. Weiner, Rev. Sci. Instrum. **71**(5), 1929 (2000)
46. G. Steinmeyer, Appl. Phys. A **79**(7), 1663 (2004)
47. A. Mar, R. Helkey, J. Bowers, D. Mehuys, D. Welch, I.E.E.E. Photon, Technol. Lett. **6**(9), 1067 (1994)
48. A.M. Fox, D.A.B. Miller, G. Livescu, J.E. Cunningham, W.Y. Jan, IEEE J. Quantum Electron. **27**(10), 2281 (1991)
49. D.A.B. Miller, D.S. Chemla, T.C. Dames, A. Gossard, W. Wiegmann, T.H. Wood, C.A. Burrus, Phys. Rev. B **32**(2), 1043 (1985)
50. J.N. Walpole, Opt. Quantum Electron. **28**, 623 (1996)
51. C. Fiebig, G. Blume, C. Kaspari, D. Feise, J. Fricke, M. Matalla, W. John, H. Wenzel, K. Paschke, G. Erbert, Electron. Lett. **44**(21), 1253 (2009)

52. J.C. Diels, W. Rudolph, *Ultrashort Laser Pulse Phenomena*, 2nd edn. (Academic Press (Elsevier), Amsterdam, 2006)
53. O.E. Martinez, J. Opt. Soc. Am. B **3**(7), 929 (1986)
54. O.E. Martinez, IEEE J. Quantum Electron. **24**(12), 2530 (1998)
55. A. Pietrzak, H. Wenzel, P. Crump, F. Bugge, J. Fricke, M. Spreemann, G. Erbert, G. Tränkle, **48**(5), 568 (2012)
56. M. Ziegler, M. Hempel, H.E. Larsen, J.W. Tomm, P.E. Andersen, S. Clausen, S.N. Elliott, T. Elsaesser, Appl. Phys. Lett. **97**
57. U. Keller, K.J. Weingarten, F.X. Kärtner, D. Kopf, B. Braun, I.D. Jung, R. Fluck, C. Hönninger, N. Matuschek, J.A. der Au, IEEE J. Sel. Top. Quantum Electron. **2**(3), 435 (1996)
58. R. Paschotta, U. Keller, Appl. Phys. B **73**(7), 653 (2001)
59. U. Keller, Nature **424**, 831 (2003)
60. F. Schättiger, D. Bauer, J. Demsar, T. Dekorsy, J. Kleinbauer, D.H. Sutter, J. Puustinen, M. Guina, Appl. Phys. B **106**(3), 605 (2012)
61. D.J. Derickson, R.J. Helkey, A. Mar, J.R. Karin, J.G. Wasserbauer, J.E. Bowers, IEEE J. Quantum Electron. **28**(10), 2186 (1992)
62. M.A. Cataluna, Y. Ding, D.I. Nikitichev, K.A. Fedorova, E.U. Rafailov, IEEE J. Sel. Top. Quantum Electron. **17**(5), 1302–2011 (2011)
63. R. Scollo, H.J. Lohe, F. Robin, D. Erni, E. Gini, H. Jäckel, IEEE J. Quantum Electron. **45**(4), 322 (2009)
64. G. Tandoi, C.N. Ironside, J.H. Marsh, A.C. Bryce, IEEE J. Quantum Electron. **48**(3), 318 (2012)
65. T. Xu, M. Rossetti, P. Bardella, I. Montrosset, IEEE J. Quantum Electron. **48**(9), 1193 (2012)
66. A.V. Uskov, J.R. Karin, R. Nagarajan, J.E. Bowers, IEEE J. Sel. Top. Quantum **1**(2), 552 (1995)
67. P.J. Delfyett, C.H. Lee, G.A. Alphonsse, J.C. Connolly, Appl. Phys. Lett. **57**(10), 971 (1990)
68. Y.K. Chen, M.C. Wu, T. Tanbun-Ek, R.A. Logan, M.A. Chin, Appl. Phys. Lett. **58**(12), 1253 (1991)
69. Y.K. Chen, M.C. Wu, IEEE J. Quantum Electron. **28**(10), 2176 (1992)
70. J.F. Martins-Filho, E.A. Avrutin, C.N. Ironside, J.S. Roberts, IEEE J. Sel. Top. Quantum **1**(2), 539 (1995)
71. M.C. Wu, Y.K. Chen, T. Tanbun-Ek, R.A. Logan, M.A. Chin, G. Raybon, Appl. Phys. Lett. **57**(8), 759 (1990)
72. R.S. Grant, W. Sibbett, Appl. Phys. Lett. **58**(11), 1119 (1991)
73. J.A. Cavaillès, D.A.B. Miller, J.E. Cunningham, P.L.K. Wa, A. Miller, Appl. Phys. Lett. **61** (4), 426 (1992)
74. K. Hall, G. Lenz, A. Darwish, E. Ippen, Opt. Commun. **111**, 589 (1994)
75. E. Gehrig, D. Woll, M.A. Tremont, A. Robertson, R. Wallenstein, O. Hess, J. Opt. Soc. Am. B **17**(8), 1452 (2000)
76. D.C. Hutchings, M. Sheik-Bahae, D.J. Hagan, E.W. van Stryland, Opt. Quantum Electron. **24** (1), 1 (1992)
77. N.A. Olsson, G.P. Agrawal, Appl. Phys. Lett. **55**(1), 13 (1989)
78. A. Dienes, L.W. Carr, J. Appl. Phys. **69**(3), 1766 (1991)
79. M.Y. Hong, Y.H. Chang, A. Dienes, J.P. Heritage, P.J. Delfyett, IEEE J. Quantum Electron. **30**(4), 1122 (1994)
80. T. Ulm, F. Harth, H. Fuchs, J.A. L'huillier, R. Wallenstein, Appl. Phys. B **92**(4), 481 (2008)
81. E.B. Treacy, IEEE J. Quantum Electron. **5**(9), 454 (1969)
82. O.E. Martinez, IEEE J. Quantum Electron. **23**(1), 59 (1987)
83. H.A. Haus, IEEE J. Quantum Electron. **6**(6), 1173 (2000)
84. R. Trebino, *Frequency-Resolved Optical Gating: The Measurement of Ultrashort Laser Pulses*, 1st edn. (Kluwer Academic Publishers, Norwell, 2000)
85. G. Steinmeyer, J. Opt. A: Pure Appl. Opt. **5**(1), R1 (2003)

Chapter 3
Femtosecond Mode-Locked Semiconductor Disk Lasers

Uwe Griebner, Peter Klopp, Martin Zorn and Markus Weyers

Abstract The generation of ultrashort pulses with passively mode-locked semiconductor disk lasers (SDLs) incorporating only an optically-pumped surface-emitting semiconductor gain element and a semiconductor saturable absorber mirror (SESAM) is presented. The optimum parameters for nearly Fourier-limited femtosecond pulses in single- or multiple-pulse regimes are investigated. On the basis of the experience gained a harmonically mode-locked SDL emitting sub-200 fs pulses at a very high repetition rate of 92 GHz and a fundamentally mode-locked SDL generating practically chirpfree pulses with durations close to 100 fs at a rate of 5 GHz are demonstrated in the 1-μm wavelength range. The latter set a record for shortest pulse durations achieved directly from any fundamentally or harmonically mode-locked semiconductor laser. Overall, the results are a further step of modelocked SDLs in becoming useful compact and low-cost ultrashort-pulse sources.

3.1 Introduction

Semiconductor disk lasers (SDLs), also called optically-pumped vertical-external-cavity surface-emitting lasers (VECSELs) or optically pumped semiconductor lasers (OPSLs), are a relatively new laser family combining high output power,

---

U. Griebner (✉) · P. Klopp
Max-Born-Institut für Nichtlineare Optik und Kurzzeitspektroskopie,
Max-Born-Straße 2 A, 12489 Berlin, Germany
e-mail: griebner@mbi-berlin.de

Present Address:
M. Zorn
JENOPTIK Diode Lab GmbH, Berlin, Germany

M. Weyers · M. Zorn
Ferdinand-Braun-Institut, Leibniz-Institut für Höchstfrequenztechnik,
Gustav-Kirchhoff-Str. 4, 12489 Berlin, Germany

S. Nolte et al. (eds.), *Ultrashort Pulse Laser Technology*,
Springer Series in Optical Sciences 195, DOI 10.1007/978-3-319-17659-8_3

wide wavelength coverage, and high beam quality. Another attractive feature of the SDL concept is the potential of ultrashort-pulse generation at high repetition rates, which is the focus of this chapter.

In general, semiconductor heterostructures enable compact and cheap lasers operating in continuous-wave (cw) or pulsed regimes. Semiconductor bandgap engineering gives large freedom concerning the emission wavelength, which can be chosen almost anywhere from ≈340 nm to ≈30 μm [1, 2]. Femtosecond mode-locked semiconductor disk lasers have the potential to replace rather complex, expensive laser systems and to establish ultrashort-pulse applications outside of scientific laboratories. Furthermore, the typically high pulse repetition rates of ≥1 GHz make SDLs attractive, e.g., for THz time-domain spectroscopy with asynchronous optical sampling (ASOPS) [3] or for frequency comb generation [4]. The high pulse repetition rates of mode-locked SDLs are also interesting for communication applications [5].

Compared to edge-emitting semiconductor lasers [6], in vertical-cavity surface-emitting lasers (VCSELs) [7] the laser cavity axis and the light emission are perpendicular to the plane of the epitaxial laser structure, allowing the emission of a circular fundamental transverse-mode beam. However, increasing the output beam diameter to larger than 10 μm, the laser output becomes multimode, and uniform current injection over such large areas is difficult with edge injection through transparent contact layers. A solution to realize carrier excitation uniformly across such a large area is optical pumping. Efficient and simple continuous-wave semiconductor diode lasers with multi-mode emission and high output power levels can be used as pump sources.

The laser concept of optically pumped vertical-external-cavity surface-emitting lasers (VECSELs) or semiconductor disk lasers (SDLs) was developed about one decade ago [8]. An SDL pumped by a multimode laser diode can be considered as a diode-pumped solid-state laser, where the gain medium consists of a semiconductor structure instead of a classical ion-doped crystal or glass. The laser geometry resembles the well-established solid-state disk laser configuration [9], where a thin solid-state gain medium, such as an Yb-doped crystal, with a high-reflecting dielectric backside coating is used as an active mirror and placed directly on a heat sink. In the simplest SDL design, an additional spherical mirror and a highly reflective mirror (usually a Bragg mirror consisting of semiconductor layers with different refractive indices) underneath the semiconductor gain medium (usually a multi-quantum-well structure) constitute the laser cavity. Using SDLs, cw output powers of several 10 W have been demonstrated with excellent beam quality [10, 11].

Similar to diode-pumped solid-state lasers, SDLs can be passively mode-locked by introducing a semiconductor saturable absorber mirror (SESAM) into the cavity [5]. However, there is a great difference between dielectric gain materials such as ion-doped crystals or glass and semiconductor gain media. The latter exhibit several orders of magnitude larger absorption and emission cross sections, resulting in low gain saturation fluences and short upper-state lifetimes. This limits the amount of stored energy in the gain material and the minimum achievable pulse repetition rates. Due to low saturation fluences, Q-switched mode-locking instabilities are strongly reduced [12], which is an important advantage for achieving very high pulse repetition rates. The full potential of semiconductor lasers for ultrashort-pulse generation has not yet been exploited. In principle, the gain bandwidth of semiconductor lasers allows fundamental or harmonic mode-locking with sub-100-fs pulse durations [13].

For a long time it was believed that the pulse durations now achieved by mode-locked SDLs could be obtained only from oscillators using dielectric or dye media and not from semiconductor lasers, since the strong carrier-density dependence of the complex refractive index and the carrier dynamics in semiconductors introduce a strong chirp. One of the first passively mode-locked SDLs, demonstrated in 2000 by Hoogland et al., generated 22-ps pulses at a repetition rate of 4 GHz [14]. Since that time, considerable progress has been made in terms of spectral coverage, output power, pulse duration, and repetition rate [5]. Experiments with SESAM-mode-locked SDLs showed that in some operation regimes the chirp contributions approximately compensate each other. In this case, one may obtain something like a "soliton-like" pulse [15]. Nevertheless, the minimum achievable pulse duration in this case had not been much shorter than about half a picosecond [16, 17]. From monolithic edge-emitting lasers, there were demonstrations of sub-500 fs pulses. Such a laser was reported to reach a repetition rate of >1 THz with a pulse width of 260 fs [18], but in this case one had to speak rather of a sine-shaped output instead of separate pulses. While the asymmetric beam profile is a general difficulty with edge-emitting semiconductor lasers, additional problems connected with pulse shortening are the large amounts of dispersion and nonlinearities introduced due to the long interaction length. As a result, strongly chirped pulses and increased timing jitter can occur. As an example, 395-fs pulses at 21 GHz with a time-bandwidth product of ≈ 1 were achieved from a monolithic two-section quantum dot laser which is about three times larger than the Fourier limit [19].

In 2008, we demonstrated a practically chirp-free 290-fs SDL [20] and Wilcox et al. showed 260-fs pulses from a similar laser [21]. After these milestones, we were able to further reduce the emitted pulse duration of mode-locked SDLs to 190 fs at 3 GHz pulse repetition rate [22] and then to 107 fs at 5 GHz [23]. Furthermore, we observed extraordinarily high pulse repetition rates from an SDL. In [24], Lorenser et al. reported a rate of 50 GHz, obtained with a pulse duration of ≈ 3 ps. We improved this value to ≈ 92 GHz with sub-200-fs pulses.

In this chapter, we present design criteria and experimental conditions for ultrashort-pulse generation from mode-locked SDLs with an emission wavelength around 1 μm. We describe different regimes of laser operation and present almost transform-limited pulses at very high repetition rates or with a shortest duration in the 100-fs range.

3.2 Semiconductor Elements

3.2.1 Design, Growth, and Processing

For the semiconductor elements of a sub-picosecond SDL, there are specific requirements which follow from the pulse shaping process and determine our design strategy. As we will show in the following paragraphs, pulse shaping in such SDLs cannot be described with the "soliton-like mode-locking" picture applicable to ultrashort-pulse lasers with dielectric gain media or to picosecond SDLs [25]. Instead, we found fast and suitably strong amplitude modulation, in particular, a fast saturable absorber, spectral matching of the elements, and minimized group delay dispersion (GDD) to be key issues. Furthermore, a larger gain bandwidth is expected to support shorter pulses, too. The SESAMs and the SDL gain structures studied were designed for laser operation with a center wavelength around 1030 nm. The design was assisted by simulation, mainly with respect to GDD.

All epitaxial layers were grown by metalorganic vapor phase epitaxy (MOVPE) in an Aixtron 200/4 reactor in 3 × 2″ configuration. The sources used were trimethylgallium, trimethylaluminum, trimethylindium, arsine, and phosphine. For the SDL gain elements, removal of the heat from the optically pumped region is crucial. GaAs has a relatively low thermal conductivity and thus the substrate hinders the heat dissipation. Transparent heat sinks like diamond on the surface of the gain chip are technologically challenging (bonding is difficult) and affect dispersion. Therefore, we used gain structures where the Bragg mirror was attached directly to a heat sink and the substrate was removed. This made it necessary to first grow the active periodic gain structure with the quantum wells (QWs) and then the Bragg mirror. The active part was grown at 650 °C and the distributed Bragg reflector (DBR mirror) at 700 °C. The wafer was cut into 2 × 2 mm^2 or 4 × 6 mm^2 pieces, which were mounted mirror-side down ("bottom up") on CuW heat sinks of the same size using AuSn solder. Afterwards the substrate was etched off with an aqueous solution of sulphuric acid and hydrogen peroxide. A ≈300-nm-thick $In_{0.48}Ga_{0.52}P$ layer served as an etch-stopper. Finally, an antireflective dielectric coating was applied and the soldered structure was mounted onto a large copper heat sink.

For the SESAM structure the DBR mirror was grown at 770 °C, while the InGaAs QW was grown at 510 °C. After coating the wafer backside with metal, the SESAM wafers were diced and the chips were soldered onto a CuW submount [26]. These submounts were attached to a Peltier cooling/heating element that allowed for a variation of the band edge via changing the temperature.

3.2.2 Semiconductor Saturable Absorber Mirror

Our SESAM structures for the 1 μm wavelength range consisted of a single InGaAs QW sandwiched between two GaAs layers grown on an AlAs/(Al)GaAs DBR mirror. The indium content in the QW defined the operation wavelength [27]. Usually a SiN_x (or comparable) antireflection coating was applied. This also provided a protection of the GaAs surface against oxidation. To achieve very short carrier recombination times in the QW and by this a fast over-all recovery of the absorber, two approaches were studied. First, the QW was surrounded by equally thick (approximately 60–70 nm) GaAs layers. As-clusters and point defects were created in the QW by implanting arsenic (As) ions. Subsequent annealing results in the formation of As clusters in the InGaAs QW, enhancing recombination [28, 29]. The second type of SESAM structure employed a "surface-near" QW without implantation to achieve short relaxation times via surface states. "Surface-near" means that the upper GaAs layer was only a thin cap, whose thickness was varied between 1 and 5 nm. Such a "surface-near" SESAM structure is shown in Fig. 3.1a. In general, the surface-near QWs showed a shorter relaxation time compared to the ion treated ones. For the SiN_x coating two different thicknesses were tested, resulting in a non-resonant and a resonant design. Best results with respect to short pulse durations were obtained using the non-resonant design.

The resonant responses of surface-near SESAMs were characterized by pump-probe and nonlinear reflectivity measurements. For this purpose, the emission of a mode-locked Yb-laser delivering pulses with a duration of ≈250 fs at 1030 nm (*Amplitude*, model: *MIKAN*) was focused on the sample. This SESAM had a 2-nm GaAs cap layer and was the absorber device operating in the SDL with the shortest pulse duration (paragraph 5).

Pump-probe traces are shown in Fig. 3.2a for two values of the pulse fluence. In general, with a pump and probe pulse duration of a few hundred fs at low to moderate fluences, a measured SESAM response (change of reflectivity versus

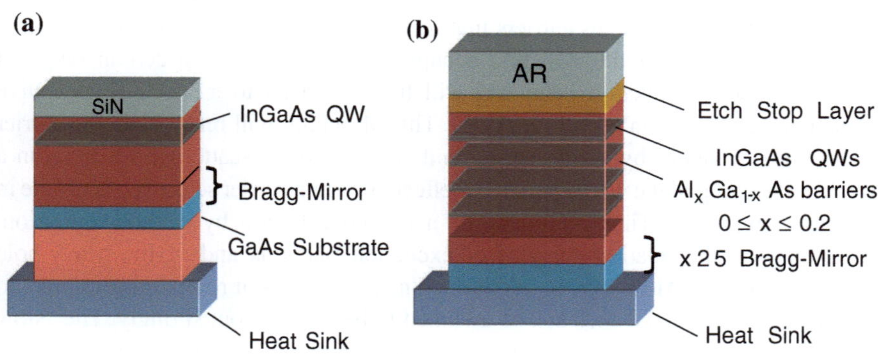

Fig. 3.1 Structure of the semiconductor elements: **a** "Surface-near" semiconductor saturable absorber mirror (SESAM). **b** Surface-emitting 4-QW gain structure

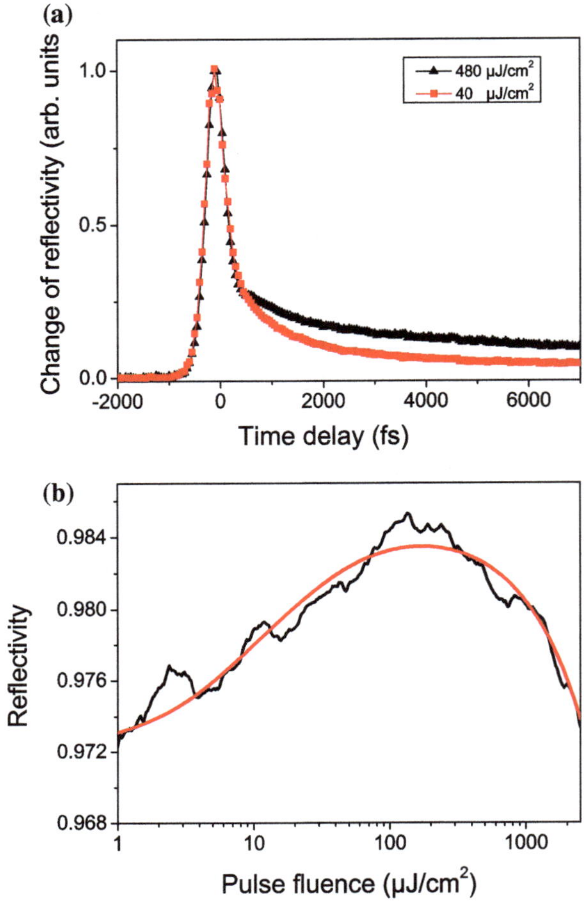

Fig. 3.2 Pump-probe response (**a**) and nonlinear reflectivity (**b**) of the surface-near QW semiconductor saturable absorber mirror. The *red curve* in (**b**) is a fit of the saturation behavior including two-photon absorption

pump-probe delay) consists of different parts. The "fast component" exhibits a decay time in the order of the pulse duration (here: <300 fs) and is related to spectral hole-burning (SHB). SHB means carriers that have been excited by a pump pulse with a limited spectral bandwidth will first occupy only states within a certain range of energies in the conduction band and will therefore lead to a spectrally selective bleaching of absorption (Pauli blocking). This bleaching will relax when the carrier distribution broadens by carrier-carrier and carrier-phonon scattering, resulting in a weaker bleaching of the whole band. The effect of carrier-carrier scattering is visible in the fast component. There may also be a fast contribution by exciton ionisation, depending on the relative energies of excitation phonons and of the heavy-hole exciton resonance. However, in our surface-near SESAMs at room temperature and above, the exciton resonance can be assumed to be smeared out strongly. The "slow component" is contributed by carrier-carrier scattering and interband relaxation, i.e., carrier recombination, with the latter more or less dominating. Spatial diffusion time of the free carriers will also play a role, if there are no recombination centers nearby.

In Fig. 3.2a, in case of the lower value of the pulse fluence (40 μJ/cm^2), the total 1/*e* absorber relaxation time was about 1 ps. This indicates an enhanced tunneling of carriers into surface states. Since these states lie between conduction and valence band in terms of energy, their presence, unaffected by the additional SiN layer, greatly accelerates the interband relaxation process. A comparison of pump-probe response measurements at 1060 nm between SESAMs with an As-ion-treated or a surface-near QW (2-nm cap) can be found in [30]. As-ion implantation yielded longer 1/*e* relaxation times (3 ps and more compared to 1 ps). Thicker GaAs caps (>2 nm), too, resulted in significantly slower recombination. All experiments presented in the following paragraphs were performed using SESAMs with a near-surface QW and a 2-nm GaAs cap. So far, we have not been able to generate any femtosecond pulses with slower types of absorbers.

When the pulse fluence in Fig. 3.2a was increased from 40 to 480 μJ/cm^2, the shape of the absorber response changed and the absorption recovered more slowly, with a 1/*e* relaxation time of >2 ps. This can be explained by a stronger saturation of the fast, SHB-related component compared to the slower one. The observation does not support the hypothesis of an optical Stark effect in the SESAM as the dominant pulse shaping mechanism in sub-picosecond SDLs, which had been proposed by Wilcox et al. [21]. This would predict a faster response for higher fluences. A more detailed discussion of this issue is included in [22].

The observed SESAM behavior has important consequences for mode-locked SDLs, which have to rely on a fast absorber for ultrashort-pulse generation: The shorter the pulse duration in the SDL cavity, the more important the fast component of the absorber response. Hence, SHB helps to shape ultrashort pulses. The higher the pulse fluence, the slower the overall relaxation of the SESAM. For shorter pulses, oversaturation of the absorber will more strongly affect the pulse shaping. Hence, SHB makes it more difficult to stabilize the mode-locking regime. The complex temporal absorption behavior also corresponds to other pulse-shaping effects, i.e., spectral dynamics (dynamic filtering) and self phase modulation (SPM). Please note that the semiconductor gain element will show, to some degree, analogue behavior (SHB in the gain spectrum, fast and slow gain relaxation components, stronger gain saturation for shorter pulses, gain filter dynamics, SPM).

The nonlinear reflectivity of the surface-near SESAM was measured with the laser mentioned above at 1030 nm in a setup similar to the one reported in [31]. The SESAM was heated to a temperature of 68 °C, which corresponded to the operating point in the SDL laser experiment delivering the shortest pulses of 107 fs at 1030 nm. The measured absolute reflection of the SESAM versus input pulse fluence is depicted in Fig. 3.2b. We extract a saturation fluence of about 10 μJ/cm^2, a modulation depth of ≈1.2 %, and a nonsaturable loss of ≈1.5 %. With increasing pulse fluence (>100 μJ/cm^2), two-photon absorption can be observed, leading to a decrease of the reflectivity and, in the worst case, to damage of the device. Additional nonlinear reflectivity measurements were performed for SESAM temperatures of 46 and 22 °C, delivering modulation depths of ≈0.8 and ≈0.6 %, respectively. The reduced modulation depth is related to the corresponding shift of exciton resonance and band gap to higher energies. This behavior is schematically illustrated in Fig. 3.7, cf. Sect. 3.4.1.

3.2.3 SDL Gain Structure

A simplified surface-emitting SDL gain structure is shown in Fig. 3.1b. Similar to the SESAM structures, the SDL gain chips consisted of an AlAs/GaAs DBR mirror followed by an active region. The active section contained InGaAs QWs and (Al) GaAs barriers. SDL gain structures were tested with different barriers and QW numbers (3–13 QWs). We investigated two different barrier configurations: The simpler one used GaAs barriers and is denoted in the following as step-index structure (STIN, see [32] and Fig. 3.3a). The second barrier design used graded $Al_xGa_{1-x}As$ layers having an aluminum content of x = 0 at the QW side and of x = 0.2 at the AlGaAsP strain compensation layer (graded-index or GRIN structure, see also [32] and Fig. 3.3b). The AlGaAsP strain compensating layers (not contained in the simplified design in Fig. 3.1b) had to be introduced since the high strain incorporated by the InGaAs QWs would otherwise cause lattice relaxation

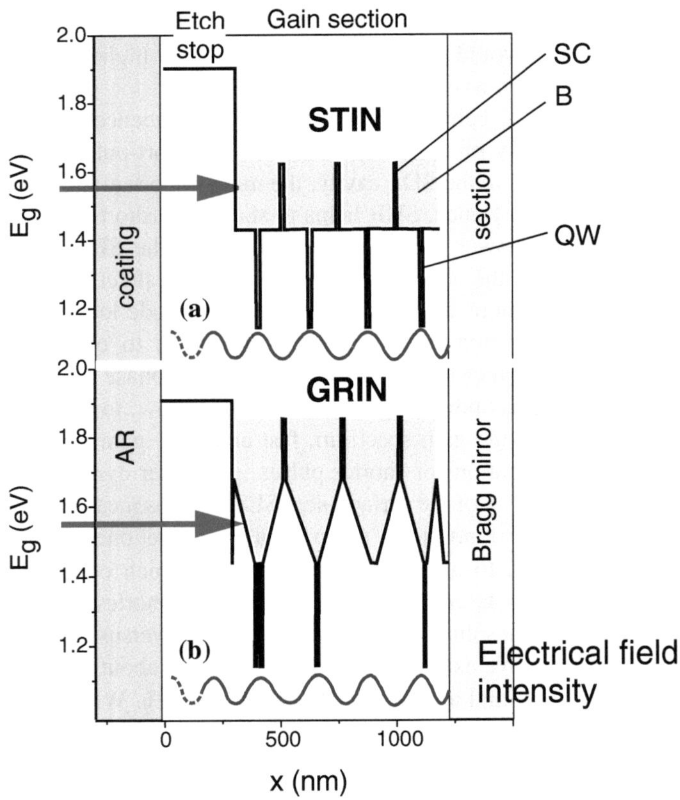

Fig. 3.3 Schematic band gap diagrams of 4-quantum well (QW) gain structures with step index (STIN) (**a**) and graded index (GRIN) design (**b**). *Arrows* mark the excitation photon energy (*SC* strain compensation, *B* barriers)

and formation of dislocation networks. Such would reduce the radiative recombination and can lead to rapid device degradation.

Figure 3.3 schematically shows the band gap diagram of the GRIN gain structure in comparison to a step index (STIN) architecture. The QWs were placed into the electric-field maxima of the interference pattern formed by the incident and reflected light beam (see red curves in Fig. 3.3), thus increasing the longitudinal confinement and consequently the gain. We varied the numbers of QWs in the antinodes of the electrical field. For example, by introduction of a double quantum well and omitting the last but one QW in our structure, the amount of pump power provided to each QW was modified (Fig. 3.3b). In our case, this meant unequal pumping of the QWs and resulted in an inhomogeneous broadening of the gain spectrum of the structure, supporting shorter pulses. The pump radiation is mainly absorbed by the spacer layers, and subsequently the generated carriers drift into the QWs. A graded Al-content of the barriers results in a graded bandgap and thereby in a quasi-electric field, which promotes the carrier drift towards the QWs. This was confirmed by time-resolved photo luminescence (PL) experiments. The rise time of the QW PL after barrier excitation was determined to be 17.7 ps for the STIN gain section, whereas the corresponding value observed for the GRIN gain section was <5 ps [32]. Because of the superior carrier collection capability, the GRIN design, which we used for most of our gain media, is expected to result in higher laser efficiency compared to the STIN-design [32].

Figure 3.4 shows the reflectance measurement of an SDL gain structure before the antireflection (AR) coating. The reflectance was measured using a tungsten white-light source. A reflectance stop band centered around 1030 nm was observed, since the element was designed to serve as a mirror of the laser cavity. Due to the missing AR coating, a resonance was formed within the gain structure, resulting in the deep reflectance decrease at 1026 nm. The corresponding PL measurement, performed with a semiconductor laser diode exciting at 797 nm, is also shown in Fig. 3.4. Comparing the spectra measured with and without AR coating, the influence of the cavity resonance effect is clearly visible. Without AR coating the peak was shifted towards the resonance wavelength and a second small peak appeared at 970 nm near the stop band edge. Only after AR coating, the undisturbed PL emission was measured, since the resonance had practically disappeared due to the minimized reflectance at the SDL/air interface. The undisturbed PL exhibited a FWHM of about 20 nm, indicating that the gain medium should principally allow laser pulse durations in the sub-100-fs range.

We obtained sub-ps laser pulses only with AR-coated SDL gain chips. The AR coating has a strong influence on the group delay dispersion (GDD) of the gain element as shown in Fig. 3.5, where an uncoated and a coated GRIN structure are compared. In the uncoated case, the element showed the spectral-phase characteristics of a Gires-Tournois interferometer (GTI). For femtosecond SDLs, the structure thickness and hence the spectral positions of the GTI resonances are chosen such that the GTI is operated in the vicinity of zero dispersion [16]. However, these positions depend on the operation temperature (Fig. 3.5). Furthermore, the dispersion varies considerably over the spectral width of a short pulse, and there are

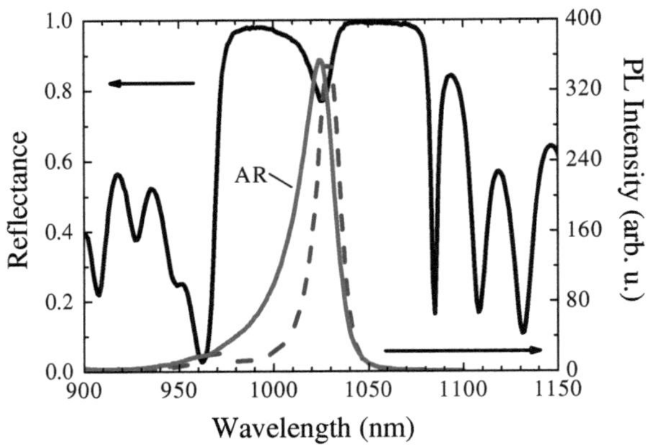

Fig. 3.4 Measured reflectance (*left axis*) and photoluminescence (*right axis*) of an SDL gain structure (STIN design with 6 QWs). The photoluminescence was additionally measured after the antireflection coating (*dashed line*)

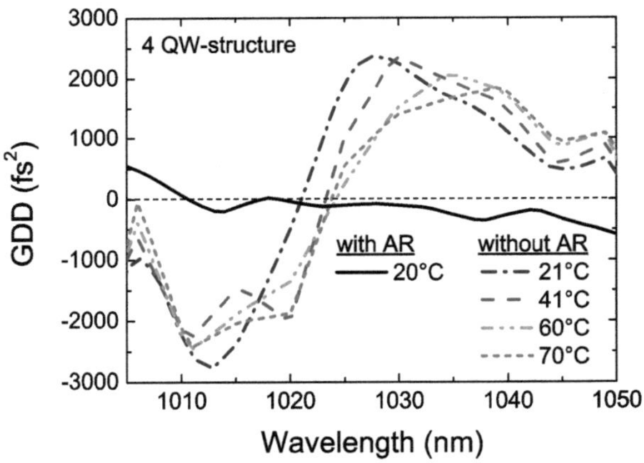

Fig. 3.5 Group delay dispersion (GDD) of a 4-QW GRIN gain structure. An AR-coated sample (*black solid line*) is compared with an uncoated sample made from the same wafer. The latter shows a significant temperature dependence as indicated by the *other curves*

also dispersion dynamics due to the interplay of GTI resonance and the gain or absorption spectrum changed by the pulse. With AR coating, the measured dispersion is low for all wavelengths, i.e., always below the resolution limit ($\approx\pm200$ fs^2) of our white-light interferometer (black solid line in Fig. 3.5). Nevertheless, we can expect a thin structure with 3 or 4 QWs to have lower residual dispersion compared to a thicker one with, e.g., 6 or 13 QWs, since the GDD is proportional to the square of the thickness of the structure [33]. This motivated the development of three- and four-QW gain media for shortest pulse operation. The above mentioned findings are also valid for the SESAM structures used here, but to a lesser degree, because their active zone is very thin (one QW only).

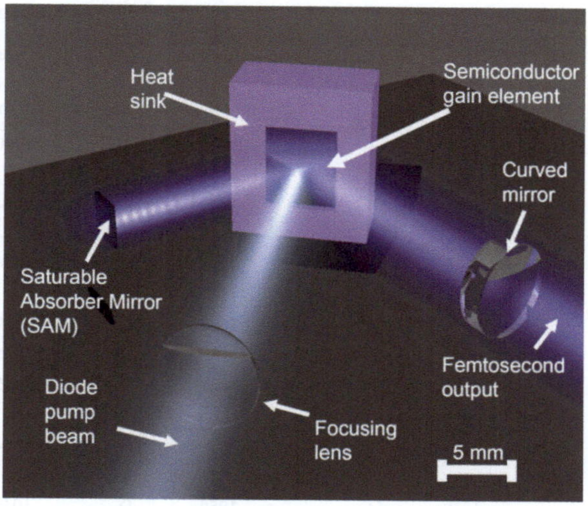

Fig. 3.6 Setup of a mode-locked semiconductor disk laser (SDL). The *V-shaped* cavity contains a saturable absorber mirror (SESAM) as the mode-locker, an InGaAs/AlGaAs disk gain element at the folding point, and a curved output coupler

3.3 Setup of the Mode-Locked Semiconductor Disk Laser

Figure 3.6 shows the scheme of a V-folded, nearly hemispherical SDL oscillator consisting of only three elements: the SDL gain structure, the SESAM and the curved output coupler. The temperature of both semiconductor elements can be independently controlled by the Peltier elements on which they are mounted. The length of the resonator was typically around 50 mm. There was a tight focus with a waist size of roughly 15–20 μm to create a high pulse fluence (several times the saturation fluence) on the SESAM. An estimation of the focus size is difficult, since solder-related deformations of the gain chip (see [26]) and a deviation of the output coupler curvature from its nominal value prevent an exact calculation. Our SDLs were optically pumped by diode lasers emitting around 800 or 840 nm.

3.4 Parameters for the Generation of Femtosecond Pulses from Mode-Locked Semiconductor Disk Lasers

3.4.1 Spectral Tuning of the SESAM

The subject of paragraph 4 is the influence of adjustable laser parameters on the ultrashort-pulse performance. At first, we investigated the relative energetic settings of the laser photon energy and of the SESAM band edge and exciton resonance.

For our initial experiments, we chose a 6-QW STIN gain chip and three different SESAMs, whose room temperature photoluminescence varied from one to the other by steps of 10 nm. With one of the absorbers, stable mode-locking with sub-ps pulse durations was achieved. The result is listed in Table 3.1 as experiment no. 1.

Table 3.1 Optimum SESAM temperatures for shortest-pulse generation from our SDLs

Experiment no.	SESAM PL at (nm)	SESAM temperature (°C)	Exciton transition at (nm)	Gain chip	Laser emission at (nm)	Minimum pulse duration (fs)	Comment/ Reference
1	1028	43	**1025**	A 6 QW	**1040**	590	[27]
2	1028	59	**1030**	B 4 QW	**1033**	350	[34]
3	1031	61	**1033**	C 4 QW	**1036**	290	[20]
4	1031	85	**1040**	C 4 QW	**1039**	300	Figure 3.9
5	1031	106	**1047**	D 4 QW	**1045**	210	Figures 3.10, 3.11 and 3.12
6	1031	111	**1048**	D 4 QW	**1044**	190	[22]
7	1050	54	**1050**	D 4 QW	**1045**	210	–
8	1028	20	**1018**	E 4 QW	**1021.5**	198	Figure 3.13
9	1028	68	**1032**	F 3 QW	**1030**	107	Figure 3.14

Three near-surface SESAMs (photoluminescence at 1028, 1031, 1050 nm) and six gain chips (*A–F*) were used. *Bold numbers* The wavelength of the SESAM exciton transition (error ±5 nm) is compared to the respective laser emission wavelength

Using the setup of Fig. 3.6 with a 0.5 % transmission output coupler, practically chirp-free pulses as short as 590 fs were generated with an average output power of 30 mW at 1040 nm [27]. For this absorber, we found that the SESAM exciton energy was slightly higher than the energy of the laser photons. Figure 3.7a illustrates this situation, where the laser is operated on the red side of the exciton resonance of the SESAM.

A way to determine the position of the exciton resonance is from a photocurrent (PC) spectrum of the SESAM, which was recorded using a Fourier-transform spectrometer. The position is approximately at the maximum of dI_{PC}/dE (see Fig. 3.7b; I_{PC}, photocurrent; E, excitation photon energy). The PL maximum of the above absorber was at 1028 nm, while the PC measurement yielded an exciton transition wavelength of 1015 nm. From further comparisons of PL spectra with PC spectra, we assume that for our type of SESAM at room temperature the heavy-hole exciton resonance is generally ≈10 nm below the PL maximum [27]. We found a variation of this shift in the range of ±5 nm, which we attribute basically to the difficulties in determining the exciton position in the PC spectrum. Band gap and exciton resonance are no sharp features, but smeared towards lower and higher energies by temperature-related and structurally induced broadening. While a sharp exciton transition should cause a significant dependence of the dynamic behavior on the excitation energy or on the absorber temperature, we did not find clear signs

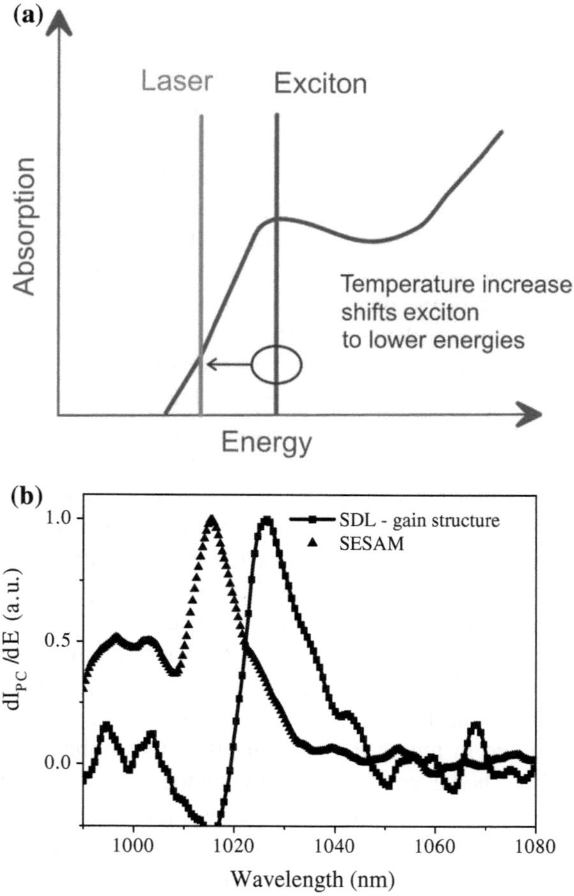

Fig. 3.7 Scheme of the energetic position of the SESAM exciton resonance (marked by *black vertical line*) relative to the laser photon energy (*red marker*). The *arrow* symbolizes that exciton resonance and band edge can be shifted by temperature tuning (**a**). Photocurrent measurements of SDL gain structure and SESAM at room temperature (**b**)

of this in pump-probe measurements of our SESAMs. Therefore, the exciton transition is to be understood mainly as a landmark to characterize the operating point on the slope of the SESAM absorption spectrum. This operating point will determine the modulation depth of the SESAM and the dynamic spectral filtering effect, which both strongly affect the mode-locked laser regime. In the above experiment, taking into account the SESAM temperature (43 °C), the laser wavelength was roughly 15 nm above the exciton transition. From the nonlinear-reflectivity measurements of this absorber in paragraph 2.2., we estimate ≈0.5 % of saturable absorption for this case.

For our next experiment (labeled no. 4 in Table 3.1), we used a similar SESAM with a PL maximum at 1031 nm. By heating of the SESAM, the exciton resonance and the band gap of the device were tuned towards the laser photon energy (see arrow in Fig. 3.7a). This increased the effective saturable absorption, providing stronger pulse shaping. Compared to the initial experiments, we reduced the number of QWs in the gain structure to minimize the dispersion; we chose a 4-QW GRIN structure containing two single QWs and a double QW as shown in Fig. 3.3b). The heat sink temperature of the gain chip was kept at 19 °C. The optical pump power at 840 nm amounted to 0.92 W, and an 0.5 % output coupler was used. The SESAM temperature was increased from room temperature to 93 °C. At room temperature, the laser photon energy was far below the energies of the band gap and the heavy-hole exciton resonance of the absorber. By heating of the SESAM, these spectroscopic features were red-shifted towards the laser photon energy by 0.25 meV/°C (0.3 nm/°C). Absorption increased from practically zero for the SESAM at room temperature, which resulted in an output power of 21 mW, to values in the order of 1 %, indicated by the considerable decrease of the output power to 10 mW around 85 °C. At the same time the laser emission wavelength was shifted towards the red, from 1034.5 to 1039 nm, since the laser minimized absorption loss as far as this was advantageous considering the filtering loss from the limited gain bandwidth.

Figure 3.8 shows the pulse duration versus the saturable absorber temperature. We observed stable single-pulse sub-ps mode-locking only around 85 °C in a range of about 10 °C. Here, the pulse had a duration around 300 fs and only very little chirp. For lower temperatures, the pulse broadened, developed a strong tail, became unstable, and the laser switched into a double-, multiple-, or long-pulse regime. This observation can be explained mainly by the decrease of saturable absorption, when tuning the temperature in this direction. Then, passive amplitude modulation became too weak to provide sufficiently strong pulse shaping. The double- or multiple-pulsing tendency was additionally promoted by the increase of the pulse energy due to the lowered loss. Countering this by decreasing the pump power resulted in pulses much longer than those obtained in the optimum range around 85 °C, however. If the SESAM temperature was driven too high, i.e., above 90 °C, the laser switched to a much longer pulse with a center wavelength of ≈1032 nm. This operation regime experienced lower loss from gain filtering, but obviously the contributions to the chirp were not balanced. Although the exact evolution of laser performance with changing SESAM temperature differed somewhat depending on the individual laser configuration and adjustment, so far for all the investigated femtosecond SDLs we observed a similar behavior as in Fig. 3.8.

The optimum SESAM temperatures for shortest-pulse generation obtained from nine experiments (including those from this chapter) are listed in Table 3.1, together with relevant parameters and characteristics of the respective laser configurations. The three saturable absorbers had the same design, but were made for different laser wavelengths. They are distinguished by the center wavelengths of their room temperature photoluminescence (PL) amounting to 1028, 1031 and 1050 nm. The "estimated exciton transition wavelengths" at the operating temperature are based on the room temperature PL, the shift of 10 nm as mentioned above, and a red-shift by about 0.3 nm/°C when increasing the SESAM temperature.

Six gain chips were investigated; chip *A* was a 6-QW structure with ungraded barriers (STIN), chips *B* to *E* were 4-QW graded-barrier designs like in Fig. 3.3b. *F* was a 3-QW graded-barrier design. Compared to the other gain structures, *E* contained no double QW; instead, inhomogeneous broadening was realised by different In-contents in the QWs (designed for a total $\Delta\lambda \approx 6$ nm). *A* to *C* were

Fig. 3.8 Pulse duration of the mode-locked SDL measured for different saturable absorber temperatures

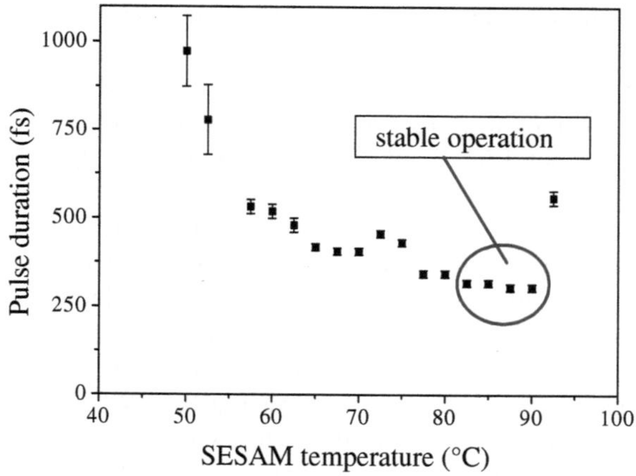

2 x 2 mm^2 samples (*B* and *C* from the same wafer); *D* to *F* were 4 x 6 mm^2 samples. In general, the laser emission wavelength was close to the exciton resonance. The estimation of the exciton transition energy is only rough; we see a trend, however: While in our first experiments, the laser photon energy was rather below the exciton energy, the laser operated energetically at or even slightly above the exciton resonance in the experiments where pulse durations around 200 fs or shorter were obtained. In any case, we observed that heating a SESAM above its optimum operation temperature led to a further decreased output power of the respective SDL and in the shortest-pulse experiments no. 5–9 lasing even ceased above some temperature, which means that effective SESAM absorption was still increasing in this direction.

3.4.2 Dependence of Mode-Locked Laser Performance on Pump-Power

As another important influence for the mode-locked performance we investigated that of the intracavity pulse energy and of the corresponding fluences [22]. In particular, we wanted to know if our SDL would show a behavior typical for soliton-like mode-locking [15, 25].

The pump power incident on the gain medium was varied from 0.77 to 1.32 W, while using an 0.5 % output coupler and keeping the heatsink of the gain chip at 16 °C and that of the SESAM at 106 °C. For the SESAM, the corresponding exciton transition wavelength was estimated at 1047 nm. The laser resonator was 50 mm long.

With femtosecond SDLs, we generally observe two values for the lasing threshold. The lower value is obtained when decreasing the pump power for the already mode-locked laser, since, in this case, the absorber is significantly bleached. When starting the laser, however, the SESAM is practically unbleached; the higher absorption loss will lead to a higher threshold. In this experiment, a pump power of 0.77 W was very close to the laser threshold in case of mode-locking and the lower limit to permit measurements. Here, the laser would cease emission within less than a minute. When increasing the pump power again, lasing restarted around 1.0 W. The output power showed a linear dependence on the incident pump power, amounting to 4–13 mW around 1045 nm for 0.77–1.24 W of 840-nm pump light. At about 1.32 W of pump radiation, the output power jumped up to 17 mW, indicating that here the onset of ps radiation significantly reduced the total loss of the laser. Probably, spectral hole-burning (SHB) in the gain chip also played a role, because fs pulses can extract only a part of the energy stored in the conduction band, while, for a longer pulse, carriers at higher states within the band become available by intraband relaxation during the course of the pulse.

Figure 3.9 displays a set of nine autocorrelation traces from our SDL. The first trace at the bottom was recorded at 0.77 W of pump power and shows the autocorrelation of a practically perfect *sech*^{2} pulse. A *sech*^{2} temporal shape is characteristic for optical solitons [15]. The pulse duration did not change much, when the pump power was increased, as can be seen also from Fig. 3.10a presenting the FWHM values obtained from the respective fits. This behavior is not soliton-like. The quality of the fits worsened while increasing the pump power up to 1.00 W. The pulses developed a tail, which was observed as a weak pedestal in the autocorrelation signal. At 1.00 W, single-pulse lasing became unstable. If the laser switched to double pulses at 1.00 W, these were again almost pedestal-free, just like the pulse at 0.77 W. In both cases the output pulse energy was approximately 1.5 pJ, as shown in Fig. 3.10b. If the pump power was increased above 1.00 W, again a pulse tail developed. The related pedestal was most pronounced just before double pulsing became unstable and the laser started producing ps pulses. The ps-pulse shape did not have much to do with *sech*^{2} and did not resemble a soliton-like pulse any more.

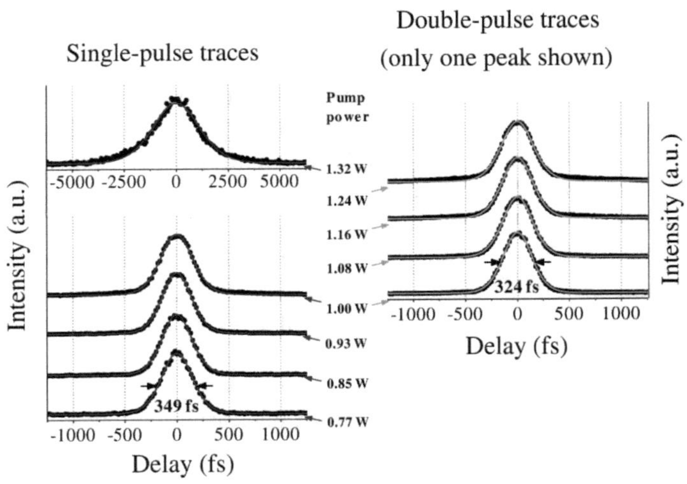

Fig. 3.9 SDL autocorrelation traces for different pump powers. Respective pump powers are indicated by *arrows*. All traces were background-free; offsets were used to arrange all traces in one graph. *Solid curves* are fits assuming a *sech*^{2} pulse shape

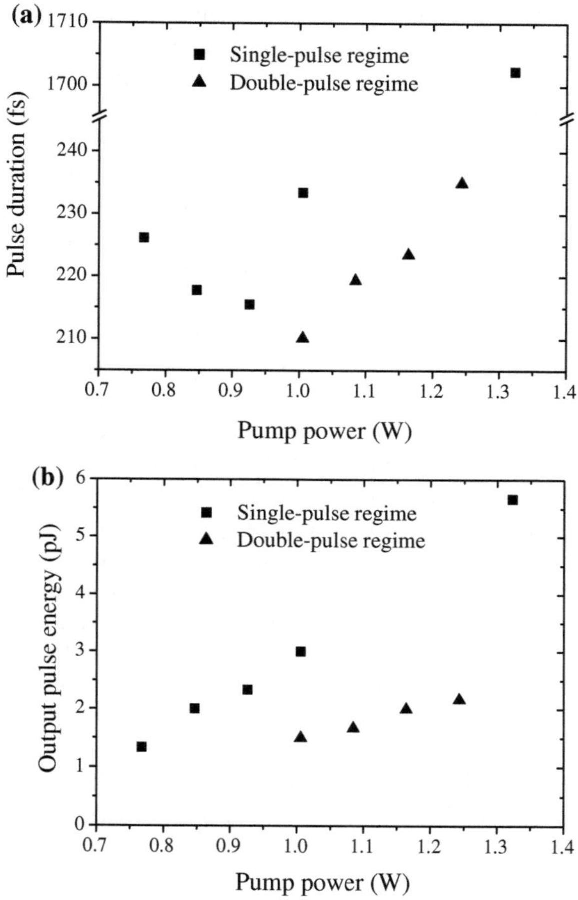

Fig. 3.10 **a** SDL pulse durations at different pump powers. At 1.00 W there was a transition from single-pulse operation to double pulsing. Slightly above 1.24 W, the laser switched back to single-pulse operation. The pulse duration was much longer, then; please note the axis break. **b** Pulse energies of the mode-locked SDL

Comparing Fig. 3.10a with b illustrates that the pulse durations did not obey the soliton area theorem, which would predict an inverse proportionality of pulse duration and pulse energy [15]. Durations are not even generally shorter for higher energies. Below 1.32 W of pump power, pulses were always around 220 fs. Minimum durations were 215 fs for the single pulse and 210 fs for double pulses. From the highest pulse energy before double pulsing started (Fig. 3.10b), 3 pJ, we estimate a fluence in the order of 100 μJ/cm^2 on the SESAM. This has to be compared to a saturation fluence of 10 μJ/cm^2 we typically determine for our surface-near SESAMs using pulses with similar duration. The onset of double or multiple pulsing in case of an oversaturated absorber is a common phenomenon with soliton-like mode-locking [28]. It is additionally promoted by increased loss from two-photon absorption, which starts to become significant around 100 μJ/cm^2 in the SESAM nonlinear reflectivity curve in Fig. 3.2b, and by SHB in the gain chip. The latter can be explained similarly as for the onset of ps pulses: Carriers at states not available for a single pulse will relax during the time in between pulses and become available for a second pulse, leading to increased extraction efficiency.

The optical spectra corresponding to the autocorrelation traces in Fig. 3.9 are shown in Fig. 3.11. Only at the respective lowest energies we find single and double pulses to be almost chirp-free, i.e., soliton-like.

For 0.77 W of pump power, we observed emission centered near 1043 nm and a spectral bandwidth (FWHM) of 5.6 nm, corresponding to a time-bandwidth product of 0.35, close to the Fourier limit for a secant hyperbolic. The development of a pulse tail was associated with the coming up of a red shoulder in the spectrum, which extended over almost 10 nm. So, the pulses acquired additional bandwidth, which would be expected for a soliton due to stronger SPM for higher pulse intensity. However, no new frequency components were generated on the blue side with respect to the spectrum at 0.77 W, and the pulse was becoming increasingly chirped.

As the laser switched to double pulses around 1.00 W of pump power, SPM and the saturation of the SESAM were reduced due to lower pulse energy and therefore the largest part of the red shoulder disappeared, growing again with increasing pump power. The evolution of spectral shape and width was very similar for single and double pulses. The shoulder growth was a little faster for the double pulses, if the pulse energies shown in Fig. 3.10b are considered. Picosecond pulses at 1.32 W of pump light corresponded to a spectrum with a FWHM of 1.1 nm.

With increasing pump power, the spectral maximum of the single pulse in Fig. 3.11a was red-shifted from 1043 to 1046 nm; the double-pulse spectral maximum went from 1045 to 1047 nm; the picosecond pulse is centered at 1048 nm. This shift documents the growing heat load. The absorption of the SESAM decreased towards the red; hence, the shift weakened the pulse shaping by the absorber. This resulted in insufficient suppression of the red shoulder and temporal

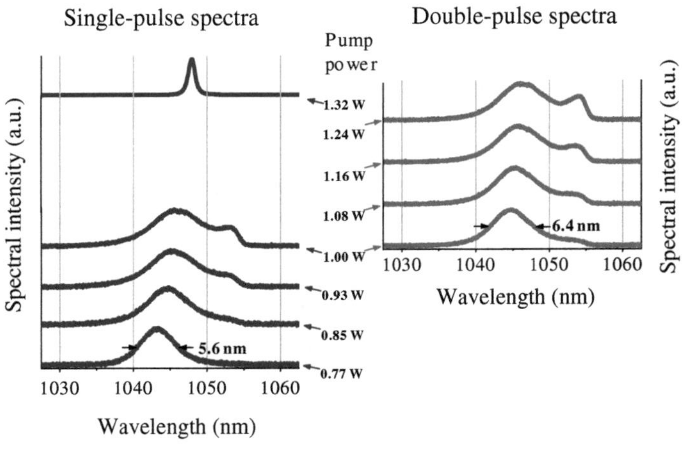

Fig. 3.11 Optical emission spectra of the mode-locked SDL at different pump powers, recorded simultaneously with the autocorrelation traces shown in Fig. 3.9. Offsets were used to present all spectra in one graph

tail; finally, the bandwidth that is required for sub-ps generation could not be mode-locked any more.

Another way to investigate the influence of the intracavity pulse energy and, hence, of the fluences on the laser performance is by variation of the output coupler transmission T_{OC}. Our results with $T_{OC} = 0.2–1.5\ \%$ (the gain was too low to permit lasing with $T_{OC} = 3\ \%$) confirmed our above findings. The minimum pulse durations varied from 190 to 230 fs [22].

3.4.3 Discussion—Mode-Locking Mechanism in SDLs

Usually, soliton-like pulse-shaping is required for sub-ps pulse generation in solid state lasers [25]. Some observations typical for soliton-like mode-locking are made with our SDLs as well as with femtosecond lasers using dielectric gain media: temporal shape of intensity close to $sech^2$, practically no chirp for appropriate operation parameters, spectral bandwidth increasing with pulse energy, multiple pulsing at high intracavity powers. However, with our SDLs, there is no inverse proportionality of pulse duration and energy, i.e., the soliton theorem is not fulfilled, and the absorber properties are crucial for SDL pulse durations, while they have little influence on durations in dielectric-media lasers. This indicates, important differences exist with the underlying pulse-shaping effects. Please note, that our sub-ps SDLs are different from the ps SDLs that contain relatively large amounts of intracavity dispersion and that were described as soliton-like in [15, 35].

In lasers based on dielectric gain elements, conditions for very short pulses are a relatively large amount of Kerr SPM, balanced in terms of chirp by appropriate GDD, and minimized spectral filtering effects (particularly, a large gain bandwidth). Passive amplitude modulation (PAM) by an absorber initiates and stabilizes mode-locking, but the absorber modulation depth and relaxation time does not affect the pulse duration much [25].

Pulses in our mode-locked SDL resonator are shaped mainly by the spectro-temporal behavior of saturable absorption and gain and the associated self-phase modulation. Very short pulses can be obtained, if PAM is sufficiently strong, the contributions to chirp largely compensate each other, filtering is minimized, and fast dynamics dominate. This is possible when using a fast SESAM and a broad-gain medium, minimizing GDD, matching SESAM and gain medium, and keeping the laser photon energy close to the excitonic resonance of the SESAM, as shown in Chap. 4 and [22].

If single-pulse operation is desired, too strong saturation of both semiconductor elements must be avoided. However, as SDL pulses become shorter, it is increasingly more difficult to fulfill the latter, since saturation will stem mainly from SHB during the pulse duration and, hence, saturation fluences will be lower than for longer pulses. For pulse durations <200 fs, in our experiments we found stabilization of the single-pulse SDL regime to be a major issue, even if operating close to the lasing threshold. In general, an oversaturated absorber will not suppress additional pulses in the resonator [28]. We also observed that this inability was more likely or pronounced when we reduced the maximum SESAM modulation depth, lowering the loss handicap for additional pulses this way.

Generally, additional pulses will be promoted by more efficient energy extraction from the gain medium and/or nonlinear losses with inverse saturation characteristics, e.g., two-photon absorption in the SESAM. For lasers with dielectric laser media and soliton-like mode-locking, increased extraction efficiency can be explained by the soliton theorem, i.e., the lower pulse energy in case of multi-pulsing corresponds to a smaller spectral pulse width, which is favourable with respect to the limited spectral bandwidth of the gain. With SDLs, the time constants of the gain chip become important: If the resonator round-trip time is similar to or longer than the recombination time in the gain chip (order of 1 ns), an SDL may produce additional pulses to reduce the amount of laser inversion decaying in between the pulse transits [36]. This will usually lead to a harmonically mode-locked regime. Spatial hole burning, too, will reduce the energy extraction per pulse, if the pulse duration is in the order of or shorter than the carrier-carrier or carrier-phonon scattering times. This was a possible reason why the maximum output pulse energy in Fig. 3.10b was limited to about 3 pJ. The transitions into a femtosecond double-pulse or a picosecond single-pulse regime are motivated by scattering processes that close or avoid the spectral hole in the gain, making more free carriers available for stimulated recombination; energy extraction is more efficient.

We often observed a third, alternative transition when increasing the intracavity pulse intensities, especially the intensity on the gain chip. In this case, multiple pulsing occured not by additional, ungrouped pulses but by emission of pulse groups with close temporal spacings (“macro pulses”, “pulse molecules”). These pulse molecules consisted of two to seven pulses. An example is shown in Fig. 3.12 (achieved with gain chip *F*, SESAM “PL at 1031 nm”). Quarterman et al. have found macro pulses with up to 70 peaks [37]. So far, we do not know applications for SDL pulse molecules. Harmonic mode-locking of SDLs enables very high

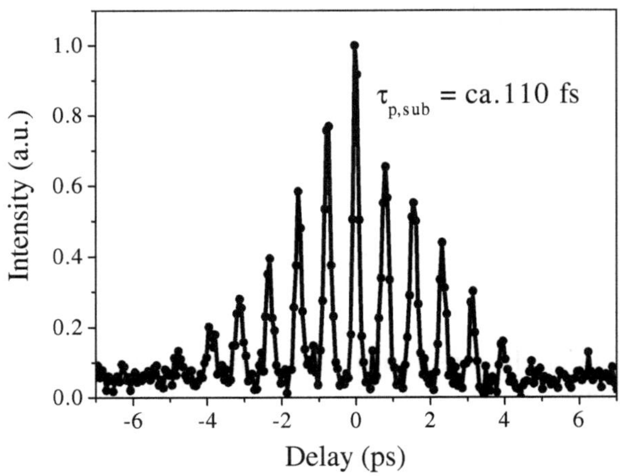

Fig. 3.12 Autocorrelation trace of a “pulse molecule” containing seven pulses from a mode-locked SDL

repetition rates at sub-picosecond pulse durations, as we will show in the next paragraph. Such a performance is attractive especially for communications with THz data rates via time multiplexing.

3.5 High Pulse Repetition Rates and Ultrashort Pulse Durations from a Mode-Locked Semiconductor Disk Laser

3.5.1 Sub-200-fs Pulses at 92 GHz Repetition Rate in the Harmonically Mode-Locked Regime

To obtain short pulses at high repetition rates, we chose a SESAM modulation depth capable of mode-locking a sufficiently broad emission bandwidth and applied a high pump power resulting in a high intracavity power. For this experiment, the four-QW-gain structure labelled E in Table 3.1 was used in a ≈29-mm-long "V-shaped" laser cavity. The absorber that matched the gain chip best was the SESAM characterized in Fig. 3.2. Its temperature was stabilized at 20 °C, where the modulation depth amounted to ≈0.6 % at 1030 nm. Taking into account the actual laser wavelength of 1022 nm, we assume a value of about 0.8 % for saturable absorption. Pumping with a power of ≈4 W from an 808-nm laser diode, we detected a trace of separate pulses with a spacing of ≈11 ps. This result is documented by the autocorrelation trace in the 50-ps range in Fig. 3.13a. It corresponds to a repetition rate of ≈92 GHz. This means 18th harmonic-order mode-locking of our laser cavity. The autocorrelation trace recorded in the 1.5-ps range can be approximated assuming a $sech^2$ shape of the pulse intensity and a pulse duration of ≈198 fs (Fig. 3.13b). The 92-GHz rate is confirmed by the optical spectrum (Fig. 3.13c), showing the corresponding modulation with a spacing of approximately 300 pm. Due to the extraordinarily high optical-pulse repetition frequency and bandwidth limitations of photodetectors and microwave analyzers, it was not possible to measure a microwave spectrum of the laser output. For an output coupler transmission of 0.2 %, we measured an output power of 31 mW.

No efforts were made for stabilizing the laser. After several minutes, the spectral comb lost contrast, which could be regained by readjustment of the SDL. In general, with an increasing number of pulses circulating in the cavity, the stability of the laser operation regime will decline. In particular, the number of pulses may vary. Stable harmonic mode-locking with a pulse rate at a demanded value will require an active stabilization of the laser resonator length and the application of a spectral filter (e.g., etalons as sub-cavity) fixing the spectral comb belonging to the desired harmonic order [38]. The filtering approach has already been demonstrated for mode-locked edge emitting laser diodes. Such lasers can reach rates >1 THz without time multiplexing, but one cannot speak of separate pulses. There is a rather sine-shape output due to uncomplete recovery of the saturable absorber and the

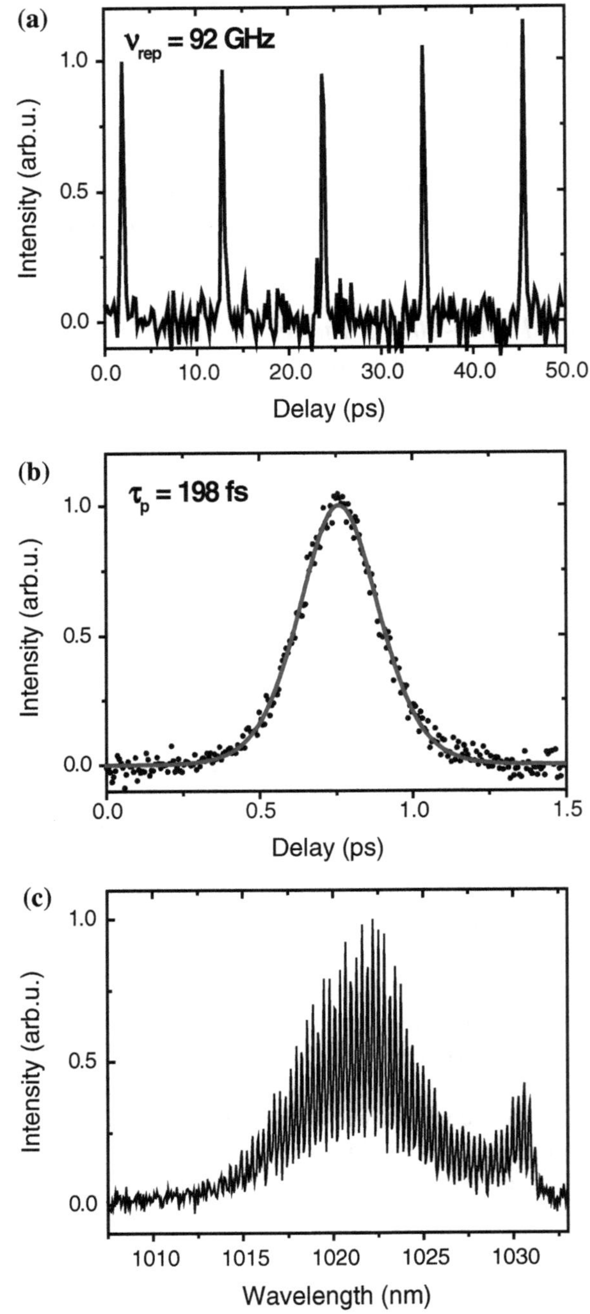

Fig. 3.13 Harmonically mode-locked semiconductor disk laser at 92 GHz pulse repetition rate:
a Autocorrelation trace of the pulse train.
b Higher-resolution autocorrelation of one pulse. The fit assuming a $sech^2$ pulse results in a pulse duration τ_p of 198 fs. **c** Optical spectrum centered near 1022 nm. The mode spacing of ≈300 pm corresponds to the pulse rate ν_{rep}

small number of resonator modes involved in the locking. For SDLs, higher orders of mode-locking and even higher pulse rates than the 92 GHz without multiplexing shown here should be possible, but carrier accumulation in the absorber will turn out as a limitation when the temporal spacing between the pulses becomes too short.

The pulse rate of 92 GHz surpassed the previous 50-GHz rate demonstrated for a mode-locked SDL with 3 ps pulses [24] and, to the best of our knowledge, at present, there is no other laser oscillator—of any kind—combining such a high repetition frequency with a pulse duration below 200 fs. The output power of 31 mW reported here was limited by the available pump power; >100 mW of output should be possible if a higher transmission of the output coupler is chosen (here, T_{OC} was only 0.2 %).

3.5.2 *107-fs Pulses at 5 GHz Repetition Rate in the Fundamentally Mode-Locked Regime*

To achieve a single-pulse regime and the shortest pulses, we increased the SESAM temperature to 68 °C and by this its modulation depth to 1.2 % and reduced the pump power to 2.8 W to keep the pulse intensities moderate on both the SESAM and the gain structure. For the results in Fig. 3.14, we used a 3-QW gain structure (chip *F*). With above 4-QW chip (*E*), minimum pulse durations were 10 fs longer. We estimated fluences of ≈100 μJ/cm^2 on the SESAM and ≈6 μJ/cm^2 on the gain structure and conclude that, in this SDL, these values were low enough to avoid oversaturation of the semiconductor elements. As shown in Fig. 3.14a, the autocorrelation displayed a pulse duration τ_p = 107 fs (*sech*^{2}-shaped intensity assumed). The optical spectrum was centered near 1030 nm and had a full width at half maximum of ≈10.2 nm (Fig. 3.14b). With a time-bandwidth product of ≈0.31, the pulse is approximately Fourier-limited. The small peak near 1046 nm indicates an only weak pulse tail, which is barely visible in the autocorrelation trace. An output power of 3 mW was measured.

The single-pulse regime was documented by a wide-range radio frequency spectrum (Fig. 3.15a), recorded with a 26.5-GHz signal analyzer. The first peak at ν_{rep} ≈ 5.136 GHz is the fundamental resonance frequency of the laser cavity. Decreased intensities of the higher-order harmonics are due to the limited bandwidth of the photodiode. Figure 3.15b is a higher-resolution scan (5 kHz bandwidth) of the fundamental. The signal-to-noise ratio was ≈70 dB.

In principle, the gain bandwidth of SDLs can support even shorter pulses (see typical PL curve in Fig. 3.4). In [37], with an SDL, pulse bunches were observed, which had an envelope of a few ps and revealed a substructure of ≈60-fs peaks. However, no single-pulse operation was achieved, which was explained by the influence of SHB in the gain chip. Therefore, to the best of our knowledge, the 107-fs pulses set a new milestone for pulse durations from fundamentally mode-locked

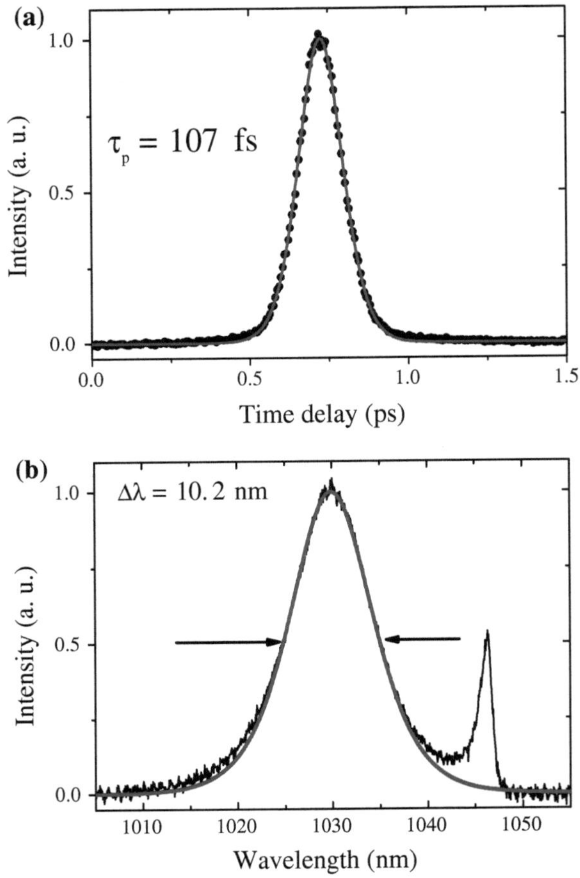

Fig. 3.14 **a** Autocorrelation trace of the semiconductor disk laser in the single-pulse regime. A *sech*^{2} fit yields a pulse duration τ_p of 107 fs. **b** Optical spectrum centered near 1030 nm

SDLs (and from any fundamentally or harmonically mode-locked semiconductor laser without external pulse compression). The result shows that SHB does not prevent single-pulse operation of SDLs at ultrashort pulse durations. Nevertheless, SHB must be considered a severe issue when trying to achieve high output powers and high pulse energies with single-pulse operation and pulse durations in the 100-fs range at the same time. While power scaling is relatively easy for cw or picosecond SDLs, the average output power of 120 mW for a pulse duration of 355 fs at 1 GHz in [39] is the highest reported so far for sub-500 fs SDLs. Moderate improvement of the value of 3 mW demonstrated here with T_{OC} = 0.2 % can be expected with a higher output coupler transmission. However, operating the SDL much further above the lasing threshold will drive the laser into multiple-pulsing.

A way to boost the ultrashort-pulse output of SDLs to power levels required for applications is by use of tapered diode amplifiers (TDAs) and/or fiber amplifiers [22, 40]. A TDA with pulsed electrical pumping is capable of acting both as an amplifier and as a fast pulse picker. This approach will give free choice of pulse repetition frequency and make SDL systems applicable also for uses requiring

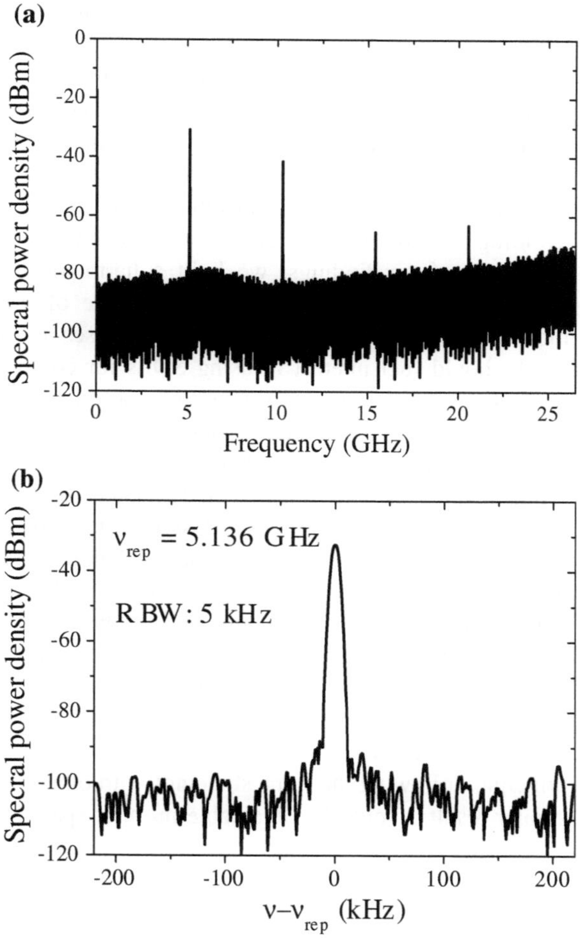

Fig. 3.15 Radio frequency spectrum of the SDL in the single-pulse regime. **a** Wide range spectrum. The fundamental laser frequency is $\nu_{rep} \approx 5.136$ GHz. **b** Vicinity of the fundamental mode

lower repetition frequencies. As a proof of principle, in [22], we examined a TDA/SDL combination and reduced the pulse repetition frequency from 3 GHz to 47 MHz.

3.6 Summary

This chapter addressed the generation of femtosecond pulses with passively mode-locked semiconductor disk lasers (SDLs) in the 1-μm wavelength range. We investigated the optimum parameters for almost chirp-free femtosecond pulses in single- or multiple-pulse regimes. On the whole, the pulse-shaping process in our SDLs is not "soliton-like" in the classical picture applied to dielectric-gain-media lasers. In particular, the pulse durations strongly depended on the relaxation behavior and the effective modulation depth of the semiconductor saturable absorber mirror (SESAM). Throughout, we used semiconductor media with minimized group delay dispersion, i.e., with antireflective coatings and thin active zones. Wavelength matching of gain chip and SESAM was essential for ultrashort-pulse generation, achieving a sufficiently deep absorber modulation. Spectral-hole burning (SHB) of both gain chip and absorber affected the stability of single-pulse operation. For fundamental mode-locking, operation close to the lasing threshold was required, avoiding too strong saturation of the semiconductor elements.

Based on our findings, we built a harmonically mode-locked SDL emitting 198-fs pulses at a very high repetition rate of ≈92 GHz and a fundamentally mode-locked SDL demonstrating 107-fs pulses at a rate of ≈5 GHz. The 92-GHz result should fuel interest in using SDLs for communications or frequency comb generation. At present, there seem to be no other laser oscillators that can generate sub-200-fs pulses at such high rates. The practically chirpfree 107-fs pulses shown here set a new record for shortest pulse durations achieved directly from any fundamentally or harmonically mode-locked semiconductor laser. With respect to pulse durations for mode-locked lasers in this emission region, InGaAs/(Al)GaAs SDL gain media now surpass typical Nd^{3+}-doped crystals and compete with Yb^{3+}-doped media. Combination of SDLs with electrically pumped tapered diode amplifiers offers an elegant approach to boost the output power and to realize variable pulse repetition rates (in principle, down to 1 Hz) [22].

Overall, the results are a further step of mode-locked SDLs in becoming useful compact and low-cost ultrashort-pulse sources. They are especially attractive for applications requiring high pulse repetition rates, like in THz spectroscopy [3]. In principle, following the same strategies with other semiconductor material systems, similar results should be obtainable in other spectral regions.

References

1. H. Yoshida, Y. Yamashita, M. Kuwabara, H. Kan, A 342-nm ultraviolet AlGaN multiple-quantum-well laser diode. Nat. Photonics **2**, 551–554 (2008)
2. K.J. Linden, Single mode, short cavity, Pb-salt diode lasers operating in the 5, 10, and 30 μm spectral regions. IEEE J. Quantum Electron. **21**, 391–394 (1985)
3. R. Gebs, P. Klopp, G. Klatt, T. Dekorsy, U. Griebner, A. Bartels, Time-domain THz spectroscopy based on asynchronous optical sampling with a femtosecond semiconductor disk laser. Electron. Lett. **46**, 75 (2010)
4. F. Quinlan, G. Ycas, S. Osterman, S.A. Diddams, A 12.5 GHz-spaced optical frequency comb spanning > 400 nm for near-infrared astronomical spectrograph calibration. Rev. Sci. Instr. **81**, 063105 (2010)
5. U. Keller, A.C. Tropper, Passively modelocked surface-emitting semiconductor lasers. Phys. Rep. **429**, 67–120 (2006)
6. S.L. Chuang, *Optoelectronic Devices* (Wiley, New York, 1995)
7. C.W. Wilmsen, H. Temkin, L.A. Coldren (eds.), *Vertical-Cavity Surface-Emitting Lasers: Design, Fabrication, Characterization, and Applications* (Cambridge University Press, Cambridge, 1999)

8. M. Kuznetsov, F. Hakimi, R. Sprague, A. Mooradian, Design and characteristics of high-power (>0.5-W CW) diode-pumped vertical-external-cavity surface-emitting semiconductor lasers with circular TEM00 beams. IEEE J. Sel. Top. Quantum Electron. **5**, 561–573 (1999)
9. C. Steven, K. Contag, M. Larionov, A. Giesen, H. Hügel, A 1-kW CW thin disk laser. J. Sel. Top. Quantum Electron. **6**, 650–657 (2000)
10. J.L. Chilla, S.D. Butterworth, A. Zeitschel, J.P. Charles, A.L. Caprara, M.K. Reed, L. Spinelli, in *High power optically pumped semiconductor lasers,* ed. by R. Scheps, H.J. Hoffman. Solid State Lasers XIII: Technology and Devices. Proc. SPIE **5332**, 143–150 (2004)
11. B. Rudin, A. Rutz, M. Hoffmann, D.J.H.C. Maas, A.-R. Bellancourt, E. Gini, T. Südmeyer, U. Keller, Highly efficient optically pumped vertical-emitting semiconductor laser with more than 20 W average output power in a fundamental transverse mode. Opt. Lett. **33**, 2719–2721 (2008)
12. C. Hönninger, R. Paschotta, F. Morier-Genoud, M. Moser, U. Keller, Q-switching stability limits of continuous-wave passive mode locking. J. Opt. Soc. Am. B **16**, 46–56 (1999)
13. M.E. Barnes, Z. Mihoubi, K.G. Wilcox, A.H. Quarterman, I. Farrer, D.A. Ritchie, A. Garnache, S. Hoogland, V. Apostolopoulos, A.C. Tropper, Gain bandwidth characterization of surface-emitting quantum well laser gain structures for femtosecond operation. Opt. Express **18**, 21330–21341 (2010)
14. S. Hoogland, S. Dhanjal, A.C. Tropper, J.S. Roberts, R. Häring, R. Paschotta, F. Morier-Genoud, U. Keller, Passively mode-locked diode-pumped surface-emitting semiconductor laser. IEEE Photon. Technol. Lett. **12**, 1135–1137 (2000)
15. R. Paschotta, R. Häring, A. Garnache, S. Hoogland, A.C. Tropper, U. Keller, Soliton-like pulseshaping mechanism in passively mode-locked surface-emitting semiconductor lasers. Appl. Phys. B **75**, 445–451 (2002)
16. A. Garnache, S. Hoogland, A.C. Tropper, I. Sagnes, G. Saint-Girons, J.S. Roberts, Sub-500-fs soliton-like pulse in a passively mode-locked broadband surface-emitting laser with 100 mW average power. Appl. Phys. Lett. **80**, 3892–3894 (2002)
17. S. Hoogland, A. Garnache, I. Sagnes, J.S. Roberts, A.C. Tropper, 10-GHz Train of Sub-500-fs optical soliton-like pulses from a surface-emitting semiconductor laser. IEEE Photon. Technol. Lett. **17**, 267–269 (2005)
18. S. Arahira, S. Oshiba, Y. Matsui, T. Kunii, Y. Ogawa, Terahertz-rate optical pulse generation from a passively mode-locked semiconductor laser diode. Opt. Lett. **19**, 834–836 (1994)
19. E.U. Rafailov, M.A. Cataluna, W. Sibbett, N.D. Il'inskaya, Yu. M. Zadiranov, A.E. Zhukov, V.M. Ustinov, D.A. Livshits, A.R. Kovsh, N.N. Ledentsov, High-power picosecond and femtosecond pulse generation from a two-section mode-locked quantum-dot laser. Appl. Phys. Lett. **87**, 081107 (2005)
20. P. Klopp, F. Saas, M. Zorn, M. Weyers, U. Griebner, 290-fs pulses from a semiconductor disk laser. Opt. Express **16**, 5770–5775 (2008)
21. K.G. Wilcox, Z. Mihoubi, G.J. Daniell, S. Elsmere, A. Quarterman, I. Farrer, D.A. Ritchie, A. Tropper, Ultrafast optical Stark mode-locked semiconductor laser. Opt. Lett. **33**, 2797 (2008)
22. P. Klopp, U. Griebner, M. Zorn, A. Klehr, A. Liero, M. Weyers, G. Erbert, Mode-locked InGaAs-AlGaAs disk laser generating sub-200-fs pulses, pulse picking and amplification by a tapered diode amplifier. Opt. Express **17**, 10820 (2009)
23. P. Klopp, U. Griebner, M. Zorn, M. Weyers, Pulse repetition rate up to 92 GHz or pulse duration shorter than 110 fs from a mode-locked semiconductor disk laser. Appl. Phys. Lett. **98**, 071103 (2011)
24. D. Lorenser, J.H.C. Maas, H.J. Unold, A.-R. Bellancourt, B. Rudin, E. Gini, D. Ebeling, U. Keller, 50-GHz passively mode-locked surface-emitting semiconductor laser with 100-mW average output power. IEEE J. Quantum Electron. **42**, 838–847 (2006)
25. F.X. Kärtner, I.D. Jung, U. Keller, Soliton mode-locking with saturable absorbers. IEEE J. Selected Topics Quantum Electron. **2**, 540–556 (1996)
26. U. Zeimer, J. Grenzer, D. Korn, S. Döring, M. Zorn, W. Pittroff, U. Pietsch, F. Saas, M. Weyers, X-ray diffraction spot mapping—a tool to study structural properties of semiconductor disk laser devices. Phys. Stat. Sol. (a) **204**, 2753–2759 (2007)

27. F. Saas, G. Steinmeyer, U. Griebner, M. Zorn, M. Weyers, Exciton resonance tuning for the generation of sub-picosecond pulses from a mode-locked semiconductor disk laser. Appl. Phys. Lett. **89**, 141107 (2006)
28. M.J. Lederer, B. Luther-Davies, H.H. Tan, C. Jagadish, N.N. Akhmediev, J.M. Soto-Crespo, Multipulse operation of a Ti:sapphire laser mode locked by an ion-implanted semiconductor saturable-absorber mirror. J. Opt. Soc. Am. B **16**, 895–904 (1999)
29. J.W. Tomm, V. Strelchuk, A. Gerhardt, U. Zeimer, M. Zorn, H. Kissel, M. Weyers, J. Jimenez, Properties of As^{+}-implanted and annealed GaAs and InGaAs quantum wells: Structural and band-structure modifications. J. Appl. Phys. **95**, 1122–1126 (2004)
30. M. Zorn, P. Klopp, F. Saas, A. Ginolas, O. Krüger, U. Griebner, M. Weyers, Semiconductor components for femtosecond semiconductor disk lasers grown by MOVPE. J. Crystal Growth **310**, 5187–5190 (2008)
31. M. Haiml, R. Grange, U. Keller, Optical characterization of semiconductor saturable absorbers. Appl. Phys. B **79**, 331–339 (2004)
32. F. Saas, V. Talalaev, U. Griebner, J.W. Tomm, M. Zorn, A. Knigge, M. Weyers, Optically pumped semiconductor disk laser with graded and step indices. Appl. Phys. Lett. **89**, 151120 (2006)
33. M. Moenster, U. Griebner, W. Richter, G. Steinmeyer, Resonant saturable absorber mirrors for dispersion control in ultrafast lasers. IEEE J. Quantum Electron. **43**, 174–181 (2007)
34. P. Klopp, F. Saas, U. Griebner M. Zorn, M. Weyers, Passively mode-locked semiconductor disk laser generating sub-300-fs pulses. CLEO/QELS 2008, CThF6
35. M. Hoffmann, O.D. Sieber, D.J.H.C. Maas, V.J. Wittwer, M. Golling, T. Südmeyer, U. Keller, Experimental verification of soliton-like pulseshaping mechanisms in passively mode-locked VECSELs. Opt. Express **18**, 10143–10153 (2010)
36. E.J. Saarinen, R. Herda, O.G. Okhotnikov, Dynamics of pulse formation in mode-locked semiconductor disk lasers. J. Opt. Soc. Am. B **24**, 2784–2790 (2007)
37. A.H. Quarterman, K.G. Wilcox, V. Apostolopoulos, Z. Mihoubi, S.P. Elsmere, I. Farrer, D.A. Ritchie, A. Tropper, A passively mode-locked external-cavity semiconductor laser emitting 60-fs pulses. Nat. Photonics **3**, 729 (2009)
38. S. Arahira. S. Oshiba, Y. Matsui, T. Kunii, Y. Ogawa, 500 GHz optical short pulse generation from a monolithic passively mode-locked distributed Bragg reflector laser. Appl. Phys. Lett. **64**, 1917–1919 (1994)
39. K.G. Wilcox, A.H. Quarterman, H. Beere, D.A. Ritchie, A.C. Tropper, High peak power femtosecond pulse passively mode-locked vertical-external-cavity surface-emitting laser. IEEE Photon. Technol. Lett. **22**, 1021–1023 (2010)
40. P. Dupriez, C. Finot, A. Malinowski, J.K. Sahu, J. Nilsson, D.J. Richardson, K.G. Wilcox, H. D. Foreman, A.C. Tropper, High-power, high repetition rate picosecond and femtosecond sources based on Yb-doped fiber amplification of VECSELs. Opt. Express **14**, 9611–9616 (2006)

Chapter 4
Compact Ultrafast Oscillators and High Performance Ultrafast Amplifiers Based on Ytterbium-Doped Fibers

J Limpert, T. Eidam, M. Baumgartl, F. Röser, M. Plötner, B. Ortaç, S. Nolte and A. Tünnermann

Abstract This chapter reviews the fundamentals and achievements of ultrashort pulse generation and amplification in ytterbium-doped fibers. Compact and ultrastable passively mode-locked fiber oscillators represent an ideal seed source for high performance femtosecond fiber amplification systems, which have been scaled towards kW-level average power and pulse energies well above the mJ-level. These laser systems will have significant impact in numerous scientific and industrial applications.

4.1 Introduction and Motivation

A number of important practical as well as fundamental research applications of ultrafast lasers appeared over the last decades, a trend initiated by the step from old dye-laser technology towards solid-state lasers. These high-power ultrafast solid-state lasers use small rods as the amplifier media, for instance, Titanium-doped sapphire as the most widespread one and have the potential to generate significantly higher pulse energies, higher powers and shorter pulse durations in combination

J. Limpert (✉) · T. Eidam · M. Baumgartl · S. Nolte · A. Tünnermann
Friedrich-Schiller-Universität Jena, Institut Für Angewandte Physik,
Max-Wien-Platz 1, 07743 Jena, Germany
e-mail: jens.limpert@uni-jena.de

F. Röser
Helmholtz-Zentrum Dresden-Rossendorf, Bautzner Landstraße 400,
01328 Dresden, Germany

M. Plötner
Fraunhofer IOF, Albert-Einstein-Straße 7, 07745 Jena, Germany

B. Ortaç
Bilkent Univ, UNAM Inst Mat Sci & Nanotechnol, TR-06800, Ankara, Turkey

S. Nolte et al. (eds.), *Ultrashort Pulse Laser Technology*,
Springer Series in Optical Sciences 195, DOI 10.1007/978-3-319-17659-8_4

with increased reliability than dye-lasers. However, these systems are difficult to scale in average power and suffer from low efficiencies, because direct diode pumping is not possible. Furthermore, the complexity of short pulse high energy Ti: sapphire lasers still constrain the employment of ultrafast laser technology in industrial environments. Material processing as one of the main driver applications relies on reliable, compact and cost effective pulsed sources.

In order to overcome thermo-optical effects, which limit the power scaling capability of these systems, several novel gain media designs, such as thin disk or slab, have been introduced [1, 2]. However, due to the low single pass gain of these amplifier materials, very complex systems, such as regenerative amplification schemes, are required to obtain a reasonable output. Therefore the robustness, compactness and long-term stability are restricted in short pulse bulk solid-state laser systems.

Alternatively, forming the gain medium long and thin not only leads to outstanding thermo-optical properties, but also to a very high single pass gain. Fiber-based laser systems have the reputation to be immune against any thermo-optical problems due to their special geometry. The excellent heat dissipation is due to the large ratio of surface-to-active volume of such a fiber. The beam quality of the guided mode is determined by the fiber core design and is therefore power-independent.

Due to the confinement of both the laser and pump radiation the intensity is maintained over the entire fiber length and is not limited to the Rayleigh length as it is the case in longitudinally pumped bulk lasers. The gain of the laser medium is determined by the product of pump light intensity and interaction length with the laser radiation in the gain medium. Therefore, the decisive product can be orders of magnitude higher in fibers than in other bulk solid-state lasers. This results in a very efficient operation of fiber laser systems exhibiting very high gain and low pump threshold values. Additionally, complete integration of the laser process in a waveguide provides an inherent compactness and long-term stability of fiber lasers.

In particular Ytterbium-doped glass fibers, which have a quantum defect of less than 10 %, can provide optical-to-optical efficiencies well above 80 % and, therefore, low thermal load. These fiber laser systems are especially interesting for high power ultrashort pulse generation and amplification because of several unique properties [3]: Firstly, a broad emission spectrum allows for short pulse amplification. In ytterbium-doped glass fibers the amplification bandwidth of approximately 40 nm supports in principle pulses of durations as short as $\sim$30 fs. Furthermore, the absorption spectrum covers a wavelength range in which powerful diode lasers are commercially available. An additional point to note is that the long fluorescence lifetime ($\sim$1 ms) results in a high-energy storage capability. Excited-state absorption of pump or signal radiation, or concentration quenching by ion-ion energy transfer processes does not occur with ytterbium, because only two energy-level manifolds are relevant for all optical wavelengths.

High power fiber lasers usually use the double-clad fiber concept (Fig. 4.1), invented in 1988 by Snitzer [4]. Such a double-clad fiber is characterized by a second waveguide, which is highly multimode, surrounding the active core. Into

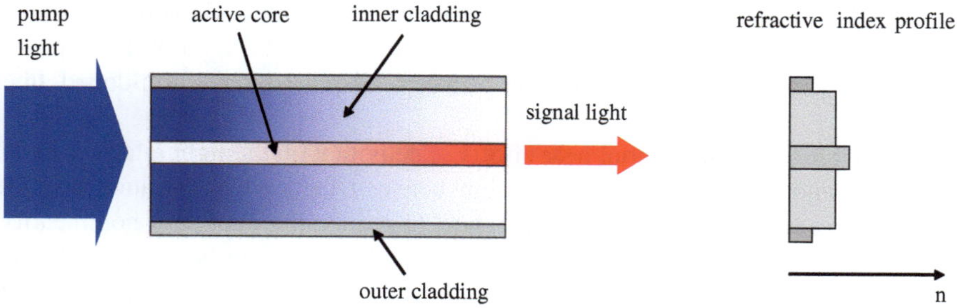

Fig. 4.1 The double-clad fiber concept

this second waveguide, also called inner cladding or pump core, low brightness high power diode laser radiation can be launched. This pump light is gradually absorbed over the entire fiber length and is converted into high brightness high power laser radiation. Thus, double-clad rare-earth doped fibers can provide a highly efficient brightness improvement by pump-to-laser radiation conversion by the laser process itself.

All the mentioned properties make rare-earth-doped fibers superior to other solid-state laser concepts in a variety of performance categories. This has become obvious following several recent demonstrations of continuous-wave fiber laser systems exhibiting more than 10 kW of average power while maintaining an excellent beam quality [5].

4.2 Fundamentals of Short Pulse Propagation in Passive and Active Fibers

The fiber geometry itself is responsible for most of the outstanding properties of rare-earth-doped fibers that make them attractive gain media. However, this geometry also promotes nonlinear effects by making the light propagate under tight confinement over considerably long lengths. In fact, in the context of ultrashort pulse amplification, nonlinearity is mostly harmful and imposes performance limitations in fiber laser systems.

Nonlinear effects in fibers can be manifold [6]. The lowest-order nonlinear effects in standard optical fibers originate from the third-order susceptibility $\chi^{(3)}$. These effects can be divided into those related to an intensity-dependent refractive index and those resulting from stimulated inelastic scattering. Self-phase modulation (SPM), four-wave-mixing (FWM), and self-focusing all fall into the first category, whereas stimulated Raman scattering (SRS) and stimulated Brillouin scattering (SBS) are effects of the second category.

In general, the nonlinearity coefficients in silica glass fibers are intrinsically small. Both the nonlinear index coefficient n_2 and the gain coefficients of SRS and SBS are at least two orders of magnitude smaller than in other common nonlinear media [7]. Nevertheless, due to the large product of intensity and interaction length inside the fiber core, nonlinear effects can be observed at very low peak power levels and can basically limit the performance of pulsed rare-earth-doped fiber systems.

All the aforementioned nonlinear effects scale both with the light intensity in the fiber core and with the interaction length between the optical radiation and the nonlinear medium—that is, the fiber. Hence, to reduce the impact of nonlinearity, temporal and spatial scaling is required. Temporal scaling can be achieved by the well-known technique of chirped pulse amplification (CPA), which is schematically depicted in Fig. 4.2.

In this technique, ultrashort optical pulses from a mode-locked oscillator are stretched in time by a certain factor, which can be as large as 10,000, by passing them through a dispersive delay line. Therefore, during amplification, the peak power of the pulses is considerably reduced, as are the nonlinear effects. After amplification, the stretched amplified pulses travel through a second dispersive delay line with the opposite sign of the stretcher dispersion, resulting in a recompression back to ultrashort pulse duration.

On the other hand, spatial scaling requires advanced fiber designs that present a large mode area of the actively doped core and an absorption length that is as short as possible, thus reducing the nonlinear interaction length. Consequently, there has been a pursuit of novel fiber designs with increased core dimensions that are still able to emit a stable fundamental mode. Conventional active step-index fibers with reduced numerical aperture, assisted by coiled or tapered sections to achieve single-mode operation, allow for core diameters of up to 40 μm. Rare-earth-doped photonic crystal fibers (PCFs) allow for significantly larger core sizes of up to 100 μm [8–10] due to the significantly better control of the index step between a nano-structured core and the holey photonic crystal cladding. Due to the PCF cladding's design freedom, an additional functionality, such as polarizing or polarization maintaining properties, can be added [11]. An example of a large core ytterbium-doped fiber is known as rod-type PCF. A cross section of the ultra-large mode area fiber is shown in Fig. 4.3. The 200 μm inner cladding (pump waveguide) is

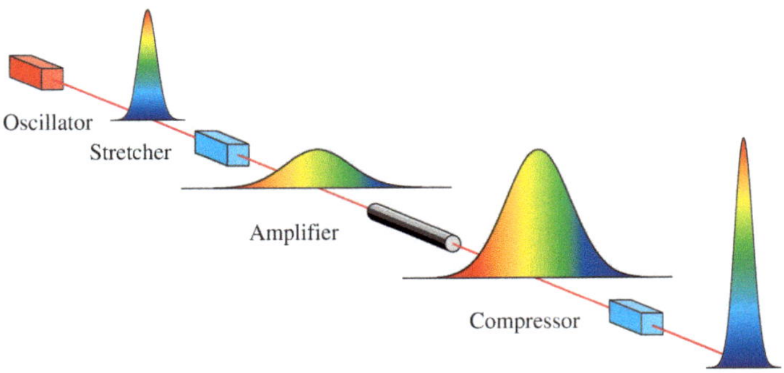

Fig. 4.2 Principle of chirped pulse amplification (CPA)

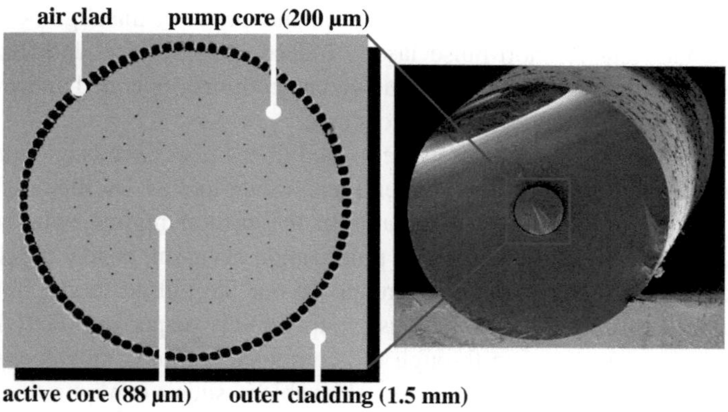

Fig. 4.3 Microscope image of a rod type photonic crystal fiber

surrounded by an air clad that consists of ninety 400 nm thick and ~10 μm long silica bridges. This structure provides a numerical aperture of ~0.6 at 976 nm, allowing for an efficient coupling of multimode pump radiation into the fiber. Nineteen missing holes in the center of the fiber form the Yb/Al codoped active core region, which has a corner-to-corner distance of 88 μm. Three rings of small, carefully dimensioned airholes (pitch $\Lambda = \sim 14.9$ μm; relative hole size $d/\Lambda = \sim 0.1$) around this core provide the confinement of the radiation in the doped core. The mode area of the fundamental mode is as large as 4000 μm^2. The small ratio of pump core area to active core area results in an enhanced pump light absorption of about 30 dB/m at 976 nm. The entire inner structure is surrounded by a stiff 1.5-mm diameter fused silica outer cladding. This large outer cladding is primarily introduced to keep the fiber straight, thus preventing bend-induced losses or distortions of the weakly guided fundamental mode. Thus, in such a straight structure, the large mode area is accessible over the entire fiber length, which is not the case in conventional LMA fibers. In addition, the outer cladding makes the fiber mechanically robust on its own, so that no extra coating material is required; this allows for straightforward high-power extraction. By applying this kind of fiber, novel performance levels have been achieved in various operation regimes, ranging from femtosecond to nanosecond pulses [12–14].

4.3 Compact All-Fiber Femtosecond Mode-Locked Lasers

Sources of ultra-short laser pulses have become a versatile tool for an increasing number of industrial and scientific applications. Despite of the outstanding performance of passively mode-locked solid-state lasers, such systems have difficulties to step out the laboratory environment. This is mainly due to the bulky elements and free-space propagation inside the cavity. In contrast, fiber based sources give the possibility to generate laser radiation in a completely integrated manner, hence, they allow for the development of compact, inexpensive, misalignment-free and environmentally stable short-pulse lasers. Indeed, passively mode-locked single-mode rare-earth-doped fiber lasers are nowadays routinely operated and are entering the market to address real world applications.

Different operation regimes of mode-locked fiber lasers have been reported over the recent decades. Following conventional mode-locked oscillators, the most intuitive way is to run a fiber oscillator in the soliton regime, whereby mode-locking is obtained using nonlinear polarization rotation inside the fiber or a semiconductor saturable absorber mirror to favour the pulsed mode. Soliton fiber lasers are built entirely from anomalous group-velocity dispersion (GVD) fiber and the pulse maintains its shape through the combined action of negative GVD and Kerr nonlinearity. However, the energy achievable in such a configuration is limited by the soliton area theorem to some tens of picojoules [15, 16]. Disobeying this theorem is leading to an excess of nonlinearity and finally is causing a pulse break-up. An increase of pulse energy from fiber oscillators can be achieved by stretching the pulse during its propagation. Thereby, the reduction of peak power keeps the nonlinear phenomena under control. This approach is referred to as stretched-pulse operation, in which the fiber laser comprises segments of large normal and anomalous GVD in the cavity [17]. The pulse width experiences large variations per cavity round trip, with a change in the chirp sign from positive at the end of the normal GVD segment to negative at the end of the anomalous GVD segment. In particular, with small net positive cavity dispersion, the pulse energy is larger than that produced in the soliton regime. Stretched pulse lasers with output energies from tens of picojoules to some nanojoules have been reported [18–22] before nonlinearity provokes instabilities. However, under certain circumstances nonlinearity can be advantageous. It has been shown that in fibers possessing normal GVD a pulse can propagate in a wave-breaking free manner despite strong nonlinearity [23]. In a fiber oscillator operated in the self-similar regime, the pulse accumulates a linear, or at least a monotonic, chirp, which is partially compensated at points in the cavity using a linear process (e.g. diffraction gratings). In addition the spectral bandwidth has to be filtered by the gain medium to obtain periodic self-consistency [24, 25].

For real world applications self-starting and environmentally stable ultrafast all-fiber sources are of particular interest. Hence, a linear cavity fiber laser generating self-similar pulses has been developed, which comprises only polarization maintaining (PM) fibers. Two different cavity configurations have been investigated and self-starting self-similar pulses could be generated over a wide range of parameters such as output coupling, pump power and net cavity dispersion.

In Fig. 4.4 the two different cavity configurations can be seen. In both cases the mode-locking mechanism is based on a semiconductor saturable absorber mirror (SAM). Highly efficient transmission gratings are used for intra cavity dispersion compensation. The fiber length inside the cavity was chosen to be the same in the two cases. The only difference is the choice of output coupling. In one case (Fig. 4.4a) the output coupler is a fixed fiber pigtailed coupler allowing for a fiber based output, in the other case (Fig. 4.4b) the output coupler is based on a bulk polarizer and a quarter-wave plate allowing for a tunable coupling ratio.

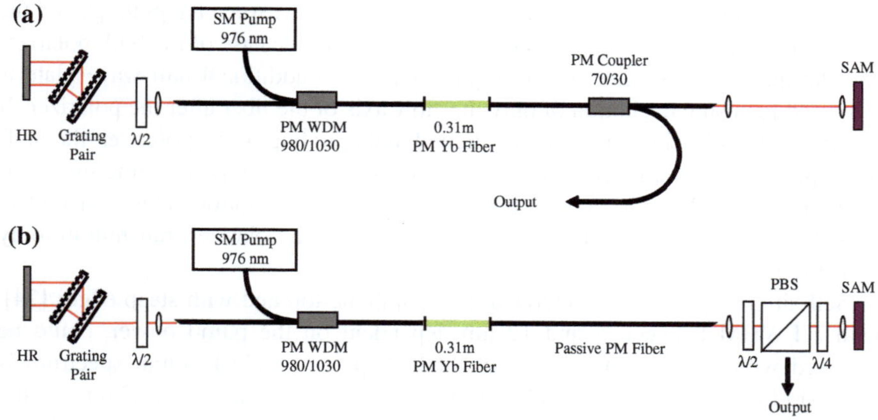

Fig. 4.4 The two different cavity designs: **a** with fiber output coupler and **b** with variable bulk output coupler. *PM* polarization maintaining, *HR* high reflection mirror, *SAM* saturable absorber mirror, *PBS* polarization beam splitter

The gain medium is a 31 cm long highly ytterbium doped (~300 dB/m absorption @ 976 nm) PM fiber with a mode-field diameter of 4.8 μm. The utilization of the minimal length of gain fiber allows to decouple gain bandwidth filtering from the nonlinear evolution in the undoped fiber because the effect of GVD and nonlinearity can be neglected during the amplification [24]. This fiber is pumped through a thin-film PM wavelength division multiplexer (WDM) by a single mode diode providing a maximum output power of 400 mW at a wavelength of 976 nm. The passive fibers used in the setup are Panda 980 PM fibers with mode field diameters of 7 μm @ 1035 nm and a dispersion of 0.024 ps^2/m. The total fiber length inside the cavity is 5.6 m for both cavity designs and in both cases the length of the passive fiber on both sides of the gain medium is equal in order to create symmetric conditions for the pulse evolution after the gain medium in both directions. All the PM fibers were fusion spliced together with an estimated polarization extinction ratio above 37.5 dB and all fiber ends were either angle polished or angle cleaved. The SAM is commercially available [26] and is based on a non-resonant design, using a GaAs/AlAs Bragg mirror with 27 layer pairs and 26 low temperature molecular beam epitaxy grown InGaAs quantum wells in front of the mirror. The AR coated device has a low-intensity absorption of 45 %, a modulation depth of 30 % and a saturation fluence of ~100 mJ/cm^2. In a pump-probe experiment using 200 fs pulses, the recovery dynamics of the optical excitation has been measured. The SAM shows a bi-temporal impulse response with a short relaxation time of <200 fs and a slower part of 500 fs. The ratio of the fast and slow parts has been determined to 3:2. To achieve the saturation threshold a telescope is used to image the output of the fiber onto the SAM.

The transmission gratings used for intra cavity dispersion compensation are 1250 lines/mm gratings made of fused silica [27] with a high transmission into the first order (>95 % @ 1035 nm). The gratings are set up in Lithrow angle (40°) with a grating separation of 16 mm. A half-wave plate is used between the gratings and the PM fiber to ensure excitation of only the slow axis. In the case of the bulk polarizer and the quarter-wave plate as an output coupler, an additional half-wave plate is also used to ensure excitation of only the slow axis of the fiber after the polarizer. If the axis of the fiber itself is properly aligned to the grating or the polarizer, the half-wave plate can be removed resulting in an even simpler setup. With the fiber pigtailed thin-film 30:70 PM coupler used in the other configuration, a half-wave plate is no longer necessary, as the coupler itself works as a polarizer transmitting only light in the slow axis.

Self-similar spectra, characterized by a parabolic top and with steep edges [24], with a FWHM between 8 and 12 nm dependent on the pump power, could be obtained with a net cavity dispersion of 0.03 ps^2. A typical output spectrum is shown in Fig. 4.5, where the FWHM of the self-similar spectrum is 11.3 nm. As the pump power is increased the spectral width increases monotonically, until the threshold for double-pulsing is reached (i.e. two pulses per round trip). The relatively large value of the net cavity dispersion of ($\geq$0.03 ps^2) is chosen to obtain a symmetric spectrum.

In contrast, for lower but still positive values of net cavity dispersion broader, but asymmetric and more structured spectra could be observed, which correspond to the stretched-pulse regime [25]. In the following we will only focus on the self-similar regime. The highest output pulse energy is obtained from the setup with the bulk output coupler, where the output coupling could be tuned. For a high output coupling coefficient the highest pulse energies of 1 nJ are obtained with a repetition rate of 17 MHz.

The chirped self-similar output pulse has a duration of 7.2 ps (8.2 ps FWHM on the background free autocorrelator) (see Fig. 4.6), but can be externally compressed to an autocorrelation FWHM of 280 fs. The pulse duration can be calculated from the width of the autocorrelation by assuming a transform limited self-similar

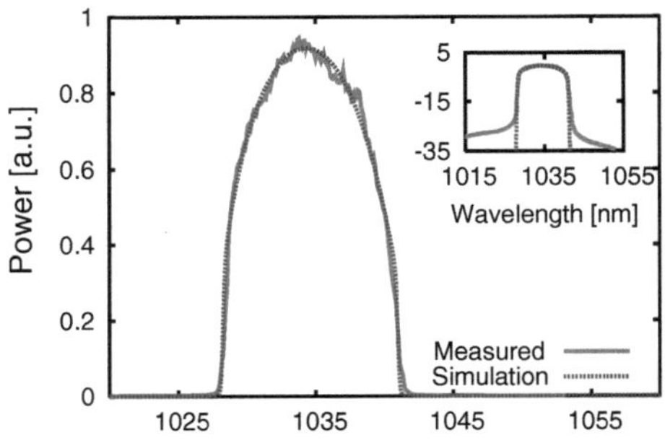

Fig. 4.5 *Solid curve* Typical output spectrum on a linear scale of the laser operated in the self-similar regime (net cavity dispersion 0.03 ps^2). *Dotted curve* Numerical simulation. *Inset* Log. scale

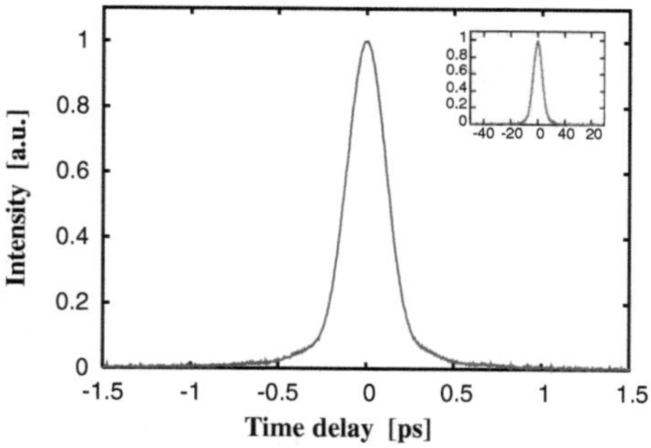

Fig. 4.6 Autocorrelation trace of the externally compressed pulse at a pulse energy of 1 nJ with an autocorrelation FWHM of 280 fs. *Inset* Uncompressed output pulse with an autocorrelation FWHM of 8.2 ps

spectrum of the compressed pulse (deconvolution factor 1.33) and is evaluated to be 210 fs. From the actual measured spectrum a deconvolution factor of 1.65 is numerically calculated assuming no chirp. It shows that even if the experimental spectrum only slightly deviates from the parabolic shape as in our case, the deconvolution factor changes significantly. Thus, an even more optimistic pulse duration of 170 fs can be evaluated. The scan range of the background free autocorrelator is 150 ps, but to check for multiple pulsing and to ensure that there is only a single pulse per round trip, a 25 GHz photo diode in combination with a 50 GHz sampling oscilloscope providing a scan range from 30 to 60 ns has been used.

Self-similar spectra similar to the spectrum shown in Fig. 4.5 are also obtained from the setup with the fiber pig-tailed coupler. The output pulse energy is 0.12 nJ, the pulse can be externally re-compressed to an autocorrelation FWHM of 350 fs. Due to the fixed output coupling of 30 %, which is lower than in the other setup, no flexibility is given to find the regime of higher pulse energy and shorter pulses. Additionally, an extra loss of 30 % is introduced inside the cavity due to the fact that the coupler is passed in both directions. Nevertheless, by optimizing the outcoupling ratio it should be possible to obtain better results with the advantage of an alignment-free fiber output.

To make sure the laser operates in the self similar regime some inspections have to be made. Firstly, the spectrum should of course exhibit a self-similar shape, which is fulfilled in our case. Secondly, an important condition for self-similar evolution is that the pulse is always positively chirped inside the cavity, possessing a minimum pulse duration after the intra cavity gratings just before entering the fiber [25]. This is also true since the negative dispersion given by the grating pair used for external compression ($-0.36\ ps^2$ (double pass)) is higher than of the pair used inside the cavity.

In addition, only half the length of intra cavity fiber is passed before the pulse is coupled out, indicating that the pulse is still highly positively chirped before entering the fiber after the intra cavity grating pair. As the dispersion from this point on inside the cavity is positive until the pulse again reaches the intra cavity gratings, this point must be a minimum point. Hence, the pulse is not completely compressed during propagation inside the cavity and, therefore, maintains its shape. As an additional verification, a numerical simulation of the cavity was carried out. Each segment was treated separately by solving the nonlinear Schrödinger equation with the parameters of our experimental setup [24]. In Fig. 4.5 the simulated spectrum can be seen to be in good agreement with the measured spectrum. In the simulation the pulse exhibits self-similar pulse propagation [25] and is always positively chirped.

In both configurations the laser is self-starting and immediately jumps back into the same mode-locked state without any external perturbations if, for instance, switching the laser off and on again. It has been shown, that SAMs exhibiting a bi-temporal impulse response can fulfill both requirements of self-starting and symmetric spectra [28]. Furthermore, the use of PM fibers makes the laser stable toward environmentally induced changes to the birefringence of the fiber. The fibers could be twisted and moved around while maintaining a stable modelocked output. This is verified by observing a uniform train of pulses using a fast photo diode and an analogue scope.

The presented approach yields a practical femtosecond laser, which is a perfect seed source for high-power amplifier systems. The parabolic pulse shape is a highly desirable initial condition for further amplification as it helps to avoid excessive accumulation of higher order phase terms along the amplifier chain, which usually degrades the pulse quality and, hence, limits the overall system performance.

The ideal femtosecond laser source for applications outside a laboratory environment would be a completely alignment- and maintenance-free system. Hence, the above laser was developed further to obtain complete fiber integration. The resulting passively mode-locked all-fiber laser is shown schematically in Fig. 4.7. The dispersion compensation is implemented in fiber integrated form as a chirped fiber bragg grating, the saturable absorber mirror is directly glued to the fiber end face. This results in a self-starting, alignment-free turn-key device, which is environmentally stable. Such a compact and robust femtosecond laser is ideally suitable as seed source for compact master oscillator power amplifier systems to be used in rough environments.

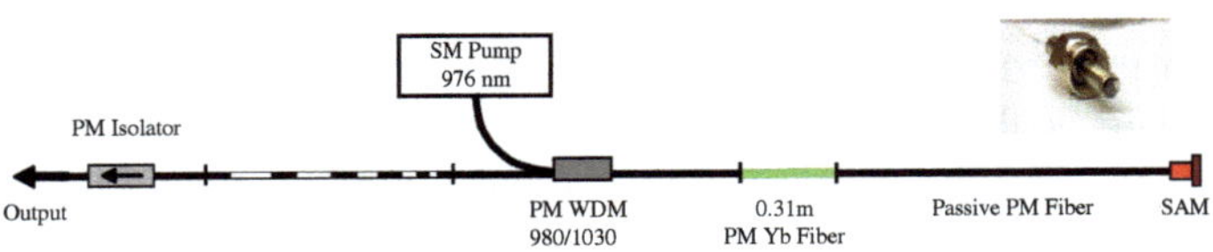

Fig. 4.7 Schematic setup of a passively mode-locker all-fiber oscillator

4.4 High Average Power Femtosecond Fiber Amplifier

In a fiber based CPA system, sufficient pulse stretching and the enlargement of the mode-field diameter of the fiber to reduce the peak power and therefore nonlinear effects such as SRS and SPM are the key points to scale the output parameters. The setup of the state-of-the-art high average power femtosecond fiber laser system is shown in Fig. 4.8 [29]. It consists of a passively mode-locked solid-state laser oscillator, a grating stretcher, a three-stage ytterbium-doped single-mode photonic crystal amplifier and a dielectric reflection-type multi-layer dielectric grating compressor.

The front end oscillator produces 200 fs pulses at 1042 nm signal wavelength with 78 MHz pulse repetition frequency and 150 mW average power. The pulses are stretched to 800 ps duration. Afterwards, the remaining signal (120 mW) is amplified to 50 W using two amplifier stages that comprise 1.2- and 1.5-m-long double-clad PCFs, both with 40 μm core and 170 μm pump cladding diameters. All fibers are pumped at 976 nm wavelength. The main amplifier fiber is a water-cooled 8-m-long double-clad fiber with 26 μm mode field diameter and 500 μm air clad. This step index fiber has no PCF structure, but the core is nanostructured and consists of Yb- and F-doped glass rods with sub-wavelength diameters arranged in such a way that they allow tuning the refractive index and, at the same time, reducing the NA. This large mode area fiber was especially designed for high average power operation, i.e. it possesses a large air clad and a signal core that guarantees single mode operation even at this average power levels. The resulting amplifier characteristic is depicted in Fig. 4.9.

At a launched pump power of 1450 W the signal output power is 950 W corresponding to 12.2 μJ pulse energy. The beam quality of the main amplifier was measured to be $M^2 = 1.3$ at this power level. The resulting maximum compressed signal power is 830 W with 10.6 μJ pulse energy. The compression efficiency drops slightly from 95 to 88 % at the maximum signal power because of depolarization taking place inside the fiber. Owing to the acquired nonlinear phase (mainly in the main amplifier) with a calculated B-integral of 11 rad, the autocorrelation width

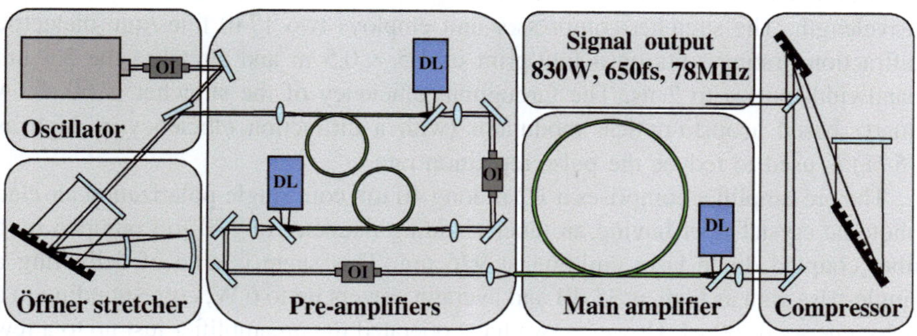

Fig. 4.8 Schematic setup of a high-average power femtosecond CPA system. *DL* Diode laser, *OI* optical isolator

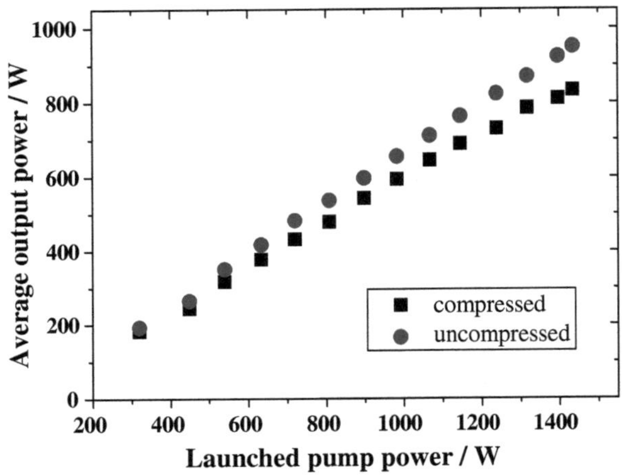

Fig. 4.9 Uncompressed (*circles*) and compressed (*squares*) signal average powers depending on the launched pump power

increases from initially 750 to 880 fs, the corresponding peak power is approximately 12 MW. The setup represents the highest average power among all femtosecond fiber laser systems with close to diffraction-limited beam quality.

4.5 High Pulse Energy Ultrafast Fiber Amplifier

Based on the considerations of nonlinearity in fiber chirped pulse amplification systems, an ultrafast high peak power setup has been developed. The schematic setup of the high energy high average power fiber CPA system is shown in Fig. 4.10 [30]. It consists of a passively mode-locked Yb:KGW oscillator, a dielectric grating stretcher-compressor unit, an acousto-optical modulator used as pulse selector and two ytterbium-doped photonic crystal fibers both used in single-pass configuration as amplification stages providing an overall gain factor of approximately 25,000.

The long-cavity Yb:KGW oscillator delivers transform-limited 400 fs pulses at a repetition rate of 9.7 MHz with an average power of 1.6 W at 1030 nm center wavelength. The stretcher-compressor-unit employs two 1740 lines/mm dielectric diffraction gratings, fits on a foot-print of 1.5 × 0.5 m and stretches the 3.3 nm bandwidth pulses to 2 ns. The throughput efficiency of the stretcher is 60 %. A quartz based acousto-optical modulator (with a diffraction efficiency as high as 75 %) is used to reduce the pulse repetition rate.

The pre-amplifier comprises a 1.2 m long 40 μm core single-polarization air-clad photonic crystal fiber having an inner cladding diameter of 170 μm pumped by a fiber coupled diode laser emitting at 976 nm. This stage is able of delivering a single-pass gain as high as 35 dB and average powers up to 6 W, corresponding to a pulse energy of 60 μJ. However, we have operated the preamplifier just up to a few μJ of pulse energy to avoid excessive accumulation of nonlinear phase in this stage. The main amplifier is constructed using a 1.2 m long low-nonlinearity air-cladding

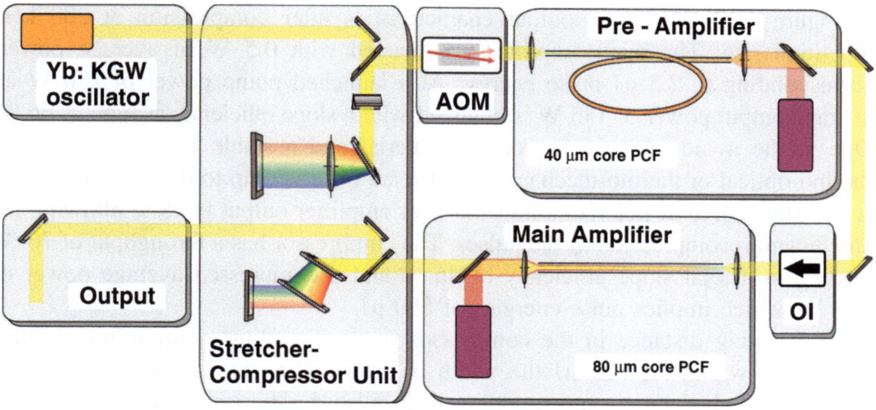

Fig. 4.10 Schematic setup of the mJ level high repetition rate fiber CPA system; *OI* optical isolator, *AOM* acousto-optical modulator

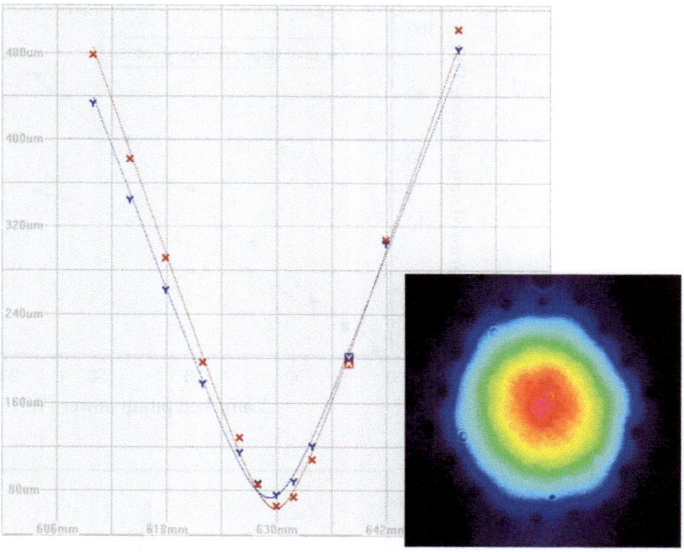

Fig. 4.11 Near-field intensity profile of the 88 μm core emission and caustic of the beam quality measurement at 100 W output power

photonic crystal rod-type fiber as described above. The core supports very few transverse modes; however, stable excitation of the fundamental mode only is achieved by seed mode matching. Figure 4.11 shows a near-field intensity profile of the large-mode area photonic crystal fiber output together with the beam quality characterization measurement. Result is a power independent beam quality characterized by a M^2-value of less than 1.2.

Figure 4.12 shows the output characteristics after compression at 200 kHz repetition rate. The main amplifier was seeded with 0.5 W of average power, corresponding to 2.5 μJ pulse energy. At a launched pump power of 230 W an average output power of 145 W is obtained with a slope efficiency as high as 66 %. Due to the avoidance of any coating material and a stable fiber mounting, no thermo-optical or thermo-mechanical issues are observed up to this average power level. The degree of polarization of the fiber amplifier output is 98 % allowing for an efficient recompression of the pulses. The compressor has a throughput of 70 %, leading to a total slope efficiency of 46 % and a compressed average power of 100 W, which implies pulse energies of 500 μJ.

The grating distance in the compressor is always adjusted for minimal autocorrelation width. At 500 μJ the width is measured to be 1.2 ps, as shown in Fig. 4.13 (dashed line), corresponding to a pulse duration of 780 fs assuming a sech^2 pulse shape. For comparison the autocorrelation trace at very low pulse energy is also shown (Fig. 4.13, dotted line). The autocorrelation traces reveal a

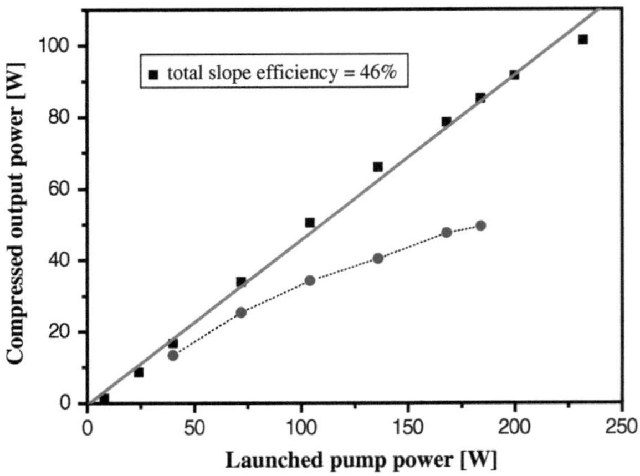

Fig. 4.12 Compressed average power of the high-energy fiber CPA system at 200 kHz (*black squares*) and at 50 kHz (*red circles*)

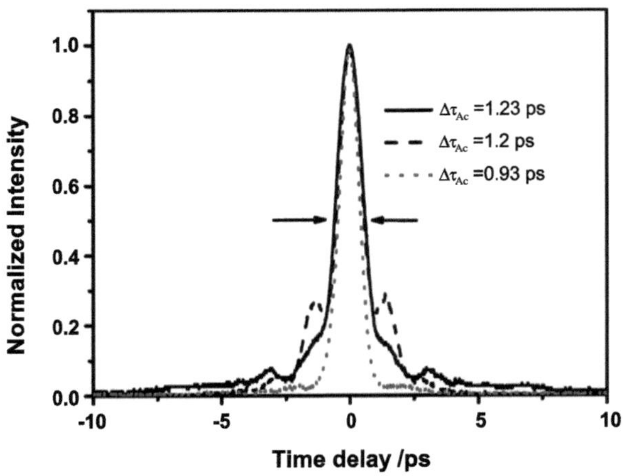

Fig. 4.13 Measured autocorrelation traces of the compressed pulses; *dotted* at low pulse energy, *dashed* 200 kHz and 500 μJ, *solid* 50 kHz and 1 mJ

wing structure growing with pulse energy, which can be attributed to the imposed nonlinear phase. The total B-integral is calculated to be 4.7 rad in the case of the 500 μJ pulses.

Following the discussion above, a further increase of pulse energy without further degradation of pulse quality was possible by decreasing the seed power launched into the main amplifier. On the one hand this reduces the B-integral of the pre-amplifier and, on the other hand the effective nonlinearity of the main amplifier, and therefore the accumulated nonlinear phase per pulse energy, is reduced as well. However, the amplification efficiency is also reduced due to increased pump light absorption saturation, as revealed in Fig. 4.12. At 100 kHz repetition rate, a launched pump power of 220 W and a seed average power of 100 mW were necessary to extract 100 W of average power corresponding to 1 mJ of pulse energy, leading to 700 μJ compressed pulse energy. At 50 kHz repetition rate and 70 mW seed power we achieved 71 W of average power with a pump power of 180 W, corresponding to 1.45 mJ energy. In this case the B-integral was 7 rad. A further increase in pump power was avoided due to an increased risk of end facet damage and pump light absorption saturation. The compressed pulses exhibited an autocorrelation width of 1.23 ps (equivalent to 800 fs pulse duration), as also shown in Fig. 4.13, and a pulse energy as high as 1 mJ. The stronger nonlinearity is indicated by the wider spread in time of the pedestal. The corresponding pulse peak power is approximately 1 GW.

The output spectrum of the high performance fiber CPA system at the highest extracted pulse energy is shown in Fig. 4.14. It is characterized by a 2.9 nm width (FWHM) and a hard cut of 7.5 nm bandwidth defined by the stretcher-compressor-unit. Only minor gain narrowing is observed for the power levels reported herein. Amplified spontaneous emission and intermediated (non-selected) pulses are suppressed better than 35 dB. The inset of Fig. 4.14 displays a larger wavelength scan of the spectrum showing no evidence of Raman scattering.

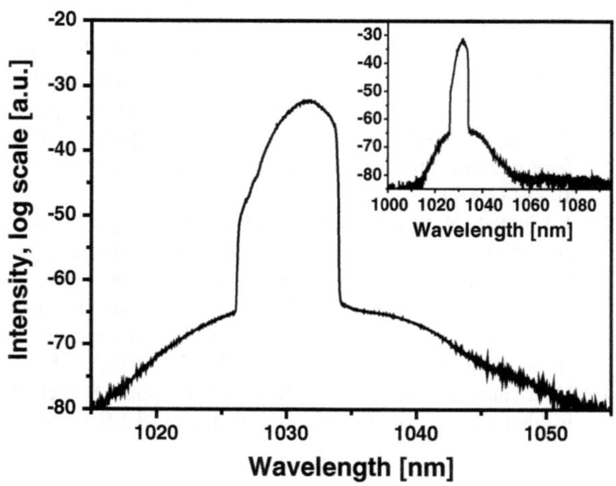

Fig. 4.14 Spectral characteristics of the output at 1.45 mJ pulse energy; *inset* broader wavelength scan

4.6 Conclusion and Outlook

Fiber based ultrashort pulse oscillators and amplifiers have reached impressive parameters regarding average power, pulse energy, peak-power and beam quality. Due to this immense progress and their robustness against environmental perturbations they are now ideal sources for fundamental science and industrial production. The further scaling potential is promising and fiber systems delivering femtosecond pulses with average powers beyond 1 kW and pulse energies of several mJ will be realized in the near future.

References

1. A. Giesen, H. Hügel, A. Voss, K. Wittig, U. Brauch, H. Opower, Scalable concept for diode-pumped high power solid-state lasers. Appl. Phys. B **58**, 365–372 (1994)
2. W.B. Jones, L.M. Goldman, J.P. Chernoch, W.S. Martin, The mini-FPL—a face-pumped laser: concept and implementation. IEEE J. Quantum Electron. **8**, 534 (1972)
3. R. Paschotta, J. Nillson, A.C. Tropper, D.C. Hanna, Ytterbium-doped fiber amplifiers. IEEE J. Quantum Electron. **33**, 1049 (1997)
4. E. Snitzer, H. Po, F. Hakimi, R. Tumminelli, B.C. McCollum, in *Double-clad, offset core Nd fiber laser" in Optical Fiber Sensors*, ed. by 1988 OSA Technical Digest Series, vol. 2 (Optical Society of America, Washington, D.C., 1988) postdeadline paper PD5
5. www.ipgphotonics.com
6. G.P. Agrawal, *Nonlinear Fiber Optics*, 4th edn. (Academic Press, San Diego, 2007)
7. R.R. Alfano, *The Supercontinuum Laser Source* (Springer-Verlag, New York, 1989)
8. J. Limpert, A. Liem, M. Reich, T. Schreiber, S. Nolte, H. Zellmer, A. Tünnermann, J. Broeng, A. Petersson, C. Jakobsen, Low-nonlinearity single-transverse-mode ytterbium-doped photonic crystal fiber amplifier. Opt. Express **12**, 1313–1319 (2004)
9. J. Limpert, O. Schmidt, J. Rothhardt, F. Röser, T. Schreiber, A. Tünnermann, S. Ermeneux, P. Yvernault, F. Salin, Extended single-mode photonic crystal fiber lasers. Opt. Express **14**, 2715–2720 (2006)
10. C.D. Brooks, F. Di Teodoro, Multimegawatt peak-power, single-transverse-mode operation of a 100 μm core diameter, Yb-doped rodlike photonic crystal fiber amplifier. Appl. Phys. Lett. **89**, 111119 (2006)
11. T. Schreiber, F. Röser, O. Schmidt, J. Limpert, R. Iliew, F. Lederer, A. Petersson, C. Jacobsen, K. Hansen, J. Broeng, A. Tünnermann, Stress-induced single-polarization single-transverse mode photonic crystal fiber with low nonlinearity. Opt. Express **13**, 7621–7630 (2005)
12. B. Ortaç, O. Schmidt, T. Schreiber, J. Limpert, A. Tünnermann, A. Hideur, High-energy femtosecond Yb-doped dispersion compensation free fiber laser. Opt. Express **15**, 10725–10732 (2007)
13. J. Limpert, N. Deguil-Robin, I. Manek-Hönninger, F. Salin, T. Schreiber, A. Liem, F. Röser, H. Zellmer, A. Tünnermann, A. Courjaud, C. Hönninger, E. Mottay, High-power picosecond fiber amplifier based on nonlinear spectral compression. Opt. Lett. **30**, 714–716 (2005)
14. O. Schmidt, J. Rothhardt, F. Röser, S. Linke, T. Schreiber, K. Rademaker, J. Limpert, S. Ermeneux, P. Yvernault, F. Salin, A. Tünnermann, Millijoule pulse energy Q-switched short-length fiber laser. Opt. Lett. **32**, 1551–1553 (2007)
15. H.A. Haus, K. Tamura, L.E. Nelson, E.P. Ippen, Stretched-pulse additive pulse mode-locking in fiber ring lasers: theory and experiment. IEEE J. Quantum Electron. **31**, 591–598 (1995)

16. K. Tamura, L.E. Nelson, H.A. Haus, E.P. Ippen, Soliton versus non-soliton operation of fibre ring lasers. Appl. Phys. Lett. **64**, 149–151 (1994)
17. K. Tamura, E. Ippen, H. Haus, L. Nelson, 77-fs pulse generation from a stretched-pulse mode-locked all-fiber ring laser. Opt. Lett. **18**, 1080–1082 (1993)
18. G. Lenz, K. Tamura, H.A. Haus, E.P. Ippen, All-solid-state femtosecond source at 1.55 μm. Opt. Lett. **20**, 1289–1291 (1995)
19. L. Nelson, S. Fleischer, G. Lenz, E. Ippen, Efficient frequency doubling of a femtosecond fiber laser. Opt. Lett. **21**, 1759–1761 (1996)
20. L. Lefort, J. Price, D. Richardson, G. Spühler, R. Paschotta, U. Keller, A. Fry, J. Weston, Practical lownoise stretched-pulse Yb3+-doped fiber laser. Opt. Lett. **27**, 291–293 (2002)
21. B. Ortaç, A. Hideur, T. Chartier, M. Brunel, C. Özkul, F. Sanchez, 90 fs generation from a stretched-pulse ytterbium doped fiber laser. Opt. Lett. **28**, 1305 (2003)
22. A. Albert, V. Couderc, L. Lefort, A. Barthélémy, High energy femtosecond pulses from an ytterbium doped fiber laser with a new cavity design. IEEE Photon. Technol. Lett. **16**, 416–418 (2004)
23. F.Ö. Ilday, J. Buckley, H. Lim, F.W. Wise, W. Clark, Generation of 50-fs, 5-nJ pulses at 1.03 μm from a wave-breaking-free fiber laser. Opt. Lett. **28**, 1365 (2003)
24. F.Ö. Ilday, J.R. Buckley, F.W. Wise, Self-similar evolution of parabolic pulses in a laser. Phys. Rev. Lett. **92**, 213902 (2004)
25. J. Buckley, F.Ö. Ilday, F.W. Wise, T. Sosnowski, Femtosecond fiber lasers with pulse energies above 10 nJ. Opt. Lett. **30**, 1888 (2005)
26. www.batop.de
27. T. Clausnitzer, J. Limpert, K. Zöllner, H. Zellmer, H.-J. Fuchs, E.-B. Kley, A. Tünnermann, M. Jupé, D. Ristau, Highly efficient transmission gratings in fused silica for chirped-pulse amplification systems. Appl. Opt. **42**, 6934–6938 (2003)
28. M. Guina, N. Xiang, A. Vainionpää, O.G. Okhotnikov, T. Sajavaara, J. Keinonen, Self-starting stretched pulse fiber laser mode locked and stabilized with slow and fast semiconductor saturable absorbers. Opt. Lett. **26**, 1809–1811 (2001)
29. T. Eidam, S. Hanf, E. Seise, T.V. Andersen, T. Gabler, C. Wirth, T. Schreiber, J. Limpert, A. Tünnermann, Femtosecond fiber CPA system emitting 830 W average output power. Opt. Lett. **35**, 94–96 (2010)
30. F. Röser, T. Eidam, J. Rothhardt, O. Schmidt, D.N. Schimpf, J. Limpert, A. Tünnermann, Millijoule pulse energy high repetition rate femtosecond fiber chirped-pulse amplification system. Opt. Lett. **32**, 3495–3497 (2007)

Chapter 5
Ultrafast Thin-Disk Lasers

Oliver H. Heckl, Jochen Kleinbauer, Dominik Bauer, Sascha Weiler, Thomas Metzger and Dirk H. Sutter

Abstract Thin-disk laser technology [1, 2] combines the inherent advantages of allowing extremely high average power, excellent beam quality, and high peak power levels [3]. In this chapter the benefits and limitations of the thin-disk concept with respect to ultrafast operation will be discussed. Different aspects of pulse duration, energy, peak- and average-power will be presented, reflecting the state of the art, with latest updates as of 2015.

5.1 Introduction

During the last decade, three solid state laser geometries have proven to support average power levels well above a kilowatt with nearly diffraction-limited transverse mode quality, all based on ytterbium as the active laser ion [4]: Thin-disk laser oscillators, fiber laser amplifier and slab laser amplifier systems, compare [5, 6] as well as the chapters on fiber and slab amplifiers of this book. All those geometries exhibit excellent heat transport capabilities compared to traditional solid state laser rods, since their ratio between cooling surface and active volume is very large. This chapter will focus on thin-disk technology in ultrafast operation.

O.H. Heckl
TRUMPF Laser-und Systemtechnik GmbH, Johann-Maus-Str. 2, 71254 Ditzingen, Germany

Present Address:
O.H. Heckl
JILA—University of Colorado, Boulder, USA

J. Kleinbauer · D. Bauer · D.H. Sutter (✉)
TRUMPF Laser GmbH, Aichhalder Str. 39, 78713 Schramberg, Germany
e-mail: dirk.sutter@de.trumpf.com

S. Weiler
TRUMPF Inc., 111 Hyde Road, Farmington, CT 06032, USA

T. Metzger
TRUMPF Scientific Lasers GmbH + Co, KG, Feringastr. 10a, 85774 Unterföhring, Germany

S. Nolte et al. (eds.), *Ultrashort Pulse Laser Technology*,
Springer Series in Optical Sciences 195, DOI 10.1007/978-3-319-17659-8_5

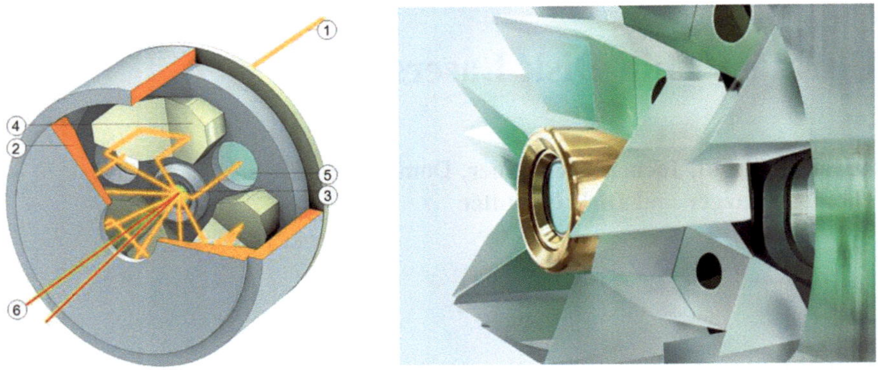

Fig. 5.1 *Left* Schematic setup of the thin-disk pump configuration with the disk in the focal plane of a parabolic mirror. The collimated pump light (*1*) is focused by a parabolic mirror (2) onto the thin-disk (*3*), re-collimated by the parabolic mirror, routed by a pair of bending mirrors (*4*) via another position on the parabolic mirror back to the disk, and ultimately the remaining unabsorbed pump light is retro-reflected back by a rear mirror (*5*). The laser beam to be amplified (*6*) reaches the disk through a hole in the parabolic mirror. *Right* Inside of a quasi-monolithic industrial implementation of a pump beam folding geometry based on multiple reflecting prisms around the central laser disk. The disk is glued to a water-cooled diamond heat sink. This efficient cooling scheme leads to an almost one dimensional heat flow and strongly reduced thermal lensing. (*Images* TRUMPF GmbH + Co. KG)

In thin-disk lasers, the active medium of the laser has the shape of a thin-disk with a highly reflective coating on one side and an antireflective coating on the other side for both pump and laser wavelength [1]. The reflective side is mounted directly onto a heat sink and the disk is used in reflection. The thin-disk thus acts as an active mirror within the laser resonator. The efficient heat flow is mainly one-dimensional towards the heat sink, i.e. collinear to the direction of the laser beam. The thermal gradients in the lateral direction are weak, and thermal lensing is therefore strongly reduced compared to a rod or slab design. The pump absorption in a single pass through the thin-disk is low, but highly efficient operation is easily obtained by arranging multiple passes of the pump light through the disk. Typically this is achieved by using a parabolic mirror to focus the incoming pump light on the disk, and several bending mirror pairs to route unabsorbed pump light back to the disks in many passes, compare Fig. 5.1. Similar schemes exist that use only two bending mirror pairs but nevertheless achieve many passes of the pump beam. The multiplication of the incident pump-beam's intensity on the disk is particularly useful for pumping (quasi-)three-level gain materials. Industrial pump cavities are nowadays fabricated as compact, quasi-monolithic blocks without any adjustable parts.

Disk lasers can provide high beam quality even at high output powers, and due to the significantly lower thermal lensing the laser resonators can be designed to be dynamically stable over a much larger range of output powers than comparable slab resonators. Compared to a fiber laser, the large diameter of the pumped spot on the disk and the small thickness of the disk allow for much lower pump beam brightness, enabling simple, cost-effective power scaling of the pump diodes. Consequently, disk lasers enable very high output power levels directly out of the laser resonator. For high-brightness continuous wave (CW) laser resonators output powers in the range of several tens of kW have been demonstrated [5, 7], and even higher powers are certainly conceivable with multi-pass thin-disk amplifiers. In short, thin-disk lasers are ideal brightness converters.

An important difference between disk and fiber or slab lasers is that disk lasers are typically operated with a highly saturated gain at low output-coupling rates. Therefore thin-disk lasers and amplifiers are very insensitive to back reflections from a work piece or an experimental setup. In contrast to fiber or slab amplifier systems, thin-disk lasers therefore have little to no need for optical isolators to prevent reflections from being amplified to destructive powers.

In addition, thin-disk lasers are ideal candidates for high-energy femtosecond (fs) or picosecond (ps) operation due to their short length of transmissive materials within the resonator in combination with large beam diameters. This results in almost negligible nonlinear effects, which benefits stable pulse formation. In contrast to a fiber or slab amplifier system, the active medium in a thin-disk oscillator or amplifier is only a fraction of a millimeter thick, while the pumped spot size is typically one to two orders of magnitude larger. Hence intensities on the thin-disk and potentially resulting nonlinear effects are minimized even at high peak powers.

For the development of ultrafast lasers, fundamental transverse mode operation is crucial. The presence of multiple higher order transverse modes can lead to mode competition and instabilities. Fundamental mode operation is achieved by choosing a mirror configuration that provides a good overlap of the TEM_{00} resonator mode with the pumped laser crystal. In rod-lasers, achieving fundamental transverse mode operation has been challenging for high average powers because of the high thermal load. This load induces thermal lensing on the intracavity optical components, particularly the gain medium, in which a substantial amount of heat is generated because of the quantum defect between pump and laser wavelength and other additional parasitic processes. The resulting lateral thermal gradients can introduce wave front distortions leading to thermal lensing and/or aberrations, which can strongly degrade the transverse beam quality. Such affects are greatly reduced in the thin-disk geometry.

A very successful way to produce pulsed output directly out of a laser oscillator is the mode-locking technique. In this technique, several longitudinal laser modes are “locked” in their relative phases to each other. These modes add up and form a laser pulse [8–10]. The lower limit of the pulse duration is ultimately given by the spectral width that can be added together and hence by the number of modes of the laser pulse. In order to force the laser into mode-locked operation one has to introduce higher losses for CW than for mode-locked operation. This can be achieved by active modulation of the losses for example with an acousto-optic or an electro-optic modulator or by passive loss modulation. Mainly two schemes have successfully been used to obtain passive mode-locking from ultrafast thin-disk lasers, i.e. Semiconductor Saturable Absorber Mirror (SESAM) [11, 12] and Kerr-lens mode-locking (KLM) [13, 14].

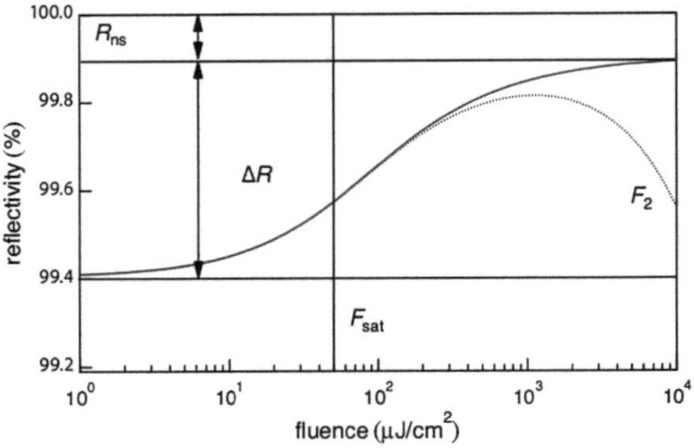

Fig. 5.2 Schematic reflectivity curves for two SESAMs (*solid blue* and *dotted green*) with the same values for modulation depth ΔR, saturation fluence F_{sat} and nonsaturable losses R_{ns}. The reflectivity of a SESAM increases with the fluence on the SESAM up to a point where two-photon absorption accounts for a "roll over", here at a fluence of 1000 μJ/cm^2 for the *green curve*. The SESAM depicted with the *blue curve* does not show a rollover in this schematic

In the approach of KLM the Kerr-nonlinearity is used to generate intensity dependent lensing to introduce losses for lower intensities, hence favoring higher intensities (pulsed operation) over CW operation. In the case of SESAM mode-locking, the saturable mirror accounts for an intensity dependent loss modulation, see Fig. 5.2. Up to the so called rollover the reflectivity of the SESAM increases with the fluence leading to reduced losses for pulsed operation. The roll over is caused by nonlinear absorption processes, such as two photon absorption. For a more detailed introduction to passive mode-locking see [10] or [15].

The concept of the thin-disk laser has the unique advantage of power scalability: The output power can be scaled by increasing the pump power and the mode areas on the gain medium. In case of mode-locked operation the mode area on the SESAM is increased by the same factor as on the gain medium. The temperature of the thin-disk will not rise if the cooling system is capable of removing the additional heat. This scaling procedure avoids damage on the key components, because the intensities on the disk and on the saturable absorber remain unchanged [16]. Furthermore, the saturation of the gain and the SESAM remains constant, which is important for preventing mode-locking instabilities [17].

5.2 Overview of Disk Materials and Pulse Durations

By far the highest CW laser output powers have been achieved from ytterbium doped yttrium aluminum garnet ($Y_3Al_5O_{12}$; "YAG") disks. The Yb^{3+} ion offers the advantage of a simple energy level structure with low quantum defect (i.e. the majority of the pump energy contributes to stimulated emission while only a small

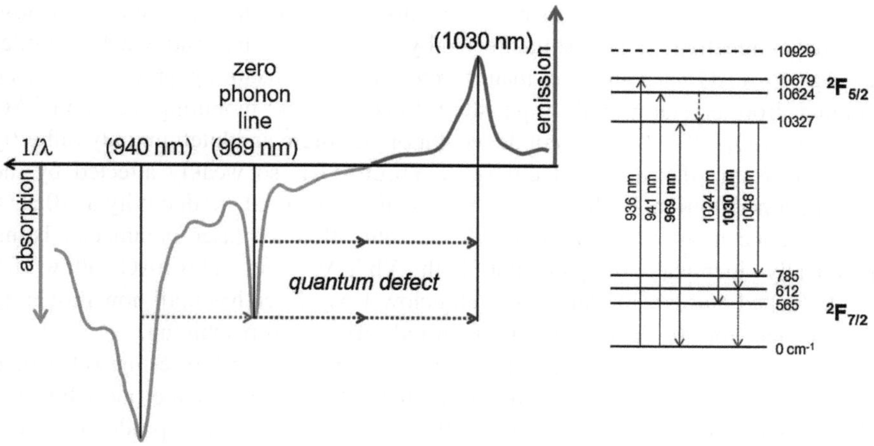

Fig. 5.3 Gain cross section of Yb:YAG (*left* shown for 25 % inversion), resulting from the $^2F_{5/2}$–$^2F_{5/2}$ transitions (energy levels on the *right*), on a wave number respective energy scale, indicating the reduction of the residual quantum defect by a third by switching to pumping at the zero phonon line

fraction decays non-radiatively) [4]. The correspondingly low heat generation is an essential prerequisite for high average power generation at good beam quality. For 938 nm pumping, the quantum defect amounts to only ~9 %, whereas traditional Nd:YAG or Nd:YVO_4 (yttrium orthovanadate) lasers pumped at 808 nm exhibit a 2.7 times larger quantum defect of 24 %. Zero phonon line pumping of Yb:YAG at 969 nm further reduces the quantum defect by a third [18, 19], compare Fig. 5.3. Although the absorption peak at 969 nm is narrower than at 938 nm, high power laser diodes can efficiently be locked to a suitably narrow spectral width (typically $\Delta\lambda_{\text{diode}} < 0.5$ nm) to reliably pump the zero-phonon transition, e.g. with a volume Bragg grating [20–22].

The 9 nm bandwidth of the main gain peak at 1030 nm is broad enough to achieve pulse durations of a few hundred fs. If desired, it is straight forward to use pulse compression schemes based on nonlinear spectral broadening and subsequent dispersive pulse compression to obtain shorter pulses, e.g. by pulse propagation in gas filled hollow-core photonic fibers [23–27]. Besides its spectroscopic properties, Yb:YAG also exhibits excellent thermal and mechanical properties of the YAG host such as excellent heat conductivity, good hardness and stiffness, and an isotropic thermal expansion coefficient, which are all important for the fabrication of thin laser disks. However, in order to increase absorption and gain of the disk, significant doping is desired, and the thermal conductivity of crystals host usually drops somewhat with rising doping concentration of the active ion. This effect is caused by the different atomic weights of the substituted ion and the doping ion: The change in atomic weight across the crystal structure increases phonon scattering and thus decreases thermal conductivity. While the high thermal conductive of YAG of 9.4 W/m K drops to about 6 W/m K at only 10 at.% of ytterbium doping (ytterbium being about twice as heavy as yttrium), an alternative host exists whose thermal conductivity is almost unaffected by ytterbium doping, and which is similar enough to YAG such that the mature Yb:YAG manufacturing processes can be applied, from crystal growth to polishing to coating and mounting, i.e. Yb:LuAG (Yb:$Lu_3A_{15}O_{12}$) [28]. As the atomic weight of ytterbium and lutetium only differ by about 1 %, the thermal conductivity of Yb:LuAG is so weakly affected by the doping concentration, leading e.g. to a 20 % higher thermal conductivity at 10 at.% doping concentrations then in Yb:YAG, with all other laser parameters being practically identical. In particular both Yb:YAG and Yb:LuAG allow for mode-locked pulses with durations well below 1 ps, which has until now proven to be sufficient for all major applications in industrial micromachining.

Nevertheless, significant academic work has been devoted to testing other gain materials in the thin-disk geometry, in particular motivated by the quest for ever shorter pulse durations, despite the challenge of developing new production techniques. The key requirement for supporting a large amount of locked longitudinal modes for fs pulses is a sufficiently large emission bandwidth, but typically this comes at the expense of reduced gain coefficients as well as inferior thermal properties. Nevertheless, some attractive host materials next to YAG have been identified for ytterbium doping, partially providing broader emission bands and at least comparable thermal conductivity, notably from the group of sesquioxides [29]. They were among the first materials to be tested as an alternative to Yb:YAG in the thin-disk laser geometry [30], and they have been recognized as potentially more efficient gain materials than Yb:YAG [31–34]. A high mechanical strength in combination with excellent thermal properties and high absorption cross sections make them ideally suited for the thin-disk laser geometry. They are also very suitable for high average power fs ultrafast sources due to their larger emission bandwidths in comparison to Yb:YAG [35, 36]. However, the growth of sesquioxide single crystals with their high melting points around 2400 °C is challenging, compare [37], which resulted in a poor crystal quality and optical-to-optical efficiency of less than 50 % in earlier experiments [33, 38]. During the last years, these difficulties have partly been overcome in laboratory environments for the growth of high quality, large scale sesquioxide single crystals [39, 40], with ceramics still being a possible alternative to single crystals [36, 41]. In particular, Ytterbium doped lutetia, Yb:Lu_2O_3, achieved 500 W average output power in CW operation [42], and the zero-phonon absorption line at 975 nm is somewhat broader than that of Yb:YAG at 969 nm. Yb:Lu_2O_3, might thus have a potential to become a second major gain material for future ultrafast disk lasers, provided it can reliably be fabricated at sufficient quality and size.

5.3 Ultrafast Disk Oscillators

The first mode-locked disk oscillator was demonstrated in a collaboration of the groups of Ursula Keller at ETH Zurich and Adolf Giesen from the University of Stuttgart, which pushed the average power level obtained directly from a mode-locked oscillator from typically less than 4–16 W [43]. It was followed by many experiments that successfully scaled the average output power [44], extended the wavelength range [45], increased the pulse energy and intensity [14, 46–52], or reduced the pulse duration, partly by employing other laser materials than Yb:YAG that offer a broader gain bandwidth [35, 53–59]. The straightforward approach to generating higher laser pulse energies has always been to reduce the pulse repetition rate, leading to a concentration of more power in fewer pulses. However, an important limitation for high energy, ultrafast oscillators stems from the nonlinearities that eventually occur at high intensities inside the resonator. While the nonlinearity of a thin-disk can typically be neglected (because of its small thickness and the large beam radius at the disk), the propagation through air during the long beam path of traditional low repetition rate oscillators results in significant nonlinear self-phase modulation at high intra-cavity intensities.

Two potentially complementary approaches have evolved to overcome this limitation, the first being to evacuate the resonator or to flush it with a gas of low nonlinearity such as helium, the other to increase the number of reflections of the disk within a resonator round-trip and to increase the output coupling ratio accordingly, giving access to a larger portion of the intracavity pulse energy. With helium flooding, energies as high as 11 μJ have been generated directly from a disk oscillator [47]. Subsequently the introduction of a mode-locked thin-disk laser employing an active multi-pass cell (AMC) by TRUMPF enabled operation at much higher output-coupling rates and lead to a substantial increase in pulse energy again based on Yb:YAG [48–51].

Before we discuss the AMC concept, we will first give an overview of pulse durations and pulse energies achieved with mode-locked thin-disk lasers. For some time it has been a common understanding that lasers providing shorter pulse durations generally tend to deliver less output power. However, KLM Yb:YAG oscillators have also been demonstrated with extremely short pulses at very high average powers [14, 60, 61], even going down to sub-50 fs, corresponding to a spectral width exceeding the fluorescence bandwidth of the gain material [62]. An overview of achieved pulse durations and average output powers in mode-locked thin-disk lasers based on different disk materials is shown on the left side of Fig. 5.4. By its nature KLM requires a somewhat stringent cavity design but it offers the advantage of an almost immediate intensity dependent response of the cavity losses. These fast dynamics can help to stabilize shorter pulse durations than achieved in purely SESAM mode-locked operation [63]. A more detailed discussion about materials for mode-locked thin-disk lasers can be found in [35]. The right side of Fig. 5.4 shows the evolution of mode-locked Yb:YAG thin-disk pulse energy records over time.

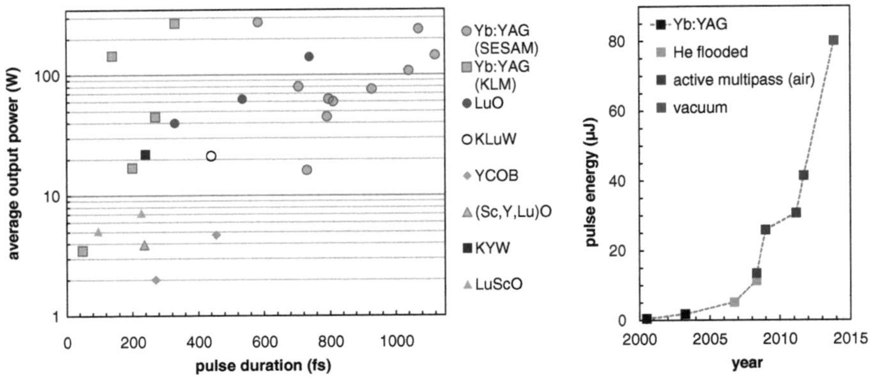

Fig. 5.4 Non-exhaustive overview of pulse duration versus average output power of Yb-doped thin-disk lasers. In the case of Yb:YAG there are two different symbols. The *dots* refer to SESAM mode-locked thin-disk lasers and the *squares* to Kerr lens mode-locking (*left*). Pulse energy records of mode-locked Yb:YAG thin-disk lasers (*right*) over time. The original publications can be found in the references cited in this chapter

By 2012 the average output power of a SESAM mode-locked thin-disk laser at the ETH Zurich was further increased to a value of 275 W by operating the laser in a vacuum environment [64]. Later in 2012 the average power record was matched with 270 W from a KLM thin-disk laser operating in air [14]. With another SESAM-based oscillator operating in vacuum the pulse energy was increased up to 80 μJ with 66 MW peak power at 1.07 ps duration in 2014 [52]. In the following year a similar peak power of 56 MW was reached with KLM at 140 fs duration at 145 W average power [65]. Overviews of SESAM-mode-locked thin-disk lasers can be found in [66–68].

In the following we address the active multi-pass cell, which allows for accessing a larger portion of the intra-cavity power and is therefore beneficial for high-intensity operation. In an active multi-pass cell, the laser disk is used inside the multi-pass cell to increase the roundtrip gain, a concept that has originally been introduced for thin-disk laser amplifiers [69, 70]. In parallel, the multi-pass increases the resonator length as desired in order to reduce the repetition rate [48–51]. Due to the higher round-trip gain, a large output coupling ratio can be used. As a consequence, the external pulse energy is much closer to the intra-cavity energy.

Figure 5.5 shows a schematic layout of an AMC disk oscillator with angular multiplexing as used in [51], which resulted in the highest average output power of and the highest pulse energy of at the time. Such schemes obviously rely on a good surface quality of the disk, as wavefront distortions add up for each pass through the AMC. TRUMPF's proprietary disk mounting is the technology of choice for fabricating disk modules with suitably low distortions [71]. To balance the nonlinear dispersion due to self-phase modulation originating mostly from the ambient air, dielectric Gires-Tournois mirrors provided negative dispersion in the resonator. A SESAM self-started and stabilized soliton mode-locking. A quarter wave plate and a thin film polarizer were used to set the desired output coupling. The portions

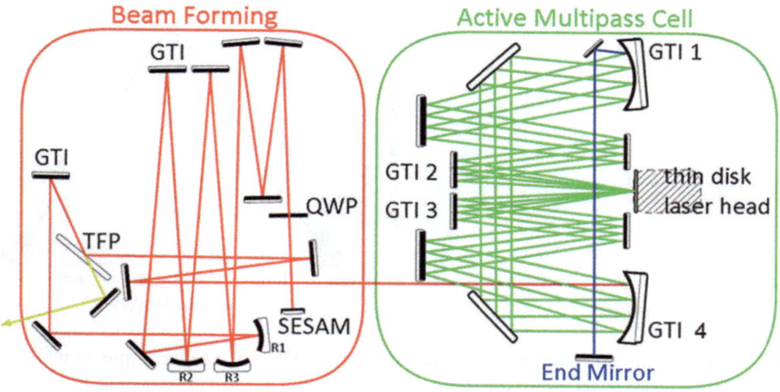

Fig. 5.5 Schematic set-up of a mode-locked oscillator with an active multi-pass cell, shown in *green*, that images the disk onto itself (for simplicity, only 4 passes in the cell are shown). *GTI* Gires-Tournois mirror; *SESAM* semiconductor saturable absorber mirror; *QWP* quarter wave plate; *TFP* thin film polarizer

of the beam path shown in red and blue were used to form a TEM_{00} resonator with the appropriate spot sizes at the thin-disk and the SESAM. Alternative to the Thin Film Polarizer (TFP) and Quarter Wave Plate (QWP), a suitable output coupler could have be used as an end mirror. Disk lasers employing an AMC can tolerate an output coupling of >70 % in contrast to the typical output coupling of <10 % for disk resonators with passive multi-pass cells. Obviously a much larger portion of the intracavity energy can be utilized at the higher output coupling. A detailed numerical study of the soliton formation in such lasers with high output coupling was given in [50].

As presented in [51] more than 40 μJ were reached at an output coupling of roughly 70 %. This pulse energy is still the highest ever achieved from an oscillator operated in ambient air. At a repetition rate of 3.51 MHz, corresponding to a resonator length of 42.7 m employing 11 passes over the thin-disk, the average power of 145 W was of the same order as the 141 W reported in [72]. The latter result was achieved by the ETH group with an $Yb:Lu_2O_3$ disk oscillator at a much higher repetition rate of 60 MHz and hence significantly lower pulse energies. Although the width of the dynamic stability zone shrinks proportional to the number of passes through the AMC, the laser reported in [51] operated in fundamental mode over the whole pump power range, indicating negligible thermal deformation of the disk's properties. The left plot of Fig. 5.6 shows the clean second harmonic autocorrelation of the obtained 41.3 μJ pulses together with a close to perfect fit based on the theoretically expected hyperbolic secant pulse shape, indicating a pulse duration of 1.12 ps. The right plot of Fig. 5.6 displays the $sech^2$ like spectrum on a linear as well as on a logarithmic scale, showing the suppression of Kelly sidebands to more than 20 dB below the main peak. This indicates a good potential for even higher pulse energies, provided that CW instabilities and double pulses can be avoided. The time-bandwidth product of

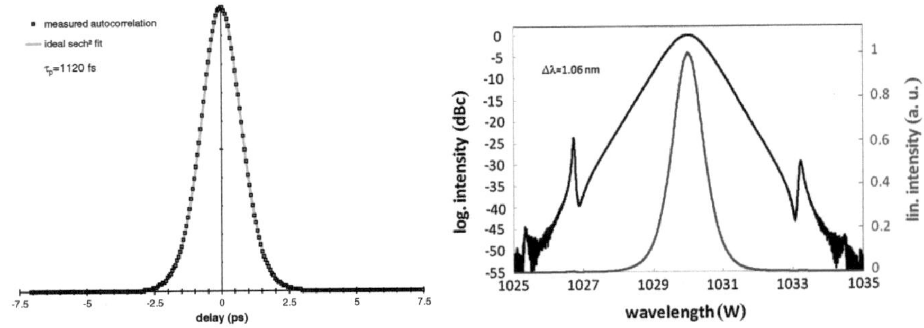

Fig. 5.6 Measured autocorrelation (*left*) and spectrum (*right*), on a logarithmic as well as a linear scale of unamplified 41.3 μJ oscillator pulses

0.336 was within 7 % of the transform limit for an ideal $sech^2$ pulse. The peak power corresponding to the measured pulse duration amounted to 32.5 MW, which was quite comparable to that of typical amplified systems and certainly sufficient for many interesting applications in micro machining. Figure 5.7 shows the spectral width and the pulse duration as a function of output power. Above 75 W of average output power stable CW mode-locking was obtained. The reciprocal reduction of pulse duration at increasing energy fitted well to the modified soliton theorem accounting for the high output coupling [49, 50]. We expect that by combining the active multi-pass approach with a helium flooded or even evacuated resonator [68], mJ level pulse energies are certainly conceivable. The evolution of mode-locked oscillators has thus led to energies that were previously only accessible with amplified systems.

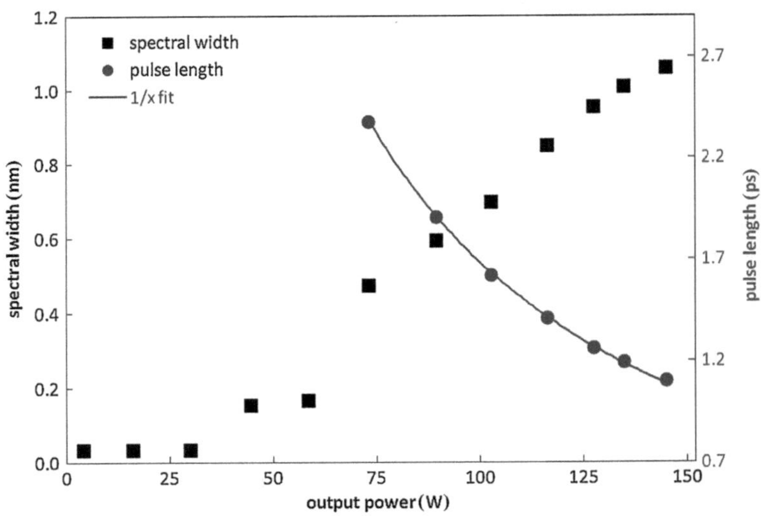

Fig. 5.7 Spectral bandwidth and pulse duration (both FWHM full width at half maximum) as a function of output power at 3.51 MHz repetition rate

5.4 Ultrafast Regenerative Thin-Disk Amplifiers

The first regenerative amplifier ("Regen") based on the thin-disk geometry was set up at ETH Zurich using an early gain module from the Giesen group in 1997, a few years prior to the first mode-locked disk oscillator demonstration. While this experiment did not reach more than 1.35 W of average power, it already achieved a peak power close to 100 MW without any need for chirped pulse amplification (CPA) [73].

A certain benefit of using a Regen is the inherently guaranteed perfect beam quality close to $M^2 = 1$ imposed by the TEM_{00} resonator. As one example Fig. 5.8 shows the final test beam characteristics of a commercial product. Coupling seed pulses into and out of this resonator is obtained by polarization switching using an intracavity Pockels cells in conjunction with a polarizing beam splitter. Beta-barium-borate (BBO) Pockels cells, e.g., can operate at standing average powers of several kW with negligible thermal lensing, due to very low absorption. In 2012, using a 25 mm long BBO-Pockels cell with 6 mm free aperture, the first 1 MHz thin-disk Regen was demonstrated [74]. That system reached sub-ps pulses of 770 fs duration at an average power of 185 W after compression. In fact, much of the average power scaling in disk Regens was only possible due to parallel improvements in high voltage switches with fast switching [75].

At the turn of the millennium, TRUMPF began its research on ultrafast lasers based on Yb:YAG disk modules originally designed for kW level CW products, the TruDisk lasers. Detailed application studies done in a federally funded collaborative project showed that few-ps pulses were ideal for precise drilling of metals, avoiding the heat affected zone and post processing known from nanosecond machining, while eliminating unwanted nonlinear interaction with the ablated plasma that occurred at shorter pulses [76]. Product development began in parallel to a second research project aiming towards increased output powers. This resulted in the world's first industrial grade 50 W ps infrared laser, which was presented in 2008 and has since been widely adopted in 24/7 production, e.g. for the drilling of fuel injection nozzles or engraving of metals and dielectrics at companies like Bosch [77]. At that time, lab results reached 80 W of average power from a TEM_{00} disk

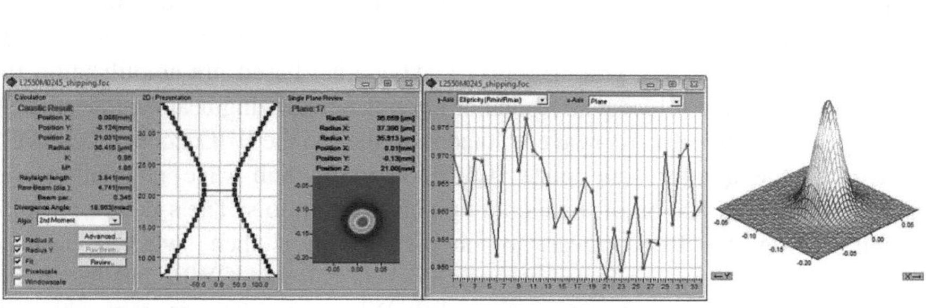

Fig. 5.8 Caustic (*left*), ellipticity (*middle*) and focus intensity profile (*right*) of a typical TruMicro 5070 with a measured $M^2 = 1.05$ and a roundness of >94 % over ±4 Rayleigh lengths (final test data at nominal power of 100 W)

Fig. 5.9 TruMicro Series 5000 laser, delivering up to 150 W of <10 ps pulses (sub-ps with CPA) at base repetition rates between 100 kHz and 1 MHz. A flexible external modulator allows for individual pulse triggering (pulse-on-demand) via TTL input as well as fast energy modulation

Regen [78]. In 2009 a more compact second generation was introduced, and in 2011, the third generation of the TruMicro Series 5000 became available, providing 100 W at 1030 nm from 200 to 800 kHz, and 60 W at 515 nm. The higher power allowed for increased throughput enabling new highly cost efficient applications e.g. for the processing of dielectrics and transparent media such as cutting of display glass in smartphone manufacturing [79]. In 2013, the technological and economic success of this innovative laser technology, which had originated from federally funded projects, was recognized by the German Future Prize, awarded by the German Federal President (Deutscher Zukunftspreis 2013 [80]).

The laser head of the fourth generation TruMicro Series 5000 [81], unveiled in 2015 with up to 150 W of average power at 1030 nm, is shown in Fig. 5.9. On a footprint of 0.6 m^2, it contains a low power, mode-locked fiber seed oscillator, followed by the disk Regen, a telescope for precise adjustment of the output beam diameter, an acousto-optic modulator for precise and highly dynamic setting of the output energy, an optional, very compact grating based pulse compressor (in the CPA version only), an optional frequency doubling or tripling module (to 515 nm resp. 343 nm), and a mechanical safety shutter, as well as the required sensors, control boards, and drivers.

Figure 5.10 shows a schematic setup of such a thin-disk Regen in ring resonator configuration. The low power fiber-based SESAM mode-locked seed oscillator generates a p-polarized train of nJ-level pulses of either picosecond duration or, in the CPA-based Femto Edition, femtosecond pulses stretched to picosecond duration by a chirped fiber Bragg grating (cFBG). After fiber-based pre-amplification the seed pulse enters the folded disk resonator through a polarizing beam splitter (PBS). As long as a positive quarter-wave voltage is applied to the BBO Pockels cell (PC), it adds to the quarter-wave plate (QWP) resulting half-wave retardation, i.e. switching the polarization of the incoming pulse from p to s. The s-polarized pulse is reflected by the PBS, travels to the disk again, is then switched back to p-polarization in its second pass through Pockels cell and QWP, and leaves the resonator after just two reflections off the disk. When the voltage is reversed before the pulse's second pass through the PC, the inverted retardation of the Pockels cell is cancelled by the QWP, such that the pulse stays s-polarized and takes many

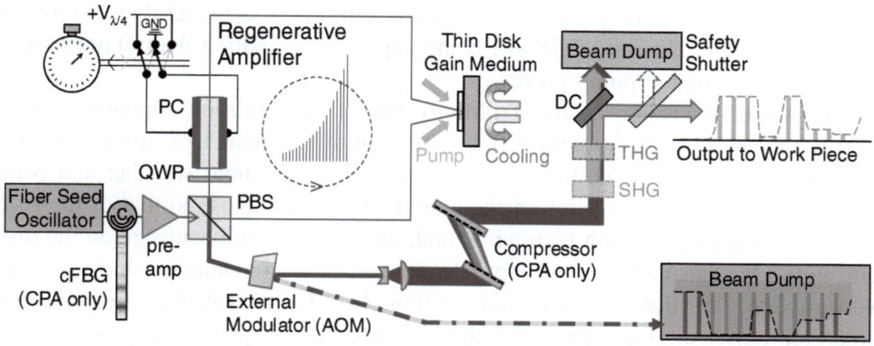

Fig. 5.10 Schematic of a regenerative thin-disk laser amplifier (*c* circulator, *cFBG* chirped fiber Bragg grating—for chirped pulse amplification (CPA) only, *PBS* polarizing beam splitter, *PC* Pockels cell, *QWP* quarter wave plate; $V_{\lambda/4}$ quarter-wave voltage; *GND* ground voltage; *AOM* acousto-optic modulator; *SHG/THG* second/third harmonic generation; *DC* dichroic mirror)

round-trips of amplification inside the Regen. The switches can apply the control voltage for the PC with sub-ns timing accuracy and <10 ns rise time, suitable for the round-trip time of the Regen resonator. Further pulses from the seed laser that may arrive during the amplification phase will only be reflected off the disk once and leave the resonator with unchanged p-polarization immediately. After the circulating s-polarized pulse has gained enough energy the Pockels cell is switched back to positive voltage, and the pulse's polarization is switched back to p such that the amplified pulse leaves the resonator. Due to the low single-pass gain of the disk, only the amplified pulse extracts significant energy from the disk, and depending on the number of round-trips the exponential amplification reaches a total gain of up to 60 dB. The disk module, compare Fig. 5.11, had already been redesigned in the

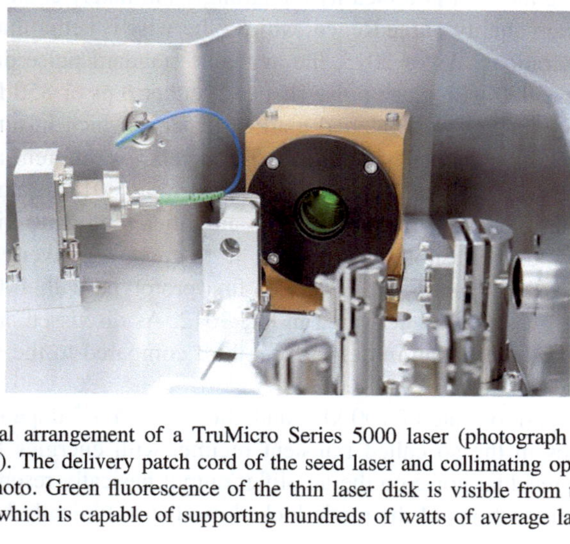

Fig. 5.11 Optical arrangement of a TruMicro Series 5000 laser (photograph from the second generation, 2009). The delivery patch cord of the seed laser and collimating optics are visible in the *left* of the photo. Green fluorescence of the thin laser disk is visible from the middle of the pump chamber, which is capable of supporting hundreds of watts of average laser power

second generation of the product to be significantly more compact than that of the kW class TruDisk lasers, while still being capable of handling many hundreds of watts of fiber-coupled pump power.

The standard version of the TruMicro 5080 Femto Edition supports a pulse energy of 200 μJ at a compressed pulse duration of less than 1 ps, the picosecond version goes up to 500 μJ. In the picosecond version (without stretcher and compressor) the sub-10 ps duration of the amplified pulses is close to that of the seed pulses, with negligible gain narrowing, and, due to the large beam size on the disk and in the Pockels cell, without any relevant self-phase modulation. In the sub-ps version, the TruMicro 5080 Femto Edition, based on CPA the stretched and amplified pulses are recompressed with a compact, single pass Treacy compressor through two transmission gratings with few-cm separation, allowing well above 90 % throughput at perfect beam quality, as only a little stretching is needed to suppress nonlinearities in the amplifier. The optical efficiency after compression and pulse picking by the first order diffraction of the acousto-optic modulator is on the order of 50 % with respect to launched pump power, and even higher for the non-CPA version without the compressor.

Clean >2 nm wide spectra had been demonstrated from a standard TruMicro 5000 Femto Edition laser head as early as 2013, with pulse compression to 650 fs [82]. Shorter pulse durations on the order of 400 fs can be reached from the same platform without using any stretcher after the fs seed, simply making use of intentional—but moderate—nonlinear spectral broadening in the amplifier, to provide the required bandwidth for shorter pulses, compare Fig. 5.12. In such a configuration, the dispersion from several reflections of strongly chirped mirror coatings can already be sufficient for pulse compression. Operating the amplifier at even higher energies allows for spectral broadening to tens of nanometers, with the temporal pulse shape becoming rather rectangular before compression [83]. In this case the spectrum supports transform-limited pulse durations below 100 fs. However, driving the Kerr nonlinearity too strongly ultimately could begin to affect the excellent beam quality of the Regen, which otherwise is better than $M^2 < 1.2$ for standard configurations. As of 2015, the available standard pulse durations of the TruMicro Series 5000 with clean pulse shapes at either 6 ps or 850 fs are therefore mostly based on practical reasons and market demand, which in turn results from extensive comparative studies of optimum parameters for different applications.

Approximately 250 W had been reached as early as 2012 at close to 70 % of optical-to-optical efficiency, maintaining the amplifier foot-print with only minor scaling of the pumped spot, resulting in a frequency converted power well exceeding 150 W, compare Fig. 5.13, and a first prototype with 100 W of second harmonic power was already delivered in late 2012. As discussed above, pumping at 969 nm reduced the quantum defect by a third compared to the standard pump wavelength of 938 nm.

At a pulse repetition rate of 100 kHz and above, the Yb^{3+} doped thin-disk gain element operates in the so called transient regime, with energies far below gain saturation. The residual noise on the amplifier output basically resembles the small pulse to pulse fluctuations of the mode-locked oscillator, typically below 1 % rms.

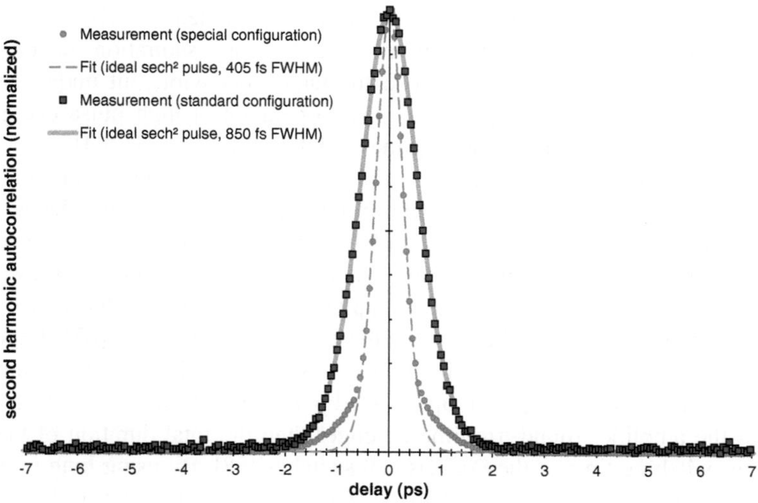

Fig. 5.12 Autocorrelations of a TruMicro 5070 Femto Edition with a standard pulse duration of 850 fs compared to a special IR version producing 405 fs pulses (*markers* measured data, *lines* fit based on sech^2 model)

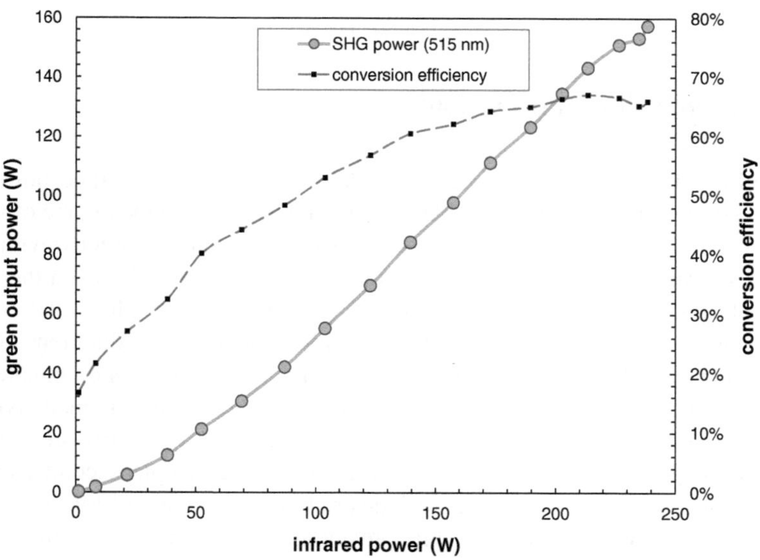

Fig. 5.13 Second harmonic conversion of a modified TruMicro Series 5000 laser head with scaled pump spot but on the same footprint of the amplifier as in the serial product TruMicro 5270, delivering 157 W of green average power at 238 W of average IR power at a repetition rate of 800 kHz (TRUMPF, 2012)

Only at much lower amplifier repetition rates of a few tens of kHz, i.e. close to resonant operation, energy fluctuations due to gain saturation effects arise. Nonlinear gain dynamics can then lead to chaotic behavior, but under carefully chosen conditions also to surprisingly stable operation at high pulse energy in a broadly intermittent single-energy regime of deterministic chaos [84].

While in almost all practical cases, the so-called single-pulse operation of the amplifier at up to a few MHz of repetition rate optimizes ablation efficiency [85], so-called burst operation, with groups of two or more pulses with pulse-to-pulse separation below tens of nanoseconds can in some cases improve surface morphology by controlled heating of the workpiece [86]. Bursts of reduced energy can also be beneficial for controlled processing of transparent media [87]. Regens do support such burst operation in multiple ways: Firstly, like for linear amplifier chains, it is possible to couple groups of seed pulses into the amplifier resonator, as long as the amplifier round-trip-time is longer than the total duration of the burst plus the switching time of the Pockels cell switches [88]. By using long resonators such as those described in Sect. 5.2 above, even total burst durations of hundreds of nanoseconds are possible. Secondly, it is possible to couple out portions of the amplified pulse in consecutive round-trips by applying suitable voltages to the intracavity Pockels cell, resulting in bursts at the repetition rate of the Regen resonator. Using patented switching technology [89] up to three identical pulse energies can thus be coupled out within one amplification cycle, which up to now has proven to be sufficient in all applications.

5.5 Conclusion and Outlook

Mode-locked thin-disk lasers oscillators deliver the highest average power and pulse energies of any mode-locked laser technology. Pulses in the regime of several tens of μJ and pulse durations around 1 ps now compete directly with more complex amplifier systems. Nevertheless amplified systems will remain the choice for applications requiring more flexibility in repetition rate or higher pulse energies in abundant industrial applications of today and tomorrow. Pulse durations of a few ps are already sufficiently short to avoid heat conduction to surrounding material of the work piece, and hence to avoid the heat affected zone and melt, typically associated with nanosecond lasers. Moreover, picosecond systems are generally simpler compared to fs systems. Thus it is fair to assume that picosecond lasers will continue to play an important role in industrial micromachining [90], while lasers with shorter pulses based on the same proven architecture but employing reliable, compact CPA schemes address more special applications.

High field experiments in particular rely on high peak intensity levels leading to a need for short pulses with high pulse energies. It is likely that the development of mode-locked thin-disk lasers will lead to reliable light sources that deliver sub 100-fs pulses at tens of microjoule pulse energies. These lasers might also be based on broadband materials other than Yb:YAG. But even with Yb:YAG, sub-50 fs pulse durations, well below the fluorescence-bandwidth-limit of the gain medium, have been obtained in 2015, using distributed Kerr-lens mode-locking, representing the shortest pulse duration achieved from any thin-disk oscillator and any Yb:YAG oscillator (at 1030 nm central wavelength) [62]. Alternatively, well established pulse shortening schemes based on nonlinear spectral broadening of Yb: YAG-based oscillators and subsequent pulse compression have already demonstrated durations below 10 fs [26], in this case using a hollow-core Kagomé type gas filled optical fiber.

For ultrafast science with highest light intensities, suitable pump sources for optical parametric amplification are required that provide significantly higher pulse energies as well as high average powers at few ps durations. Gigawatt-class pulses from disk amplifiers at pulse durations of less than 2 ps were first presented in 2012 [74]. Shorter pulses of about 0.2 ps have been demonstrated with disk Regens based on ytterbium tungstates [91, 92]. Intensities as high as 15 GW have been obtained after compression directly from a single stage regenerative CPA system [84]. In 2013, 300 W of average power was demonstrated with 30 mJ pulse energy [93]. By 2015 the energy had been scaled to more than 220 W of average power and more than 200 mJ after compression to a duration of <2 ps [94]. In other configurations, even kW class average powers have been obtained from chirped pulse thin-disk Regens with suitable stretching to avoid excessive intracavity peak intensities [95, 96].

The highest average powers of any ultrafast system with pulse durations below 10 ps, i.e. 1.4 kW in the near IR or 820 W in the green and 234 W in the UV, have been obtained with a multi-pass thin-disk amplifier [6, 97]. As these results were pump-power limited, there is potential for further scaling, which may 1 day lead to multi-kW average power ultrafast thin-disk laser systems. Pulse energies of a joule or beyond are certainly achievable based on multi-pass thin-disk amplifiers [98]. Record average powers have been obtained at DESY, Germany, reaching multi-kW output in burst operation [99, 100]. Such multi-pass amplifiers are basically similar to Regens in the sense that the small signal gain of the disk is exponentially multiplied with many reflections over the disk, but in contrast to a Regen no switching element is needed. Moreover, in today's multi-pass systems the power load on opto-mechanical components remains moderate compared to state-of-the-art TEM_{00} CW disk oscillators that easily operate at intracavity powers of tens of kW, leaving room for further scaling of the multi-pass average powers [5]. Multi-pass schemes with many tens of roundtrips are conceivable, certainly more than employed so far in [6, 97], such that a total gain of up to two orders of magnitude and an amplified average power in the multi-kW regime should be reachable even at seed-powers below 100 W.

As of today, ultrafast thin-disk regenerative amplifiers are the most successful ultrafast laser systems for high average power industrial micro machining, but their performance can be further boosted by multi-pass thin-disk amplifiers. In scientific applications, it would be surprising if thin-disk lasers and amplifiers did not play an important role for the generation of extreme light intensities in the coming years [101]. As demonstrated by carrier-envelope-phase-locked sub-6 fs pulses obtained by optical parametric amplification pumped with a thin-disk amplifier [84, 102], such systems certainly have the potential for laying the foundation of the third generation fs technology [103].

Acknowledgments The authors wish to thank their colleagues and partners without whom the development of the highly successful ultrafast thin-disk laser technology would not have been possible.

References

1. A. Giesen, H. Hügel, A. Voss, K. Wittig, U. Brauch, H. Opower, Scalable concept for diode-pumped high-power solid-state lasers. Appl. Phys. **B58**(5), 365–372 (1994)
2. A. Giesen, J. Speiser, Fifteen years of work on thin-disk lasers: results and scaling laws. IEEE J. Sel. Top. Quantum Electron. **13**(3), 598–609 (2007)
3. S. Schad, C. Stolzenburg, K. Michel, D. Sutter, Latest advances in high brightness disk lasers: high power performance from CW to short and ultra short pulses. Laser Tech. J. **11**(2), 49–53 (2014)
4. W. Krupke, Ytterbium solid-state lasers—the first decade. IEEE J. Sel. Top. Quantum Electron. **6**(6), 1287–1296 (2000)
5. S. Schad, V. Kuhn, T. Gottwald, V. Negoita, A. Killi, K. Wallmeroth, Near fundamental mode high-power thin-disk laser, in *Proceedings of SPIE 8959, Solid State Lasers XXIII: Technology and Devices*, 89590U (February 28, 2014)
6. J. Negel, A. Loescher, A. Voss, D. Bauer, D. Sutter, A. Killi, M. Abdou Ahmed, T. Graf, Ultrafast thin-disk multi-pass laser amplifier delivering 1.4 kW (4.7 mJ, 1030 nm) average power converted to 820 W at 515 nm and 234 W at 343 nm. Opt. Express **23**(16), 21064–21077 (2015)
7. A. Ullmann, M. Curtin, G. Needham, H. Wang, L. Zeldin, Development of high-brightness thin-disk lasers. J. Directed Energy **4**(1), 110–118 (2010)
8. L.E. Hargrove, R.L. Fork, M.A. Pollack, Locking of HeNe laser modes induced by synchronous intracavity modulation. Appl. Phys. Lett. **5**(1), 4 (1964)
9. E.P. Ippen, C.V. Shank, A. Dienes, Passive mode-locking of the cw dye laser. Appl. Phys. Lett. **21**(8), 348–350 (1972)
10. R. Paschotta, U. Keller, Ultrafast solid-state lasers, in *Ultrafast Lasers—Technology and Applications*, ed. by M.E. Ferman, A. Galvanauskas, G. Sucha. Optical Engineering, Vol. 80 (Marcel Dekker, 2003). ISBN-13:978-0824743499
11. U. Keller, K.J. Weingarten, F.X. Kärtner, D. Kopf, B. Braun, I.D. Jung, R. Fluck, C. Hönninger, N. Matuschek, J. Aus der Au, Semiconductor saturable absorber mirrors (SESAM's) for femtosecond to nanosecond pulse generation in solid-state lasers. IEEE J. Sel. Top. Quantum Electron. **2**(3), 435–453 (1996)
12. U. Keller, Ultrafast solid-state laser oscillators: a success story for the last 20 years with no end in sight. Appl. Phys. B **100**(1), 15–28 (2010)
13. D.E. Spence, P.N. Kean, W. Sibbet, 60-fsec pulse generation from a self-mode-locked Ti: sapphire laser. Opt. Lett. **16**(1), 42–44 (1991)
14. J. Brons, V. Pervak, E. Fedulova, M. Seidel, D. Bauer, D. Sutter, V. Kalashnikov, A. Apolonski, O. Pronin, F. Krausz, Energy scaling of KLM thin-disk oscillators. Opt. Lett. **39** (22), 6442–6445 (2014)
15. W. Koechner, in *Solid-State Laser Engineering*. Springer Series in Optical Sciences, Vol. 1 (Springer, New York, 2006). ISBN-13:978-0-387-29094-2
16. R. Paschotta, Power scalability as a precise concept for the evaluation of laser architectures. arXiv: 0711.3987v1 (2007)

17. C. Hönninger, R. Paschotta, F. Morier-Genoud, M. Moser, U. Keller, Q-switching stability limits of continuous-wave passive mode locking. J. Opt. Soc. Am. B **16**(1), 46–56 (1999)
18. D.H. Sutter, J. Kleinbauer, D. Bauer, M. Wolf, C. Tan, R. Gebs, A. Budnicki, P. Wagenblast, S. Weiler, Ultrafast disk lasers and amplifiers, in *Proceedings of SPIE*, vol. 8235, 82350X-9 (SPIE, Bellingham, WA, 2012)
19. B. Weichelt, A. Voss, M. Abdou Ahmed, T. Graf, Enhanced performance of thin-disk lasers by pumping into the zero-phonon line. Opt. Lett. **37**(15), 3045–3047 (2012)
20. O.M. Efimov, L.B. Glebov, L.N. Glebova, K.C. Richardson, V.I. Smirnov, High-efficiency Bragg gratings in photothermorefractive glass. Appl. Opt. **38**(4), 619–627 (1999)
21. B.L. Volodin, S.V. Dolgy, E.D. Melnik, E. Downs, J. Shaw, V.S. Ban, Wavelength stabilization and spectrum narrowing of high-power multimode laser diodes and arrays by use of volume Bragg gratings. Opt. Lett. **29**(16), 1891–1893 (2004)
22. V.C. Negoita, Y. Li, T. Barnowski, J. Jiang, H. An; R. Roff, M. Shih, T. Vethake, T. Gottwald, S. Schad, G. Treusch, Wavelength stabilization of high power laser systems using volume holographic gratings, in *Proceedings of SPIE 8965, High-Power Diode Laser Technology and Applications XII*, 89650S (2014)
23. O.H. Heckl, C.J. Saraceno, C.R.E. Baer, T. Südmeyer, Y.Y. Wang, Y. Cheng, F. Benabid, U. Keller, Temporal pulse compression in a xenon-filled Kagome-type hollow-core photonic crystal fiber at high average power. Opt. Express **19**(20), 19142–19149 (2011)
24. F. Emaury, C. Dutin, C. Saraceno, M. Trant, O. Heckl, Y. Wang, C. Schriber, F. Gerome, T. Südmeyer, F. Benabid, U. Keller, Beam delivery and pulse compression to sub-50 fs of a mode-locked thin-disk laser in a gas-filled Kagome-type HC-PCF fiber. Opt. Express **21**(4), 4986–4994 (2013)
25. F. Emaury, C. Saraceno, B. Debord, D. Ghosh, A. Diebold, F. Gèrôme, T. Südmeyer, F. Benabid, U. Keller, Efficient spectral broadening in the 100-W average power regime using gas-filled kagome HC-PCF and pulse compression. Opt. Lett. **39**(24), 6843–6846 (2014)
26. K. Mak, M. Seidel, O. Pronin, M. Frosz, V. Pervak, A. Apolonskiy, F. Krausz, J. Travers, P. Russell, Compression of μJ-level pulses from 250 fs to sub-10 fs at 38 MHz repetition rate using two gas-filled hollow-core kagomé-PCF stages, in *Advanced Solid State Lasers, OSA Technical Digest* (*online*) (Optical Society of America, 2014), paper AF4A.4
27. S. Pricking, R. Gebs, R. Fleischhaker, J. Kleinbauer, A. Budnicki, D. Sutter, A. Killi, Sascha Weiler, M. Mielke, B. Beaudou, B. Debord, F. Gerome, F. Benabid, Hollow core fiber delivery of sub-ps pulses from a TruMicro 5000 Femto edition thin-disk amplifier, in *Proceedings of SPIE 9356, High-Power Laser Materials Processing: Lasers, Beam Delivery, Diagnostics, and Applications IV*, 935602 (2015)
28. K. Beil, S.T. Friedrich-Thornton, F. Tellkamp, R. Peters, C. Kränkel, K. Petermann, G. Huber, Thermal and laser properties of Yb:LuAG for kW thin-disk lasers. Opt. Express **18** (20), 20712–20722 (2010)
29. R. Peters, C. Kränkel, S.T. Fredrich-Thornton, K. Beil, K. Petermann, G. Huber, O.H. Heckl, C.R.E. Baer, C.J. Saraceno, T. Südmeyer, U. Keller, Thermal analysis and efficient high power continuous-wave and mode-locked thin-disk laser operation of Yb-doped sesquioxides. Appl. Phys. B **102**(3), 509–514 (2011)
30. M. Larionov, J. Gao, S. Erhard, A. Giesen, K. Contag, V. Peters, E. Mix, L. Fornasiero, K. Petermann, G. Huber, J. Aus der Au, G. J. Spühler, F. Brunner, R. Paschotta, U. Keller, A.A. Lagatsky, A. Abdolvand, N.V. Kuleshov, Thin-disk laser operation and spectroscopic characterization of Yb-doped sesquioxides and potassium tungstates, in *Advanced Solid-State Lasers*, ed. by C. Marshall, OSA Trends in Optics and Photonics, Vol. 50 (Optical Society of America, 2001), paper WC4
31. K. Petermann, G. Huber, L. Fornasiero, S. Kuch, E. Mix, V. Peters, S.A. Basun, Rare-earth-doped sesquioxides. J. Lumin. **87–89**, 973–975 (2000)
32. K. Petermann, L. Fornasiero, E. Mix, V. Peters, High melting sesquioxides: crystal growth, spectroscopy, and laser experiments. Opt. Mater. **19**, 67–71 (2002)

33. K. Petermann, D. Fagundes-Peters, J. Johannsen, M. Mond, V. Peters, J.J. Romero, S. Kutovoi, J. Speiser, A. Giesen, Highly Yb-doped oxides for thin-disc lasers. J. Cryst. Growth **275**, 135–140 (2005)
34. R. Peters, C. Kränkel, K. Petermann, G. Huber, Broadly tunable high-power Yb:Lu_2O_3 thin-disk laser with 80 % slope efficiency. Opt. Express **15**, 7075–7082 (2007)
35. T. Südmeyer, C. Kränkel, C.R.E. Baer, O.H. Heckl, C.J. Saraceno, M. Golling, R. Peters, K. Petermann, G. Huber, U. Keller, High-power ultrafast thin-disk laser oscillators and their potential for sub-100-femtosecond pulse generation. Appl. Phys. B **97**(2), 281–295 (2009)
36. M. Tokurakawa, A. Shirakawa, K. Ueda, H. Yagi, M. Noriyuki, T. Yanagitani, A.A. Kaminskii, Diode-pumped ultrashort-pulse generation based on Yb^{3+}:Sc_2O_3 and Yb^{3+}:Y_2O_3 ceramic multi-gain-media oscillator. Opt. Express **17**, 3353–3361 (2009)
37. R. Peters, C. Kränkel, K. Petermann, G. Huber, Crystal growth by the heat exchanger method, spectroscopic characterization and laser operation of high-purity Yb:Lu_2O_3. J. Cryst. Growth **310**, 1934–1938 (2008)
38. P. Klopp, V. Petrov, U. Griebner, K. Petermann, V. Peters, G. Erbert, Highly efficient mode-locked Yb:Sc_2O_3 laser. Opt. Lett. **29**(4), 391–393 (2004)
39. R. Peters, K. Beil, C. Kränkel, K. Schenk, K. Petermann, G. Huber, Ytterbium-doped Sesquioxides for high-power solid-state lasers: recent progress in crystal growth and laser operation, in *Conference on Lasers and Electro-Optics* (Europe, 2009), paper CA9.1
40. R. Peters, Ytterbium-dotierte Sesquioxide als hocheffiziente Lasermaterialien, in *PhD-Thesis* (University of Hamburg, Germany, 2009)
41. A. Ikesue, Y.L. Aung, V. Lupei, in *Ceramic Lasers* (Cambridge University, 2013). ISBN-13:978-0521114080
42. C. Kraenkel, K. Beil, P. Koopmann, T. Li, G. Huber, Diode-pumped sesquioxide lasers in the near- and mid-infrared range, in *International Photonics and Optoelectronics Meetings, OSA Technical Digest (online)* (Optical Society of America, 2012), paper MF1A.2
43. J. Aus der Au, G.J. Spühler, T. Südmeyer, R. Paschotta, R. Hövel, M. Moser, S. Erhard, M. Karszewski, A. Giesen, U. Keller, 16.2-W average power from a diode-pumped femtosecond Yb:YAG thin-disk laser. Opt. Lett. **25**(11), 859–861 (2000)
44. E. Innerhofer, T. Südmeyer, F. Brunner, R. Häring, A. Aschwanden, R. Paschotta, C. Hönninger, M. Kumkar, U. Keller, 60-W average power in 810-fs pulses from a thin-disk Yb:YAG laser. Opt. Lett. **28**(5), 367–369 (2003)
45. F. Brunner, E. Innerhofer, S.V. Marchese, T. Südmeyer, R. Paschotta, T. Usami, H. Ito, S. Kurimura, K. Kitamura, G. Arisholm, U. Keller, Powerful red-green-blue laser source pumped with a mode-locked thin-disk laser. Opt. Lett. **29**(16), 1921–1923 (2004)
46. S.V. Marchese, T. Südmeyer, M. Golling, R. Grange, U. Keller, Pulse energy scaling to 5 μJ from a femtosecond thin-disk laser. Opt. Lett. **31**(18), 2728–2730 (2006)
47. S.V. Marchese, C.R.E. Baer, A.G. Engqvist, S. Hashimoto, D.J.H.C. Maas, M. Golling, T. Südmeyer, U. Keller, Femtosecond thin-disk laser oscillator with pulse energy beyond the 10 microjoule level. Opt. Express **16**(9), 6397–6407 (2008)
48. J. Neuhaus, J. Kleinbauer, A. Killi, S. Weiler, D. Sutter, T. Dekorsy, Passively mode-locked Yb:YAG thin-disk laser with pulse energies exceeding 13 μJ by use of an active multi-pass geometry. Opt. Lett. **33**(7), 726–728 (2008)
49. J. Neuhaus, D. Bauer, J. Zhang, A. Killi, J. Kleinbauer, M. Kumkar, S. Weiler, M. Guina, D. H. Sutter, T. Dekorsy, Subpicosecond thin-disk laser oscillator with pulse energies of up to 25.9 microjoules by use of an active multi-pass geometry. Opt. Express **16**(25), 20530–20539 (2008)
50. J. Neuhaus, D. Bauer, J. Kleinbauer, A. Killi, D.H. Sutter, T. Dekorsy, Numerical analysis of a sub-picosecond thin-disk laser oscillator with active multi-pass geometry showing a variation of pulse duration within one round trip. J. Opt. Soc. Am. B **27**(1), 65–71 (2009)
51. D. Bauer, I. Zawischa, D.H. Sutter, A. Killi, T. Dekorsy, Mode-locked Yb:YAG thin-disk oscillator with 41 μJ pulse energy at 145 W average infrared power and high power frequency conversion. Opt. Express **20**(9), 9698–9704 (2012)

52. C. Saraceno, F. Emaury, C. Schriber, M. Hoffmann, M. Golling, T. Südmeyer, U. Keller, Ultrafast thin-disk laser with 80 μJ pulse energy and 242 W of average power. Opt. Lett. **39** (1), 9–12 (2014)
53. F. Brunner, T. Südmeyer, E. Innerhofer, R. Paschotta, F. Morier-Genoud, J. Gao, K. Contag, A. Giesen, V.E. Kisel, V.G. Shcherbitsky, N.V. Kuleshov, U. Keller, 240-fs pulses with 22-W average power from a mode-locked thin-disk Yb:KY$(WO_4)_2$ laser. Opt. Lett. **27**(13), 1162–1164 (2002)
54. S.V. Marchese, C.R.E. Baer, R. Peters, C. Kränkel, A.G. Engqvist, M. Golling, D.J.H.C. Maas, K. Petermann, T. Südmeyer, G. Huber, U. Keller, Efficient femtosecond high power Yb:Lu_2O_3 thin-disk laser. Opt. Express **15**, 16966–16971 (2007)
55. C.R.E. Baer, C. Kränkel, O.H. Heckl, M. Golling, T. Südmeyer, R. Peters, K. Petermann, G. Huber, U. Keller, 227-fs P pulses from a mode-locked Yb:$LuScO_3$ thin-disk laser. Opt. Express **17**(13), 10725–10730 (2009)
56. C.R.E. Baer, C. Kränkel, C.J. Saraceno, O.H. Heckl, M. Golling, T. Südmeyer, R. Peters, K. Petermann, G. Huber, U. Keller, Femtosecond Yb:Lu_2O_3 thin-disk laser with 63 W of average power. Opt. Lett. **34**(18), 2823–2825 (2009)
57. O.H. Heckl, C. Krankel, C.R.E. Baer, C.J. Saraceno, T. Sudmeyer, K. Petermann, G. Huber, U. Keller, Continuous-wave and mode-locked Yb:YCOB thin-disk laser: first demonstration and future prospects. Opt. Express **18**, 19201–19208 (2010)
58. C.J. Saraceno, O.H. Heckl, C.R.E. Baer, M. Golling, T. Südmeyer, K. Beil, C. Kränkel, K. Petermann, G. Huber, U. Keller, SESAMs for high-power femtosecond mode-locking: power scaling of an Yb:$LuScO_3$ thin-disk laser to 23 W and 235 fs. Opt. Express **19**(21), 20288–20300 (2011)
59. C. Saraceno, O.H. Heckl, C.R.E. Baer, C. Schriber, M. Golling, K. Beil, C. Kränkel, T. Südmeyer, G. Huber, U. Keller, Sub-100 femtosecond pulses from a SESAM mode-locked thin-disk laser. Appl. Phys. B **106**(3), 559–562 (2012)
60. O. Pronin, J. Brons, C. Grasse, V. Pervak, G. Boehm, M.C. Amann, V.L. Kalashnikov, A. Apolonski, F. Krausz, High-power 200 fs KLM Yb:YAG thin-disk oscillator. Opt. Lett. **36** (24), 4746–4748 (2011)
61. O. Pronin, J. Brons, M. Seidel, J. Zhang, M. Trubetskov, D. Bauer, D. Sutter, V. Kalashnikov, A. Apolonski, V. Pervak, F. Krausz, in *High-Power Femtosecond Thin-Disk Oscillators CLEO-Europe* (2015) invited paper CA-10.5
62. J. Zhang, J. Brons, M. Seidel, V. Pervak, V. Kalashnikov, Z. Wei, A. Apolonski, F. Krausz, O. Pronin, 49-fs Yb:YAG thin-disk oscillator with distributed Kerr-lens mode-locking, in *CLEO Europe Post Deadline* (2015) paper PD-A.1
63. F.X. Kärtner, I.D. Jung, U. Keller, Soliton mode-locking with saturable absorbers. IEEE J. Sel. Top. Quantum Electron. **2**(3), 540–556 (1996)
64. C.J. Saraceno, F. Emauri, O.H. Heckl, C.R.E. Baer, M. Hoffmann, C. Schriber, M. Golling, T. Südmeyer, U. Keller, 275 W average output power from a femtosecond thin-disk oscillator operated in a vacuum environment. Opt. Express **20**(21), 23535–23541 (2012)
65. J. Brons, V. Pervak, D. Bauer, D. Sutter, A. Apolonski, O. Pronin, F. Krausz, Enhancing the gain-bandwidth of a KLM Yb:YAG thin-disk oscillator through increased nonlinearities, in *Submitted for Publication at the Advanced Solid-State Lasers Conference* (2015)
66. C.R.E. Baer, O.H. Heckl, C.J. Saraceno, C. Schriber, C. Kränkel, T. Südmeyer, U. Keller, Frontiers in high-power passively mode-locked thin-disk laser oscillators. Opt. Express **20** (7), 7054–7065 (2012)
67. C.J. Saraceno, C. Schriber, F. Emaury, O.H. Heckl, C.R.E. Baer, M. Hoffmann, K. Beil, C. Kränkel, M. Golling, T. Südmeyer, U. Keller, Cutting-edge high-power ultrafast thin disk oscillators. Appl. Sci. **3**(2), 355–395 (2013)
68. C.J. Saraceno, F. Emaury, C. Schriber, A. Diebold, M. Hoffmann, M. Golling, T. Südmeyer, U. Keller, Toward millijoule-level high-power ultrafast thin-disk oscillators. IEEE J. Sel. Top. Quantum Electron. (JSTQE) **21**(1), 106–123 (2015)
69. M. Kumkar, Laser amplifying system. European Patent EP1,286,434B1 (2001)

70. D. Müller, S. Erhard, O. Ronsin, A. Giesen, Thin-disk multi-pass amplifier, in *Advanced Solid-State Photonics*, ed. by J. Zayhowski, OSA Trends in Optics and Photonics, Vol. 83 (Optical Society of America, 2003), paper 278
71. M. Huonker, A. Voss, Ch. Schmitz, Laser amplifying system. European patent EP1,213,801B1 (2000)
72. C.R.E. Baer, C. Kränkel, C.J. Saraceno, O.H. Heckl, M. Golling, R. Peters, K. Petermann, T. Südmeyer, G. Huber, U. Keller, Femtosecond thin-disk laser with 141 W of average power. Opt. Lett. **35**(13), 2302–2304 (2010)
73. C. Hönninger, I. Johannsen, M. Moser, G. Zhang, A. Giesen, U. Keller, Diode-pumped thin-disk Yb:YAG regenerative amplifier. Appl. Phys. B **65**(3), 423–426 (1997)
74. R. Graf, T. Metzger, M. Chyla, D. Sutter, Z. Major, A. Apolonski, F. Krausz, High average power Yb:YAG thin-disk regenerative amplifier at up to 1 MHz repetition rate, in *5th EPS-QEOD Europhoton Conference* (Stockholm, 2012), post deadline paper POSD.2
75. T. Bergmann, R. Knappe, Driver citcuit for Pockels cell and laser system with such a driver circuit and a pockels cell. European Patent EP1,418,460B1 (2003)
76. F. Dausinger, G. Schmitz, D. Sutter, Femtonische laser im maschinenbau. Laser Tech. J. **2** (4), 40–47 (2005)
77. J. König, T. Bauer, Fundamentals and industrial applications of ultrashort pulsed lasers at Bosch. in *Proceedings of SPIE 7925, Frontiers in Ultrafast Optics: Biomedical, Scientific, and Industrial Applications XI*, 792510 (February 11, 2011)
78. J. Kleinbauer, D. Eckert, S. Weiler, D.H. Sutter, 80 W ultrafast CPA-free disk laser, in *Solid State Lasers XVII: Technology and Devices*, ed. by W.A. Clarkson, N. Hodgson, R.K. Shori, Proceedings of SPIE, vol. 6871 (SPIE, Bellingham, WA, 2008), 68711B
79. S. Weiler, Ultrafast lasers: high-power pico- and femtosecond lasers enable new applications. Laser Focus World **47**(10), (2011)
80. J. König, D. Sutter, S. Nolte, Deutscher Zukunftspreis (2013), http://www.deutscher-zukunftspreis.de/en/nominated/2013/team-1 (online: 2013)
81. http://www.trumpf-laser.com/en/products/solid-state-lasers/short-and-ultra-short-pulsed-lasers/trumicro-series-5000.html (online: 2015)
82. R. Fleischhaker, R. Gebs, A. Budnicki, M. Wolf, J. Kleinbauer, D.H. Sutter, Compact gigawatt-class sub-picosecond Yb:YAG thin-disk regenerative chirped-pulse amplifier with high average power at up to 800 kHz, in *2013 Conference on Lasers and Electro-Optics—International Quantum Electronics Conference* (Optical Society of America, 2013), paper CFIE4.1
83. D. Sutter, Method for the production of temporal rectangular ultra-short pulses. European Patent EP1,775,806B1 (2005)
84. T. Metzger, A. Schwarz, C.Y. Teisset, D. Sutter, A. Killi, R. Kienberger, F. Krausz, High-repetition-rate picosecond pump laser based on a Yb:YAG disk amplifier for optical parametric amplification. Opt. Lett. **34**(14), 2123–2125 (2009)
85. B. Neuenschwander, T. Kramer, B. Lauer, B. Jaeggi, Burst mode with ps-and fs-pulses: influence on the removal rate, surface quality, and heat accumulation, in *Proceedings of SPIE 9350, Laser Applications in Microelectronic and Optoelectronic Manufacturing* (LAMOM) XX, 93500U (March 4, 2015)
86. M. Sailer, F. Bauer, J. Kleiner, M. Kaiser, Scaling of ablation rates. Ablation efficiency and quality aspects of burst-mode micromachining of metals, in *Lasers in Manufacturing Conference—LIM 2015* (München, 2015)
87. D. Flamm, D. Grossmann, M. Kaiser, J. Kleiner, M. Kumkar, K. Bergner, S. Nolte, Tuning the energy deposition of ultrashort pulses inside transparent materials for laser cutting applications, in *Lasers in Manufacturing Conference—LIM 2015* (München, 2015)
88. D. Sutter, M. Kumkar, Method for producing selectively a first laser pulse in the form of an ultra-short pulse or a temporally longer second laser pulse with another pulse form against the first laser pulse, comprises producing a single ultra-short pulse. German patent application DE 102,008,003,575 A1 (2008)

89. D. Sutter, R. Flaig, T. Bergmann, Controlling pockels cells. European Patent EP1,801,635B1 (2006)
90. O. Heckl, S. Weiler, S. Luzius, I. Zawischa, D.H. Sutter, Ultrafast disk technology enables next generation micromachining laser sources, in *High-Power Laser Materials Processing: Lasers, Beam Delivery, Diagnostics, and Applications II*, ed. by F. Dorsch, Proceedings of SPIE Vol. 8603 (SPIE, Bellingham, WA, 2013), paper 8603-10
91. M. Larionov, F. Butze, D. Nickel, A. Giesen, High-repetition-rate regenerative thin-disk amplifier with 116 μJ pulse energy and 250 fs pulse duration. Opt. Lett. **32**(5), 494–496 (2007)
92. U. Buenting, H. Sayinc, D. Wandt, U. Morgner, D. Kracht, Regenerative thin-disk amplifier with combined gain spectra producing 500 μJ sub 200 fs pulses. Opt. Express **17**(10), 8046–8050 (2009)
93. C. Yuriko Teisset, M. Schultze, R. Bessing, M. Häfner, S. Prinz, D. Sutter and T. Metzger, 300 W picosecond thin-disk regenerative amplifier at 10 kHz repetition rate, in *Advanced Solid-State Lasers Congress Postdeadline*, ed. by G. Huber, P. Moulton. OSA Postdeadline Paper Digest (online) (Optical Society of America, 2013), paper JTh5A.1
94. S. Klingebiel, M. Schultze, C. Y. Teisset, R. Bessing, M. Häfner, S. Prinz, M. Gorjan, D. Sutter, K. Michel, H. G. Barros, Z. Major, F. Krausz, T. Metzger, 220 mJ ultrafast thin-disk regenerative amplifier, in *CLEO: 2015, OSA Technical Digest* (*online*) (Optical Society of America, 2015), paper STu4O.2
95. T. Metzger, M. Gorjan, M. Ueffing, C. Teisset, M. Schultze, R. Bessing, M. Häfner, S. Prinz, D. Sutter, K. Michel, H. Barros, Z. Major, F. Krausz, Picosecond thin-disk lasers, in *Conference on Lasers and Electro-Optics CLEO* (2014), invited paper JTh4L
96. T. Nubbemeyer, M. Gorjan, A. Alismail, M. Ueffing, H.G. Barros, T. Metzger, D. Sutter, Z. Major, F. Krausz, High average power kW-scale Yb:YAG thin-disk regenerative amplifier, in *CLEO Europe* (2015), paper CA-P.1
97. J.-P. Negel, A. Voss, M.A. Ahmed, D. Bauer, D. Sutter, A. Killi, T. Graf, 1.1 kW average output power from a thin-disk multi-pass amplifier for ultrashort laser pulses. Opt. Lett. **38** (24), 5442–5445 (2013)
98. J. Tümmler, R. Jung, H. Stiel, P.V. Nickles, W. Sandner, High-repetition-rate chirped-pulse-amplification thin-disk laser system with joule-level pulse energy. Opt. Lett. **34**(9), 1378–1380 (2009)
99. M. Schulz, R. Riedel, A. Willner, S. Düsterer, M.J. Prandolini, J. Feldhaus, B. Faatz, J. Rossbach, M. Drescher, F. Tavella, Pulsed operation of a high average power Yb:YAG thin-disk multi-pass amplifier. Opt. Express **20**(5), 5038–5043 (2012)
100. M. Schulz, H. Hoeppner, M. Temme, R. Riedel, B. Faatz, M. J. Prandolini, M. Drescher, F. Tavella, 14 kilowatt burst average power from 2-stage cascaded Yb:YAG thin-disk multi-pass amplifier, in *Frontiers in Optics 2013*, ed. by I. Kang, D. Reitze, N. Alic, D. Hagan. OSA Technical Digest (online) (Optical Society of America, 2013), paper FTu4A.2
101. J.-P. Chambaret, O. Chekhlov, G. Cheriaux, J. Collier, R. Dabu, P. Dombi, A.M. Dunne, K. Ertel, P. Georges, J. Hebling, J. Hein, C. Hernandez-Gomez, C. Hooker, S. Karsch, G. Korn, F. Krausz, C. Le Blanc, Zs. Major, F. Mathieu, T. Metzger, G. Mourou, P. Nickles, K. Osvay, B. Rus, W. Sandner, G. Szabó, D. Ursescu, K. Varjú, Extreme light infrastructure: laser architecture and major challenges, in *Proceedings of SPIE 7721, Solid State Lasers and Amplifiers IV, and High-Power Lasers*, 77211D (May 19, 2010)
102. S. Prinz, M. Haefner, C.Y. Teisset, R. Bessing, K. Michel, Y. Lee, X.T. Geng, S. Kim, D.E. Kim, T. Metzger, M. Schultze, CEP-stable, sub-6 fs, 300-kHz OPCPA system with more than 15 W of average power. Opt. Express **23**, 1388–1394 (2015)
103. H. Fattahi, H.G. Barros, M. Gorjan, T. Nubbemeyer, B. Alsaif, C.Y. Teisset, M. Schultze, S. Prinz, M. Haefner, M. Ueffing, A. Alismail, L. Vámos, A. Schwarz, O. Pronin, J. Brons, X.T. Geng, G. Arisholm, M. Ciappina, V.S. Yakovlev, D.-E. Kim, A.M. Azzeer, N. Karpowicz, D. Sutter, Z. Major, T. Metzger, F. Krausz, Third-generation femtosecond technology. Optica **1**(1), 45–63 (2014)

Chapter 6
High-Average Power Ultrafast Yb: Innoslab-Amplifier

Peter Russbueldt, Torsten Mans, Dieter Hoffmann and Stefan Schippel

Abstract The state of the art of Yb:Innoslab femtosecond amplifiers is presented. Diode-pumped Innoslab amplifiers are designed for a good thermal management and low nonlinearity and enable ultrafast laser systems with high average power and almost diffraction limited beam quality. The presented compact femtosecond oscillator-Yb:Innoslab amplifier MOPA provide nearly transform and diffraction limited 640 fs pulses at 620 W average output power and 20 MHz repetition rate. By cascading two amplifiers an average output power of 1.1 kW and peak power of 80 MW is achieved in a single, linearly polarized beam. The specific properties and scaling laws of Yb:Innoslabs are discussed.

6.1 Yb:Innoslab Amplifier

6.1.1 Introduction

Today, ultrafast lasers have found widespread applications in micromachining, metrology and physics. The generation of high harmonics from femtosecond pulses focused in a gas provides a compact source of coherent radiation with wave length down to a few nanometers. Pumping of optical parametric oscillators and amplifiers permits almost any wavelength and pulse duration. Micromachining by sub-picosecond radiation is characterized by a high precision. Heat affected zones are minimized and processes are much more stable and reproducible for pulse durations shorter than the electron phonon coupling time than with longer picosecond and

P. Russbueldt (✉) · T. Mans · D. Hoffmann
Fraunhofer-Institut für Lasertechnik ILT, Steinbachstraße 15, 52074 Aachen, Germany
e-mail: peter.russbueldt@ilt.fraunhofer.de

S. Schippel
Layertec GmbH, Ernst-Abbe-Weg 1, 99441 Mellingen, Germany

S. Nolte et al. (eds.), *Ultrashort Pulse Laser Technology*,
Springer Series in Optical Sciences 195, DOI 10.1007/978-3-319-17659-8_6

nanosecond pulses [1]. Exploiting nonlinear processes, absorption becomes nearly material independent enabling precision cutting, ablation and drilling of weakly absorbing materials, multi-component and multi-layer systems. Femtosecond pulses allow local in-volume modification in transparent dielectrics for the production of waveguides, etched micro-channels or micro mechanical parts. New laser sources will establish ultrafast laser technology as a cost- and time-effective tool in an industrial environment.

For average powers beyond 100 W with pulse durations below 1 ps only ytterbium-doped media are established. Besides the sufficient amplification bandwidth the main reason is the availability of high power and high brightness diode lasers for pumping in the $\lambda_p = 940–980$ nm wavelength range. The quasi-three-level energy scheme of ytterbium^{3+} induces a high Stokes efficiency at the peak emission wavelength around $\lambda \approx 1030$ nm and a small quantum defect. However, it is accompanied by a thermally populated lower state of the laser transition, that leads to absorption of the laser radiation below a minimal pump intensity.

Ytterbium-doped laser materials require high pump intensities throughout the crystal which in turn lead to high thermal loads. For efficient operation pump intensities throughout the gain volume in the order of the saturation intensity of the pump radiation $I_{sat}(\lambda_p) \approx 5$ kW cm^{-2} and an efficient heat removal from the gain volume are essential. A good thermal management avoids failure by thermo-mechanical stress and allows nearly diffraction-limited beam quality at high average power. Today the basic designs Fiber, Thin-disk and Innoslab fulfill these requirements. Amplifiers based on Innoslab and Fiber technology have demonstrated nearly diffraction-limited 1.1 kW and 830 W average power, respectively, at ~700 fs pulse duration [2, 3]. With Thin-disk ultrafast amplifiers, average powers of 200 and 32 W have been demonstrated using a regenerative and multi-pass setup, respectively [4, 5]. With a high-power Thin-Disk oscillator an average power of 275 W has been achieved [6].

Within this chapter we will discuss the Innoslab approach in detail. The fiber based amplification is discussed in Chap. 5 while the Thin-disk geometries are evaluated in Chap. 4.

6.1.2 Setup

An Innoslab laser [7] or amplifier [8] consists of a longitudinally, partially pumped slab crystal. Grinding of the mounting surfaces of the slab suppresses parasitic oscillations across the line-shaped, homogeneously pumped cross section inside the crystal. The short distance between the pumped gain volume and the large cooled mounting surfaces, see Fig. 6.1, allows for efficient heat removal. For cw-pumping, the thickness of the slab is in the order of $d \approx 1$ mm and the length in the order of $l \approx 10$ mm. The gain volume is formed by a thin line (height 0.1–1 mm) of pump

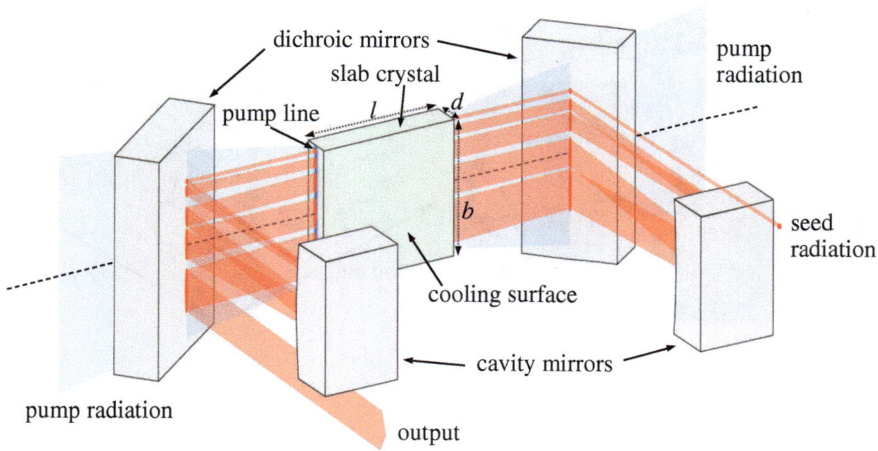

Fig. 6.1 Schematic setup of an Innoslab amplifier

radiation which is homogenized along the width *b* of the slab. This geometry is well adapted to the beam characteristics of laser diode bars. The terms fast and slow axis are therefore also used for the direction of the slab thickness and width, respectively.

One-dimensional heat flow establishes a homogeneous cylindrical thermal lens and avoids depolarization. In an amplifier a confocal arrangement of two cylindrical mirrors folds the laser radiation several times through the gain volume [2, 8–10]. Perpendicular to the pump line the thermal lens reproduces the laser mode at every round trip. In the plane of the pumped volume the beam is expanded by a constant factor at each round trip, according to the magnification *M* of the confocal cavity (Fig. 6.1). Innoslab lasers are designed for a moderate gain of about a factor of 2–10 per pass. By the confocal cavity as many as 9 passes through the slab and amplification factors of 1000 are achieved. Nevertheless, Innoslab amplifiers are single-pass amplifiers—at each passage a new section of the gain volume is saturated. In Innoslab lasers the beam expansion balances the increase of power and yields a reasonably constant intensity over the slab width. This decisively improves the efficiency by homogeneous saturation of the gain medium and constantly keeps the intensity below the damage threshold of the optical components. By a short interaction length inside the laser material and by keeping the ratio of laser intensity to saturation intensity constant the nonlinearity is minimized. The low nonlinearity of the single-pass setup is one of the key advantages of Innoslab amplifiers.

The two pump modules enclose the cavity of the Innoslab amplifier (Fig. 6.2). The setup is compact with a footprint of $500 \times 500\ \text{mm}^2$. Temperature stabilization at room temperature by water cooling of the whole setup ensures stable longtime operation.

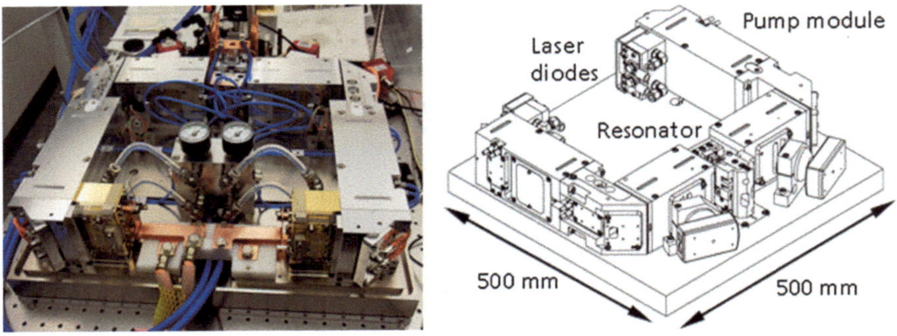

Fig. 6.2 Complete amplifier head

Fig. 6.3 Intensity profile in x-direction (*inset*) and power after a number of crystal passes

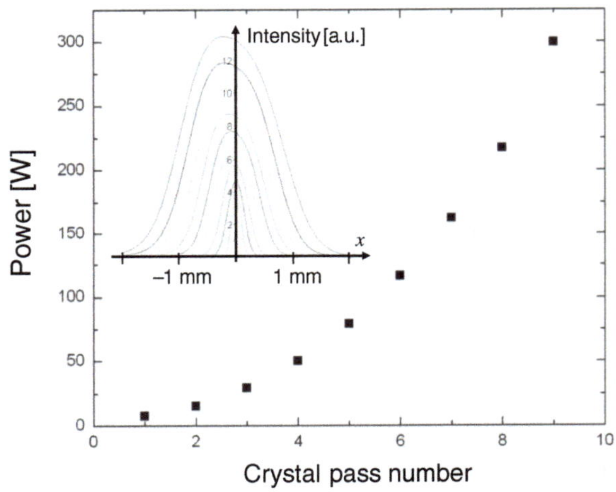

6.1.3 Simulations

Numerical simulations of the Yb:Innoslab amplifier were conducted based on two different approaches. The first approach is the expansion of the electric field according to Hermite transverse modes and the cw amplification in a quasi-three-level system (Fig. 6.3).

In order to check the amplification behavior and beam propagation with greater precision, the cw amplification for a quasi-three-level system was additionally integrated into a program which permits the rigorous propagation of a complex electric field using the Kirchhoff diffraction integral (Fig. 6.4). Also the propagation of the pump radiation in the crystal was incorporated by ray-tracing.

The Innoslab amplifier based on Yb:YAG or Yb:KGW was examined numerically using the two approaches. Taking the rate equations into account, the three-dimensional amplification distribution was calculated for a quasi-three-level system in the laser crystal with specified values for geometry, pump power and seed

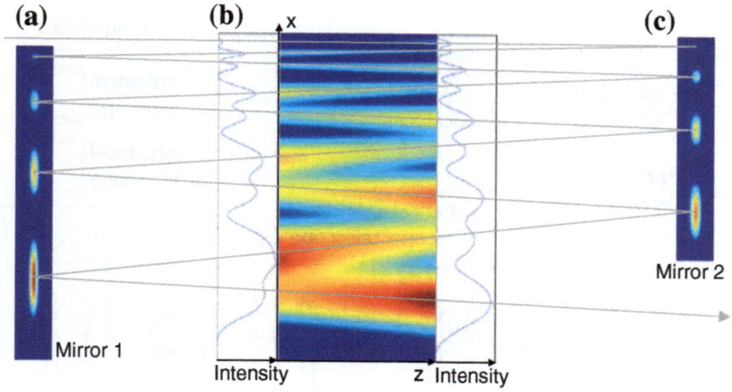

Fig. 6.4 Calculated three-dimensional intensity distribution in the crystal and on the mirrors. The beam has to pass the edges of mirror 1 (**a**), crystal (**b**) and mirror 2 (**c**) with sufficient safety distance δ (i.e. distance of the beam axis to the edge referred to the beam radius)

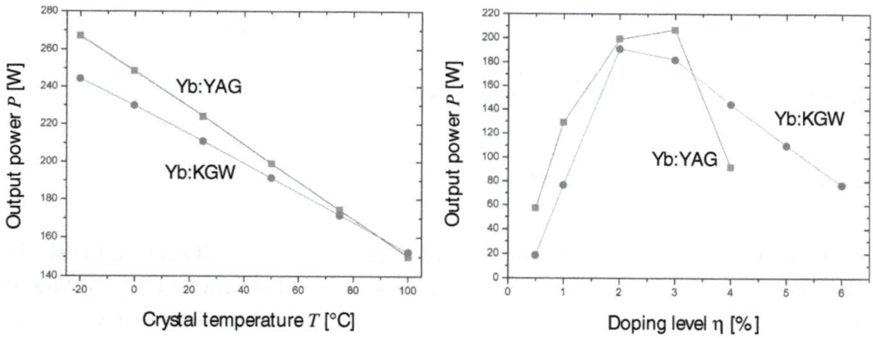

Fig. 6.5 Dependence of the amplifier output power P on crystal temperature (at a doping level of $\eta = 2$ %) and doping level (at temperature $T = 50$ °C). Parameters: magnification $M = 1.2$; pump height $2w_p = 240$ μm; pump power $P_p = 800$ W; seed power $P_s = 3$ W

power. The simulations allow evaluating different geometries, i.e. number of crystal passes, magnification of the confocal cavity and seed beam parameter as well as power of pump and seed, with respect to output power, beam quality and B-integral.

Further comparative examinations of the resonator geometries and dependences on temperature and doping level were carried out for Yb:YAG and Yb:KGW (Fig. 6.5).

The simulation data yields key requirements for the laser setup: For the required amplification a pump intensity of ~50 kW cm^{-2} must be attained in the longitudinal slab. For effective heat removal the slab should have a length of ~10 mm and should be pumped over the width of 10 mm at a height of 200–250 μm. To avoid self-absorption the slab must be pumped in the entire mode volume. The high pump intensity implies high laser intensities and therefore requires high damage thresholds of all optical surfaces. The beam path must be optimized not only with regard to high amplification but also to avoid intensity peaks.

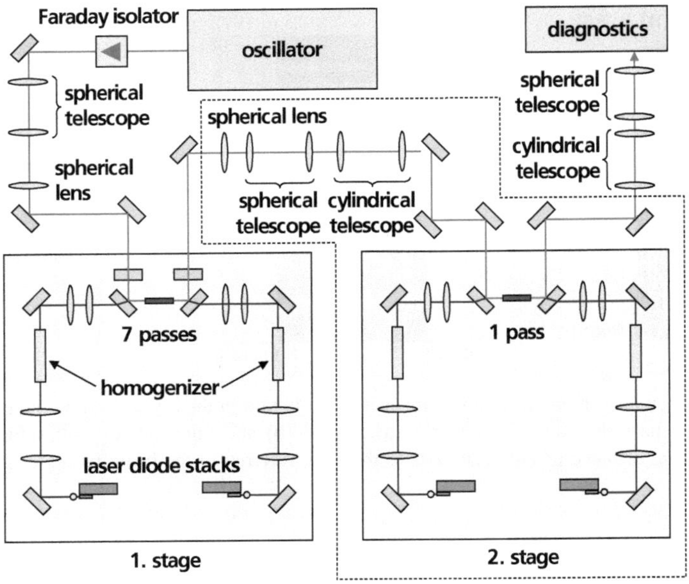

Fig. 6.6 Single-stage and dual stage amplifier setup

6.1.4 Experimental Results

The two commercially available horizontal laser diode stacks (JOLD-720-HSC-4L) used in our setup consist of four collimated laser diode bars each. They provide up to 660 W at λ_p = 940 nm. The radiation is imaged into a planar waveguide for homogenization in slow axis (Fig. 6.6). Afterwards the exit of the waveguide is imaged by a relay optic into the 1 x 10 x 10 mm^3 Yb:YAG crystal (2.5 % doping level). In fast axis the collimated laser diode bars are imaged into the middle of the slab, adapting beam parameters by two cylindrical lenses.

The amplifier was seeded by a Yb:KGW oscillator (HighQ Laser Innovation GmbH) of P_s = 2.3 W output power at ν_{rep} = 20 MHz. The almost transform limited τ = 289 fs pulses of $\Delta\lambda$ = 4.2 nm bandwidth are centered at λ_s = 1030.2 nm (Fig. 6.7).

Only a spherical telescope and a focusing lens were used for mode matching to the amplifier mode. No stretcher or cylindrical beam transformation were employed. An optical isolator was inserted to protect the oscillator against feedback by the amplifier. Up to P = 620 W of linearly polarized fs-radiation was extracted from the amplifier (Fig. 6.8) at P_s = 2.3 W seed power and 2 × 200 A diode current corresponding to P_p = 1250 W pump power. An optical-optical slope efficiency η_{opt} = 0.65 is reached above a threshold pump power of 300 W.

After seven passes inside the 10 mm long laser crystal the pulses of τ = 636 fs duration (FWHM, $sech^2$-fit) stay nearly transform limited ($\tau\ \Delta\nu$ = 0.365) without compression (Fig. 6.7). At 20 MHz repetition rate and pulse energies of E = 31 μJ

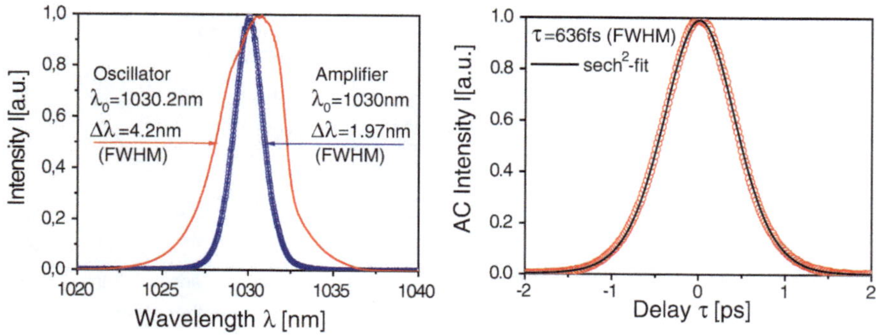

Fig. 6.7 Spectra (*left*) and autocorrelation (*right*) fitted by sech^2 of fs-oscillator (*red*) and amplifier (*blue*) at P = 620 W output power

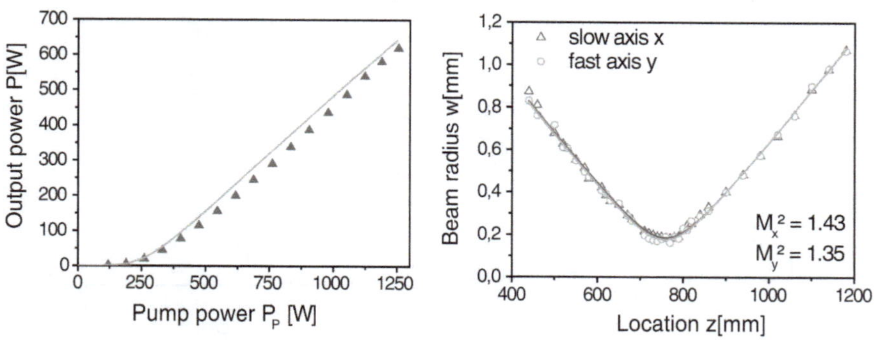

Fig. 6.8 Measured (*triangles*) and calculated (*line*) output power of the amplifier seeded by the 2.3 W femtosecond oscillator (*left*) and beam radius at P = 620 W output power (*right*)

corresponding to a B-integral B = 1.6 no sign of self phase modulation is visible. Maximum peak intensity inside the amplifier is 13.5 GW cm^{-2}. Gain narrowing reduces the spectral bandwidth $\Delta\lambda_s$ = 4.2 nm of the oscillator to $\Delta\lambda$ = 1.97 nm behind the amplifier (Fig. 6.7). After amplification a cylindrical telescope transforms the radiation into a circular beam. Beam quality was measured (Spiricon M2-200/v.4.2, 2nd momentum method) to be M_x^2 = 1.43 and M_y^2 = 1.35 respectively at P = 620 W output power (Fig. 6.8). There is no sign of background radiation or nonlinear spectral broadening. Lasing of the amplifier is completely suppressed by the seed radiation at seed powers as low as P_s = 1 W.

For even higher output power two or more amplifiers can be cascaded. In the experiment the oscillator 7-pass Innoslab MOPA setup described above is combined with a single-pass booster amplifier (Fig. 6.6). Except for the cavity, booster stage and 7-pass amplifier have the same setup. By three spherical lenses the output of the first amplifier is imaged in fast-axis telecentrically to the second amplifier in such a way, that an eighth pass is added to the first stage. In slow-axis the output of the first stage defined by the geometric margins of cavity mirrors and slab crystal is

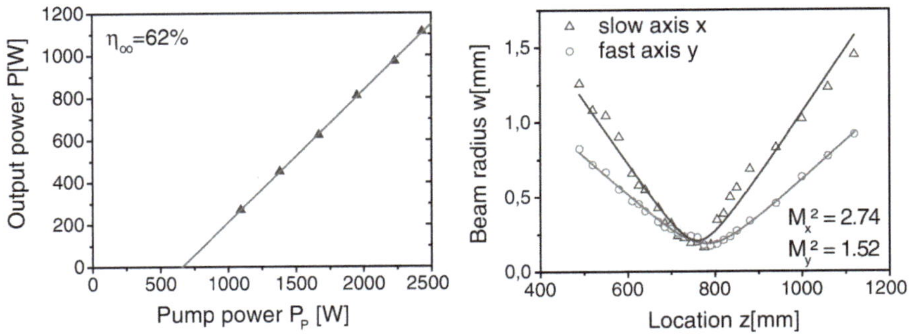

Fig. 6.9 Output power of the fs amplifier chain seeded with 2.3 W (*left*) and beam radius at P = 1.1 kW output power (*right*)

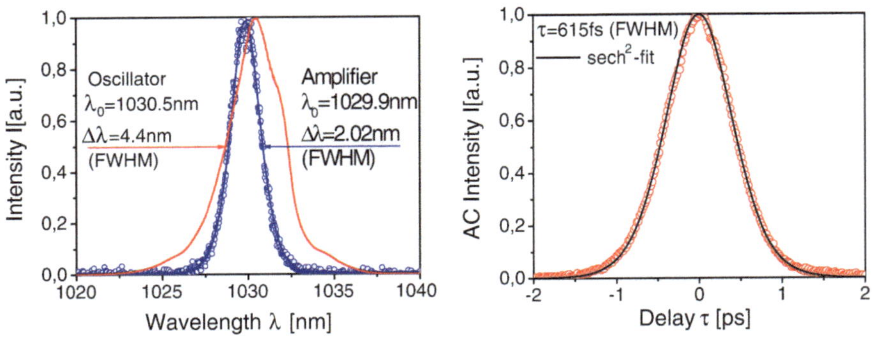

Fig. 6.10 Spectra (*left*) and autocorrelation (*right*) fitted by sech^2 (*line*) of fs-oscillator (*red*) and amplifier (*blue*) at P = 1.1 kW output power

matched by two cylindrical lenses to the 10 mm width of the slab crystal in the second stage. The imaging ensures power independent mode matching of the second amplifier stage. The high output power of the first stage allows for efficient power extraction out of the second stage by one pass. The single pass in the second stage only slightly increases the B-integral to $B = 1.8$.

Up to P = 1.1 kW of linearly polarized fs-radiation was extracted from the amplifier (Fig. 6.9) at P_s = 2.3 W seed power and 2 × 200 A in the first and 2 × 190 A diode current in the second stage corresponding to P_p = 1249 + 1180 W pump power. The corresponding peak power at the repetition rate ν_{rep} = 20 MHz is $P_{\max} = P/(\nu\ \tau)\cdot\ln(1+\sqrt{2})$ = 80 MW. Pump power in the second stage was limited by the wavelength shift of the laser diodes. After 7+1 passes the laser pulses of τ = 615 fs duration (FWHM, sech^2-fit) are nearly transform limited without compression ($\tau\ \Delta\nu = 0.362$; Fig. 6.10). At 20 MHz repetition rate and E = 55 μJ pulse energy there is no sign of self phase modulation. Gain narrowing reduces the spectral bandwidth $\Delta\lambda_s$ = 4.4 nm of the oscillator to $\Delta\lambda$ = 2.02 nm after amplification (Fig. 6.10).

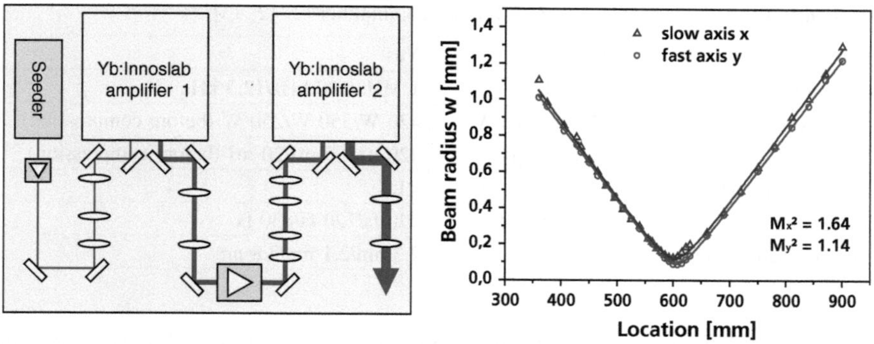

Fig. 6.11 Setup (*left*) and beam radius (*right*) of system with improved beam quality and high-power isolator between the Yb:Innoslab amplifier stages at $P = 0.95$ kW output power

After amplification a cylindrical telescope transforms the radiation into an almost circular beam. Beam quality behind transformation was measured (Spiricon M2-200/v.4.2, 2nd momentum method) to be $M_x^2 = 1.5$ and $M_y^2 = 2.7$ respectively at $P = 1.1$ kW output power (Fig. 6.9). Beam quality was mainly limited by the quality of the solder joint of slab crystal and heat sink. Again, there is no sign of nonlinear spectral broadening, background radiation or lasing of the amplifier.

More recently, an improved setup was realized (Fig. 6.11). It contains an optical isolator between the two Yb:Innoslab amplifier stages, in order to protect the seeder from feedback. The isolator consists of a magnetic field and a TGG slab crystal with dimensions $12 \times 1.5 \times 14.5\ \text{mm}^3$. The output of the first amplifier is telecentrically imaged into the isolator crystal. In order to image the margins determined by the slab crystal and the second resonator mirror to the edges of the isolator crystal, a magnification factor of 2.3 was used. This yields a complete filling of the width of the isolator crystal and ensures that the output of the first amplifier always hits the isolator crystal. The radiation is then imaged into the crystal of the second Innoslab amplifier stage, adapting the size in both transverse directions by two spherical and two cylindrical lenses. After the second stage the beam is transformed into a circular beam by a one spherical and one cylindrical lens. The slab geometry of the isolator crystal allows for an improved thermal management compared to a rod design (see 6.1.5), which is inevitable at high power and diffraction-limited beam quality due to residual absorption in the TGG crystal.

The beam quality of the setup could be improved compared to the results shown in Fig. 6.9 to $M^2 = 1.64 \times 1.14$ (Fig. 6.11). The average power is $P = 0.95$ kW at a repetition rate of $\nu = 20$ MHz and with a seed power of $P_s = 4.1$ W and 4×170 A diode current corresponding to $P_p = 1050 + 1040$ W pump power.

For frequency doubling experiments a femtosecond MOPA with a single Yb: Innoslab amplifier stage at a repetition rate of $\nu_{rep} = 50$ MHz was used. As nonlinear crystal lithium triborate (LBO) cut for type-I conversion and mounted in an oven with a temperature stability of 0.02 K was used. Up to $P_{2\omega} = 377$ W at 515 nm with a beam quality of $M^2 < 1.7$ were achieved from $P = 570$ W of infrared radiation. For

Table 6.1 Experimental results of the Yb:Innoslab amplifier [2, 12, 13]

CPA	No	Yes
Repetition rate ν_{rep}	20 MHz	1 MHz/100 kHz/12.5 kHz
Average power P	620 W/1.1 kW	420 W/350 W/250 W (before compression)
Pulse energy E	31 μJ/55 μJ	420 μJ/3.5 mJ/20 mJ (before compression)
Beam quality M^2	<1.5/< 3	<1.4
Pulse duration τ	636 fs/615 fs	720 fs/720 fs/830 fs
Spectral bandwidth $\Delta\lambda$	2.0 nm	2.3 nm/2.1 nm/2.8 nm

long term stable operation the infrared power was limited to $P = 470$ W delivering $P_{2\omega} = 287$ W at 515 nm with a conversion efficiency of 61 %. The beam quality was $M^2 < 1.4$. This is the highest average power at 515 nm and at diffraction-limited beam quality demonstrated so far. The pulse duration is $\tau = 700$ fs and the bandwidth $\Delta\nu = 0.46$ nm, yielding a time-bandwidth-product $\tau\ \Delta\nu = 0.364$ close to the transform limit [11].

The above mentioned results were achieved without chirped-pulse amplification. This makes the amplifier especially interesting for applications where a robust setup and only moderate pulse energies of some 10 μJ are needed, e.g. in micro processing. Using a chirped-pulse amplification scheme in the Innoslab amplifier allows for higher pulse energy. Up to $E = 420$ μJ at a repetition rate of $\nu_{rep} = 1$ MHz and $E = 3.5$ mJ at a repetition rate of $\nu_{rep} = 100$ kHz have been achieved by seeding the amplifier with $P_s = 5$ W of a fiber laser system, with pulse duration stretched to 2 ns [12]. A pulse energy of $E = 20$ mJ was achieved at a repetition rate of $\nu_{rep} = 12.5$ kHz [13]. Table 6.1 summarizes the results attained with the Yb: Innoslab amplifier.

6.1.5 Scaling

The good thermal management of Innoslab amplifiers is due to the small height $2w_S$ and high aspect ratio $b/(2w_S)$ of the gain volume compared to a rod (Fig. 6.12). A reasonable measure for the thermally induced aberrations is the phase difference $\Delta\phi$ between the center and the edge of the laser beam acquired at a pass through the crystal. This phase is proportional to the crystal length l and the temperature difference $\Delta T_R = P/(\lambda l)\cdot 1/(4\pi)$ or $\Delta T_S = P/(\lambda l)\cdot w_S/(4b)$. The phase difference is independent of the pumped area for the rod and it scales with the aspect ratio $b/(2w_S)$ of the pumped area for the slab. The temperature differences compare as $\Delta T_S/\Delta T_R = 1.6 \cdot 2w_S/b$. The temperature difference is determined by the distance for the heat to travel and the area for heat removal. In a rod these terms scale equally with the pump radius, while in a slab the area for heat removal is increased and the distance for the heat to travel simultaneously decreased by increasing the aspect ratio of the pump area. As a consequence a high aspect ratio $2w_S/b$ reduces thermally induced aberrations in the slab geometry.

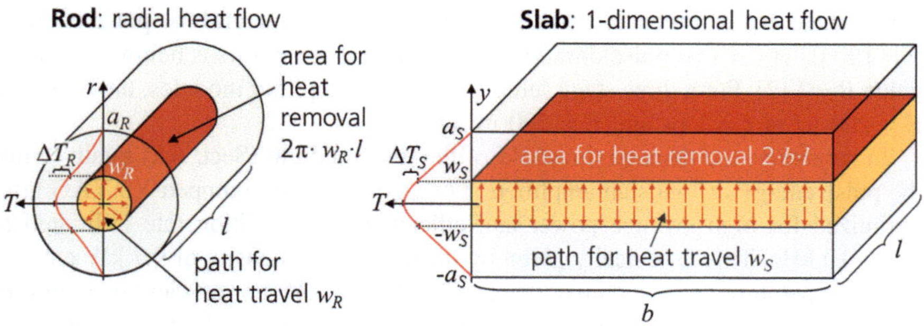

Fig. 6.12 Heat conduction in rod and slab geometry

For the realized Yb:YAG Innoslab amplifier, thermal aberrations are reduced compared to rod type amplifiers by about a factor of 30, allowing kW average power at diffraction-limited beam quality. The basic amplifier module of all realized setups consists of a 1 × 1 × 10 mm^3 slab crystal, pumped by up to 1250 W at 940 nm in a 0.2 × 10 mm^2 cross section. It enables an average power of up to 700 W and pulse energies of up to 100 μJ at 700 fs-pulses and 100 mJ for stretched ns-pulses (Table 6.1). To scale the output power, two or three amplifiers can be cascaded doubling or tripling the output power. Adding even more amplifier stages is unpractical not only due to complexity, but also due to thermal aberrations degrading the beam quality.

However, increasing the slab width b at constant pump height $2w_s$ and pump intensity keeps the thermal management inside the amplifier constant and increases the output power without degrading the beam quality. An Yb:YAG Innoslab amplifier of a slab width b = 30 mm is capable of 1.5–2 kW average power at diffraction-limited beam quality, cascading three amplifiers enables up to 5 kW. Up to 30 mm slab width, the technology of the existing setup can be used: slightly modified pump modules, the resonator setup and crystal mounting technology. Beyond a slab width of ~30 mm the size and complexity of the pumping optics increases disproportionately, mounting and cooling of the slab crystal becomes difficult. For even higher power thin-disk booster amplifiers may be favorable, perhaps enabling 10 kW ultrafast lasers 1 day.

The maximum pulse energy of laser systems is limited by three constraints. First of all, the maximum pump power and the storage time of the laser medium determine the stored energy, a few Joules in the case of Yb:YAG Innoslabs. The second limit is the damage threshold of the antireflection-coated gain medium in the order of 10 J cm^{-2}. The mode area limits the pulse energy to E < 40 mJ for the realized b = 10 mm 7-pass Innoslab amplifier, to E < 100 mJ for the multi-stage setup or $E < 0.3$ J for b = 30 mm. Finally, the B-integral defines the threshold of possible damage by self-focusing and degrading beam quality by filamentation. By beam expansion and a short interaction length inside the laser material the B-integral of Innoslab amplifiers is minimized. In case of the realized b = 10 mm single-stage 7-pass Innoslab amplifier (gain 500) the B-integral for pulse energies $E = 100$ μJ at $\tau = 1$ ps pulse duration and $E = 10$ mJ at $\tau = 1$ ns is below the critical value $B = 3$ [2]. For a three stage setup of the same amplifier modules, the limits are $E = 200$ μJ at $\tau = 1$ ps and $E = 200$ mJ at $\tau = 1$ ns.

In summary, the damage threshold of the gain medium (Sect. 6.2) finally limits the pulse energy of Innoslab amplifiers. For a continuously pumped Yb:YAG laser optimized for high average power and efficiency the repetition rate is limited to $\nu_{rep} > 10$ kHz. This is basically given by the saturation intensity of 9.7 kW cm^{-2} at room temperature and the damage threshold of Yb:YAG. For efficient operation of a single-pass amplifier like the Innoslab the laser intensity inside the laser medium should exceed the saturation intensity several times. Hence, fluences below 10 J cm^{-2} to avoid optical damage require repetition rates of multi-kHz in the best case of a constant intensity inside the gain medium. The best case is realized by beam expansion in case of the Innoslab. At the dispense of efficiency and average power the pulse energy can be further increased. The pump intensity has to be reduced in order to decrease the stored energy per area and the laser medium has to be pumped quasi-cw to maintain the thermal management. At 100 Hz repetition rate pulse energies of $\sim$500 mJ should be feasible.

If required, sub-picosecond pulses can be compressed to <100 fs at >100 W average power. In nonlinear post-compression fs-pulses are spectrally broadened by SPM in a nonlinear waveguide and their linear chirp removed by a compressor. With the Innoslab laser system and silica fibers pulse energies up to 1.6 μJ, average powers up to 100 W and pulse duration as short as 30 fs have been achieved at diffraction-limited beam quality. For higher pulse energies >100 μJ a gas-filled capillary can be used for the nonlinear medium. Optical parametric chirped pulse amplification (OPCPA) enables even shorter sub-10 fs pulses [14].

6.2 Optics Development

6.2.1 Optics for the fs Laser System

As described in the previous chapter the damage threshold of the optical coatings is the main limiting factor for Yb:YAG lasers and amplifiers and the Yb:Innoslab amplifiers in particular. Figure 6.13 is a schematic diagram of the laser system, including the required damage threshold of the components, which are exposed to the highest power densities. These are the laser crystal, the resonator end mirrors and the beam splitters for pump light coupling (indicated under 45° with reference to the resonator and pump light beam path).

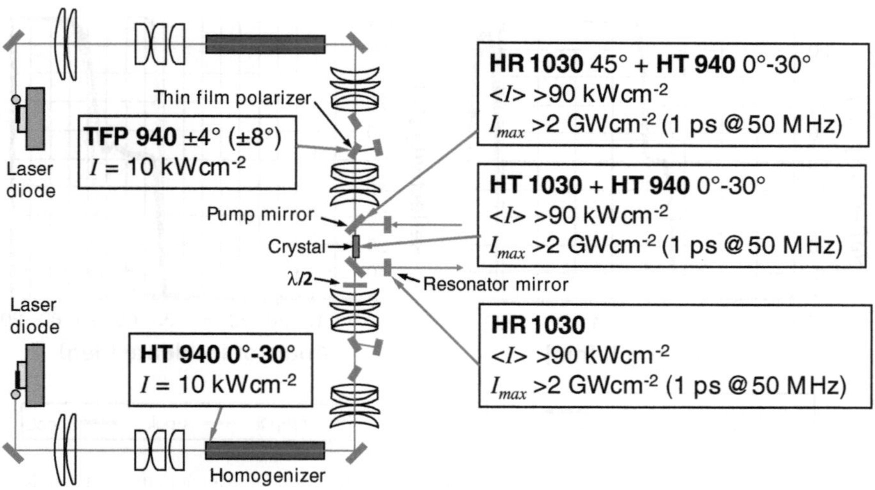

Fig. 6.13 Schematic setup of the fs laser system

6.2.2 Beam Splitter for Pump Light Coupling

The laser design provides for the crystal to be pumped from two directions. The resonator itself must therefore be folded, while the folding mirrors must at the same time be highly transmissive for the pump radiation, i.e. they have to be beam splitters or beam combiners. In addition, the pump radiation has to be focused into the crystal and therefore hit the beam splitter convergently with an angle range of ±15°. The layer system was therefore specified as highly reflecting (>99.9 %) at 45° and 1030 nm and highly transmitting (>90 %) at 45° ± 15° and 940 nm.

High-power-resistant pump mirrors, i.e. resonator mirrors with high transmission for the radiation of the pump laser diode for Yb-doped laser materials are state of the art. When the beam incidence is vertical these mirrors exhibit a very steep transition from the range of highest transmission T(0°, 980–990 nm) > 99 % to the range of highest reflection R(0°, 1020–1030 nm) > 99.9 %. The spectral gap between the two ranges only amounts to 30 nm, i.e. about 3 % of the wavelengths considered. Figure 6.14 shows the reflectivity of such coatings when used under an angle of incidence of 45°.

It can be seen that the steepness of the curve shown above is retained both for the s- and the p-polarization. The reflection band edges for s- and p-polarized light, however, have a spectral gap of approx. 50 nm. This explains the identical measurement for the transition width between the spectral ranges of high transmission and high reflection for unpolarized light at an angle of incidence of 45°. From Fig. 6.14 (right) it can be seen that the s-polarized fraction of the radiation is reflected practically completely at angles of incidence greater than 50°. The required high transmission of a convergent unpolarized pump beam thus cannot be provided by such standard layer systems.

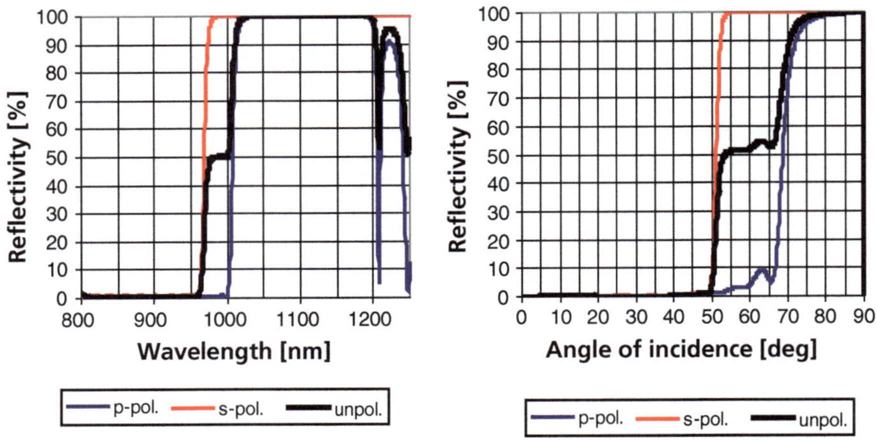

Fig. 6.14 Reflectivity of a standard beam splitter as a function of wavelength at an angle of incidence of 45° (*left*) and as a function of the angle of incidence at a wavelength of 940 nm (*right*)

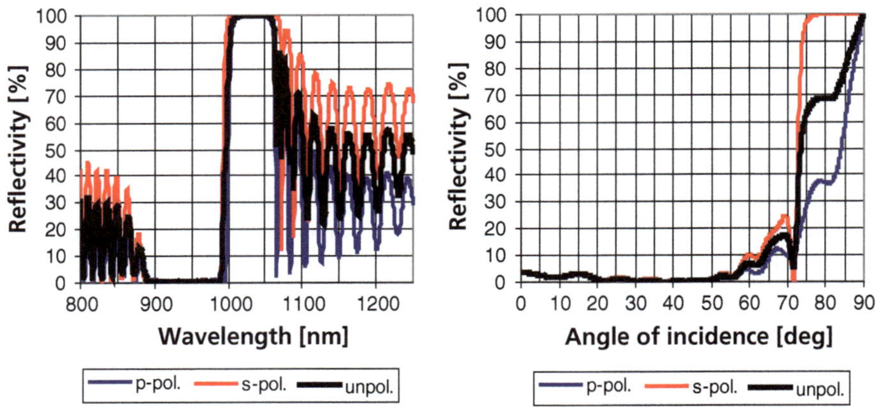

Fig. 6.15 Reflectivity of a newly developed beam splitter as a function of wavelength at an angle of incidence of 45° (*left*) and as a function of the angle of incidence at a wavelength of 940 nm (*right*)

To solve this problem, a layer design was calculated using a different layer material combination exhibiting a much narrower transition between the HR and HT ranges. Figure 6.15 shows the reflectivity of this new folding mirror.

It is clear that this layer design meets the spectral requirements stated above. The transition between the HR and HT ranges is only about 10 nm wide for unpolarized light. Even with large angles of incidence up to 60°, the reflectivity at 940 nm is only approx. 10 % for the s-component and about 8 % for unpolarized light. Owing to the changed refractive index difference of the layer materials, the reflection band is considerably narrower than with the standard beam splitter.

The standard beam splitter consists of a layer package of approx. 60 layers. For the special beam splitter about 100 individual layers are required. As the laser damage threshold generally depends to a large extent on the layer materials used, the new layer system was tested before the optics for the laser system were produced. The layer system exhibited a laser damage threshold of 1.1 J/cm^2 or 220 GW/cm^2 at a pulse duration of 5 ps, which roughly matches the damage threshold of the best standard optics. For the laser system being developed a sustained resistance to pulse peak power densities >2 GW/cm^2 was required.

6.2.3 Polishing of YAG Substrates

The laser damage threshold also depends on the substrate polishing. The faces of the slab crystals are exposed to very high energy densities and so an improvement in the quality of the polishing, in particular the avoidance of surface defects and a reduction in roughness, will raise the laser damage threshold. During the work several polishing compounds and polishing agents (pitch, various foils) as well as various polishing techniques (contact pressure, relative speed of substrates and polishing agents) were tested. Figure 6.16 shows profilometer images of the surface topography of polished YAG substrates which were polished using the previous (left) and the improved (right) technology.

The YAG polishing tests produced distinct improvements in polishing quality. The rms roughness of the polished substrates was reduced to 0.2 nm, which is comparable with a good polished finish on quartz glass. The number of surface defects and their size were also significantly reduced.

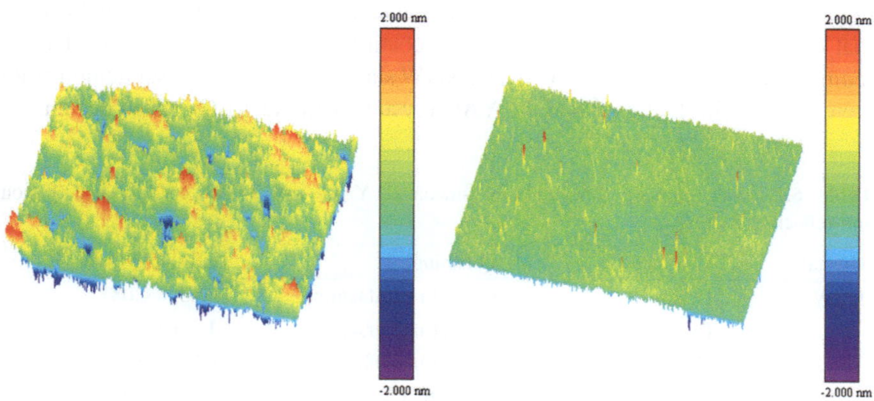

Fig. 6.16 Surface topography of a YAG substrate (profilometer measurement, image section 650 × 450 μm) polished with the previous technique (*left*, Rq = 0.49 nm) and with the improved technique (*right*, Rq = 0.20 nm)

Damage threshold measurements on uncoated YAG substrates were also conducted. The damage threshold of the uncoated substrates was improved from 0.89 to 0.98 TW cm^{-2} (λ = 1064 nm, τ = 10 ps, ν_{rep} = 10 kHz, $E_{pulse,max}$ = 200 μJ).

6.2.4 Anti-reflective Coating of YAG and KGW

Anti-reflective coatings had to be developed for both YAG and KWG crystals. The coatings were specified as anti-reflective (<0.2 %) at 0° for 940 and 1030 nm.

In order to obtain highest-possible damage thresholds, two different coating material combinations were used on the YAG anti-reflective coating tests (Ta_2O_5/SiO_2 and HfO_2/SiO_2). The coating design was not changed fundamentally (apart from slight modifications owing to the different refractive indices). For the anti-reflective coating of KGW two different coating designs were tested with the coating materials Ta_2O_5/SiO_2. Table 6.2 shows the measured damage thresholds for uncoated crystals and the various AR coatings.

As already outlined before, the damage threshold of the uncoated YAG substrates was distinctly improved by the new polishing method. On the coated YAG substrates the new polishing method performed less well than the original technique. Close inspection of the optics showed that this was due to cleaning artefacts. Ultrasonic cleaning of the YAG substrates, which is much more reliable than manual cleaning, was subsequently introduced and eradicated this effect. The problems with the cleaning also make it difficult to interpret the differences between the HfO_2/SiO_2 and the Ta_2O_5/SiO_2 systems. If very good substrate cleaning is achieved, HfO_2-based coating systems clearly have the potential to significantly raise the laser damage threshold.

In the design comparison on KGW, Design 2 was found to have the distinctly higher laser damage threshold. Despite the uncertainties mentioned, it can be stated that the laser damage threshold of all the coating designs tested is higher than the required values by a factor of 4–7. The KGW slab crystals for the laser demonstrator were coated with Design 2 and the YAG crystals with the Ta_2O_5-based design.

Table 6.2 Laser damage thresholds of uncoated YAG and KGW crystals and various anti-reflective coatings

Crystal	Coating	Polishing	LIDT (TW cm^{-2})
KGW	Design 1	Crystal manufacturer	0.83 ± 0.08
	Design 2	Crystal manufacturer	1.4 ± 0.13
YAG	Uncoated	Crystal manufacturer	0.89 ± 0.04
		LAYERTEC	0.98 ± 0.06
	Design Ta_2O_5	Crystal manufacturer	1.12 ± 0.03
		LAYERTEC	1.01 ± 0.05
	Design HfO_2	Crystal manufacturer	1.34 ± 0.05
		LAYERTEC	0.94 ± 0.10

Laser parameters: λ = 1064 nm, τ = 10 ps, ν_{rep} = 10 kHz, $E_{pulse,max}$ = 200 μJ

6.2.5 *Technology for the Production and Coating of Slab Crystals*

In addition to developing the coating systems for reflection suppression and metallization and proving their suitability in principle, a technology also had to be developed for completely coating the slab crystals. The crystals used in the project have dimensions of $10 \times 10 \times 1\ mm^3$. Two opposing surfaces measuring $10 \times 1\ mm^2$ have to be polished and anti-reflection coated for the pump and laser radiation. The $10 \times 10\ mm^2$ surfaces are metallized with gold.

The problem is that the metallization coating has to extend to the edge in order to guarantee adequate cooling over the entire surface after soldering into the heat sink, but there must not be any metal on the entry and exit surfaces for the laser radiation because even the tiniest absorption losses would immediately destroy the crystal, given the high laser power densities used here.

To achieve this reliably and reproducibly, the following technology was deployed which suitably combines the works steps of crystal polishing and coating and ensures metallization up to the edge without affecting the entry and exit surfaces.

1. Crystal is ground to the required dimensions
2. Ultrasonic cleaning
3. Metallization
4. Faces polished
5. Ultrasonic cleaning
6. AR coating

Polishing the faces after metallization ensures that no metal remains on the faces. During AR coating the metallized surfaces are protected by a special fixture which stops any dielectrics getting onto the gold layer and preventing it from being wetted by the solder.

References

1. A. Tünnermann, S. Nolte, J. Limpert, Femtosecond vs. picosecond laser material processing. Laser Techn. J **1**, 34–38 (2010)
2. P. Rußbüldt, T. Mans, J. Weitenberg, H.D. Hoffmann, R. Poprawe, Compact diode-pumped 1.1 kW Yb:YAG Innoslab femtosecond amplifier. Opt. Lett. **35**(24), 4169–4171 (2010)
3. T. Eidam, S. Hanf, E. Seise, T.V. Andersen, T. Gabler, C. Wirth, Th Schreiber, J. Limpert, A. Tünnermann, Femtosecond fiber CPA system emitting 830 W average output power. Opt. Lett. **35**(2), 94–96 (2010)
4. Trumpf Scientific SDL H3-200, http://www.trumpf-scientific-lasers.com/products/disk-lasers.html. Accessed 1 July 2013
5. J. Tümmler, R. Jung, H. Stiel, P.V. Nickles, W. Sandner, High-repetition-rate chirped-pulse-amplification thin-disk laser system with joule-level pulse energy. Opt. Lett. **34**(9), 1378–1380 (2009)

6. J. Saraceno, F. Emaury, O.H. Heckl, C.R. Baer, M. Hoffmann, C. Schriber, M. Golling, Th Südmeyer, U. Keller, 275 W average output power from a femtosecond thin disk oscillator operated in a vacuum environment. Opt. Express **20**(21), 23535–23541 (2012)
7. K. Du, N. Wu, J. Xu, J. Giesekus, P. Loosen, R. Poprawe, Partially end-pumped Nd:YAG slab laser with a hybrid resonator. Opt. Lett. **23**, 370–372 (1998)
8. P. Rußbüldt, T. Mans, G. Rotarius, J. Weitenberg, H.D. Hoffmann, R. Poprawe, 400 W Yb: YAG Innoslab fs-Amplifier. Opt. Express **17**, 12230–12245 (2009)
9. J. Giesekus, T. Mans, K.-M. Du, B. Braun, P. Loosen, R. Poprawe, in *High Power Diode End Pumped Slab MOPA System*. International Conference on Lasers and Electrooptics (CLEO), paper CThI3 (2001)
10. Luther-Davies, V.Z. Kolev, M.J. Lederer, N.R. Madsen, A.V. Rode, J. Giesekus, K.M. Du, M. Duering, Table-top 50-W laser system for ultra-fast laser ablation. Appl. Phys. A **00**, 1–5 (2004)
11. B. Gronloh, P. Russbueldt, W. Schneider, B. Jungbluth, H.-D. Hoffmann, High average power sub-picosecond pulse generation at 515 nm by extracavity frequency doubling of a mode-lockes Innoslab MOPA. Proc. of SPIE **8235**, 82351C (2012)
12. P. Russbueldt, T. Mans, G. Rotarius, H.-D. Hoffmann, R. Poprawe, T. Eidam, J. Limpert, A. Tünnermann, in *Proceedings of the Advanced Solid-State Photonics*, OSA Techn. Digest, MF4, 2009. "Hybrid 400 W Fiber-Innoslab fs-Amplifier,"
13. M. Schulz, R. Riedel, A. Willner, T. Mans, C. Schnitzler, P. Russbueldt, J. Dolkemeyer, E. Seise, T. Gottschall, S. Hädrich, S. Duesterer, H. Schlarb, J. Feldhaus, J. Limpert, B. Faatz, A. Tünnermann, J. Rossbach, M. Drescher, F. Tavella, Yb:YAG Innoslab amplifier: efficient high repetition rate subpicosecond pumping system for optical parametric chirped pulse amplification. Opt. Lett. **36**(13), 2456–2458 (2011)
14. F. Tavella, A. Marcinkevicius, F. Krausz, 90 mJ parametric chirped pulse amplification of 10 fs pulses. Opt. Express **14**, 12822–12827 (2006)

Chapter 7
Few-Cycle Mid-Infrared OPCPA System

Xun Gu, Yunpei Deng, Gilad Marcus, Thomas Metzger, Reinhard Kienberger and Ferenc Krausz

Abstract We review the principle, design and output characteristics of the MPQ/LMU LWS-1 mid-infrared few-cycle OPCPA system. With the output of CEP-stable 2.1-μm–700-μJ 15-fs (2-optical-cycle) pulses at 1 kHz, this laser is unique in the world, providing the shortest sub-mJ pulses in the mid-infrared. A system upgrade with a Yb:YAG thin-disk pump laser has recently produced 1.2-mJ 1.5-cycle pulses at 3 kHz, approaching the TW-level monocycle regime.

7.1 Introduction

Isolated attosecond pulses can now be routinely generated and utilized in a host of interesting applications requiring atomic time resolutions, creating an entire new field of attosecond physics [1]. Typically, such pulses are generated by high harmonic generation (HHG) in a noble gas, driven by intense few-cycle CEP-stable laser pulses [2, 3]. The process can be well described by the famous semiclassical three-step model [4]. First, the intense laser field ionizes the atom by tunnel ionization, releasing the electron into the continuum; second, the free electron accelerates in the laser field, and is driven back to the parent ion after the laser electric field reverses direction; finally, the electron recombines with the parent ion, releasing the energy it has acquired during acceleration as a HHG photon. Using this model, the cutoff energy of the HHG photon is derived to be [5]

$$\hbar\omega_{\mathrm{co}} = I_{\mathrm{p}} + 3.17U_{\mathrm{p}} = I_{\mathrm{p}} + \frac{3.17e^2}{32\pi^2c^3\varepsilon_0 nm} I\lambda^2 \tag{7.1}$$

X. Gu (✉) · Y. Deng · G. Marcus · T. Metzger (✉) · R. Kienberger · F. Krausz
Max-Planck Institute of Quantum Optics, Hans-Kopfermann-Str. 1, 85748 Garching, Germany
e-mail: xun.gu@ch.abb.com

T. Metzger
e-mail: Thomas.Metzger@de.trumpf.com

S. Nolte et al. (eds.), *Ultrashort Pulse Laser Technology*,
Springer Series in Optical Sciences 195, DOI 10.1007/978-3-319-17659-8_7

where I_p is the ionization potential, and the ponderomotive potential U_p scales with the driver laser intensity I and the square of the laser wavelength λ.

The typical driver laser for HHG and attosecond pulse generation is a Ti:sapphire amplifier system, generating few-cycle pulses of a few hundred microjoules at ~800 nm [2, 3]. The pulse's carrier-envelope phase (CEP) is monitored by a nonlinear interferometer, and is actively stabilized by controlling both the oscillator pumping power and a mechanical element in the amplifier. As an example, the shortest pulse ever produced was a 80-as pulse generated in such a fashion [6]. While such isolated attosecond pulses have been widely applied in many experiments, their photon energies have unfortunately been limited to ~100 eV. This is largely due to ionization depletion, which effectively limits the maximum laser intensity at the focus to 10^{14}–10^{15} W/cm^2.

To circumvent the intensity limitation, an alternate and more effective route to higher photon energies is by using a longer driver laser wavelength in the infrared. This is clearly implied by the cutoff formula above, where the ponderomotive potential is proportional to the square of the driver laser wavelength. Using a 2.1-μm driver laser, (7.1) predicts a HHG cutoff energy of 1321 eV with 1 PW/cm^2 laser intensity, in contrast to 205 eV with a 800-nm laser at the same intensity. The possibility of generating soft X-ray photons up to keV with a tabletop system is extremely attractive for many applications. For example, the generation of coherent water-window X-ray radiation (280–530 eV) is highly sought after for biomedical imaging, because of the possibility to selectively image carbon atoms without water absorption.

However, generating intense few-cycle pulses in the mid-infrared requires a different laser technology, due to the lack of a broadband laser medium in the mid-infrared similar to Ti:sapphire in the near infrared. Fortunately, there exists another very good mechanism for broadband pulse amplification, namely, optical parametric amplification (OPA).

OPA is physically the same process as difference-frequency generation (DFG), whereby one photon gets divided into two photons carrying the same total energy in a second-order nonlinear optical crystal. In OPA, the initial condition is that a strong pump beam of shorter wavelength and a weak signal beam of longer wavelength interact in a nonlinear crystal. Through the DFG process, pump photons are continuously converted into pairs of signal photons and difference-frequency "idler" photons. As a result, the pump energy flows into the signal beam and a new idler beam, whereby the weak signal can be amplified by many orders of magnitudes.

OPA can be very efficient. Combined with chirped-pulse amplification (CPA), the technology known as OPCPA [7–9] has in recent years become the most promising technique for the amplification of high-energy ultrashort laser pulses, and is so far the only method by which multi-mJ few-cycle pulses are generated [10–13]. Compared with traditional laser amplification, the advantages of OPA/OPCPA include:

- Gain bandwidth: One of the most remarkable properties of parametric amplification is broad amplification bandwidth without substantial gain narrowing.

- Single-pass gain: The single-pass gain achievable in OPA is extremely high, possible to reach even six orders of magnitude.
- Thermal load: The thermal load in OPA is almost negligible. The energy conversion from the pump into the signal and the idler is an instantaneous process. No excess energy is produced or stored in the crystal, which is different from laser amplification.
- Efficiency: High efficiencies can be reached with OPA, up to 25 % in many cases.
- Scalability to high energies: OPA can be scaled to very high pulse energies. With a broader amplification bandwidth, hence a shorter compressed pulse duration, the achievable pulse peak power can be much higher than in a conventional CPA amplifier.
- Versatility to achieve phase matching: There are several degrees of freedom that can be exploited to achieve broadband phase matching, including crystal orientation, the noncollinear angle between the pump and the signal, and the spatio-temporal properties of the beams.
- Idler wave: The idler wave can sometimes also be useful, for example, to generate a signal in spectral regions difficult to reach with conventional techniques or to use the idler wave to stabilize the CEP [14–17].

On the other hand, new challenges and disadvantages of OPA compared to the conventional laser amplification include:

- Pump-to-signal synchronization: Amplification takes place only when the pump and the signal are both present. Therefore precise synchronization is required.
- Stretching and compression: Stretching the seed pulse to a fraction of the pump pulse duration is required for efficient energy conversion. In case of a long pump pulse, recompression with a ratio of many orders of magnitude can be difficult to achieve.
- Strict phase matching conditions: Alignment of a parametric amplifier involves numerous factors including the geometry of interaction, pointing stability, pump energy fluctuations, beam collimation, etc.
- Background superfluorescence: The background emission in OPA has a different mechanism compared to the amplified spontaneous emission (ASE) in a solid-state laser amplifier. The background emission in an OPA is created by optical parametric superfluorescence due to spontaneous decay of a pump photon into an idler and a signal photon which is further parametrically amplified.
- Demanding pump beam requirements: The pump beam must have high peak power, a relatively short pulse duration ($<\sim 100$ ps), a good temporal profile and a good beam quality.

Despite the challenges, for our goal of developing an intense mid-infrared few-cycle light source, OPCPA is clearly the only technique capable of delivering the desired output parameters. The MPQ/LMU LWS-1 prototype system is based on this technique, and is the subject of discussion in this chapter.

7.2 Broadband OPA Phase-Matching Configurations

In this section, we are going to review some basic factors affecting the amplification gain bandwidth of the OPA process, which is critical for few-cycle pulses. Particularly, two phase-matching configurations will be covered.

We assume a type I configuration in a uniaxial birefringent crystal, in which the pump is an e-wave, and the signal and the idler are o-waves. This is the most common phase-matching situation for OPA applications.

The basic OPA gain formula can be derived from the coupled-wave equations between the pump, signal and idler waves. Neglecting pump depletion, and assuming plane waves and the slowing-varying-envelope approximation, we have

$$\frac{I_s(L)}{I_s(0)} = 1 + (gL)^2 \frac{\sinh^2 a}{a^2} \tag{7.2}$$

where

$$g = 4\pi d_{\mathrm{eff}} \sqrt{\frac{I_p(0)}{2\varepsilon_0 n_p n_s n_i c \lambda_s \lambda_i}} \tag{7.3}$$

$$a = \sqrt{(gL)^2 - (\Delta k L/2)^2} \tag{7.4}$$

$$\text{and} \quad \Delta k(\omega_p, \omega_s) = k_e(\omega_p) - k_o(\omega_s) - k_o(\omega_p - \omega_s) \tag{7.5}$$

Clearly, the spectral dependence of the gain, i.e., the gain bandwidth, is mostly determined by the spectral dispersion of the phase mismatch factor. While the pump frequency ω_p is basically fixed, the seed frequency ω_s varies over a large range for broadband pulses, and hence Δk can also vary. The maximum amplification is achieved at signal frequency ω_s where $\Delta k(\omega_p, \omega_s) = 0$, and the OPA gain will remain high only in the region around ω_s where Δk stays small enough. Therefore, for broadband amplification to occur, we need both

$$\Delta k(\omega_p, \omega_s) = 0 \tag{7.6}$$

$$\text{and} \quad \frac{\partial \Delta k(\omega_p, \omega_s)}{\partial \omega_s} = 0 \tag{7.7}$$

The first condition can be easily satisfied by choosing an appropriate crystal angle θ, which affects the e-wave k_p. The satisfaction of the second condition at the same time requires some more thinking.

In a collinear geometry, in which the signal, idler and pump beams propagate along the same direction, (7.7) implies

$$\left(\frac{\partial k_o}{\partial \omega}\right)_{\omega_s} - \left(\frac{\partial k_o}{\partial \omega}\right)_{\omega_p - \omega_s} = 0 \tag{7.8}$$

Obviously, this can be satisfied at $\omega_s = \omega_i = \omega_p - \omega_s$, or $\omega_p = 2\omega_s$. In this case, the signal and the idler have the same wavelength, polarization and propagation; therefore, they are in principle indistinguishable. This configuration is called *degenerate OPA*, and it ensures broadband amplification.

If the desired signal center wavelength is not half of the pump wavelength, obviously, (7.7) cannot be satisfied in a collinear geometry, and the amplification bandwidth would be narrow. An additional degree of freedom can be introduced, however, if we allow the pump and the signal beams to cross with a certain angle. In this case, $\Delta\mathbf{k}$ is defined as the vector difference between the $\mathbf{k}$-vectors of the three waves, and is dependent on both the propagation direction of the pump beam θ and the signal-pump noncollinear angle α.

Differentiating $\Delta\mathbf{k}$ against ω_s in a similar fashion will lead to

$$\left(\frac{\partial k_o}{\partial \omega}\right)_{\omega_s} \cos(\alpha + \beta) - \left(\frac{\partial k_o}{\partial \omega}\right)_{\omega_i} = 0 \tag{7.9}$$

where β is the angle between the idler and the pump, and can be calculated from α using the relation $k_o(\omega_s) \sin \alpha = k_o(\omega_i) \sin \beta$.

Therefore, by choosing appropriate angles θ and α, one can also achieve a broad OPA bandwidth. This configuration is called *noncollinear OPA* (NOPA), which is widely used in the design of many OPCPA systems [10–13].

For the MPQ/LMU LWS-1 system to be discussed in this chapter [18, 19], we have chosen to build a 1053-nm Nd:YLF pump laser [20]. Twice the pump wavelength is 2.1 μm, which is an ideal center wavelength for our infrared OPA system. Therefore, we have chosen to build a near-degenerate OPA system in order to realize a broad gain bandwidth for few-cycle pulses. The design slightly deviates from the true degenerate configuration, in that the signal and the pump are not perfectly collinear, but cross at a very small angle. The angle difference allows us to easily separate the signal and the pump beams after the OPA crystals, and avoid signal-idler interference which might disturb the signal pulse profile. The deviation from the degenerate condition is small, and the broad bandwidth is therefore not affected.

7.3 MPQ/LMU LWS-1 System Description

The MPQ/LMU LWS-1 prototype IR OPCPA system generates CEP-stable 2.1-μm–700-μJ 15-fs pulses at 1 kHz [18, 19]. With pulses spanning only about two optical cycles, this laser is unique in the world, providing the shortest sub-mJ pulses in the mid-infrared. A further upgrade to the LWS-1 system is currently underway

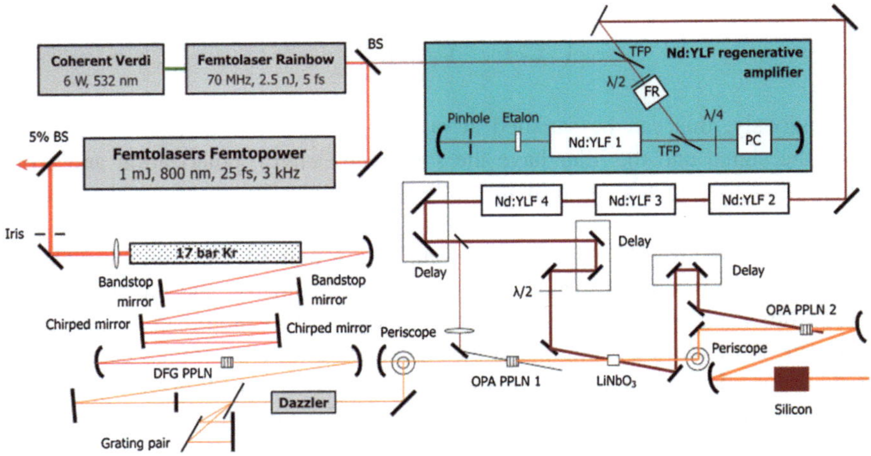

Fig. 7.1 Schematic of the MPQ/LMU LWS-1 mid-IR OPCPA system

with a new ps Yb:YAG thin-disk pump laser [21], promising to reach multi-mJ and an octave bandwidth in the coming years.

The schematic of the LWS-1 prototype system is shown in Fig. 7.1. We will discuss the design and performance of the system in detail in the following sections.

7.3.1 Infrared Seed Pulse Generation

First, we start from the generation of the broadband infrared seed pulse for our OPCPA system, by means of intrapulse DFG (optical rectification) in a PPLN. This design automatically ensures the CEP stability, which is preserved in the OPA process, and is of vital importance to many applications which rely on few-cycle pulses.

A number of options exist for the generation of mid-infrared pulses. A straightforward way is to use a broadband oscillator based on a laser medium in this spectral region. However so far the shortest pulse duration of laser oscillators operating around 2 μm has been limited to sub-ps [22].

Because of the lack of an appropriate ultrafast laser oscillator in the mid-infrared, an optical parametric oscillator [23–25] has been used as an ultrashort tunable light source. However, the bandwidth and the pulse duration are also not sufficient for our few-cycle OPCPA system. Another approach to obtain the infrared radiation is by optical rectification (intrapulse DFG) of a broadband optical pulse in the visible and near-infrared region. This scheme is frequently used in ultrafast terahertz (THz) optics and spectroscopy [26–28]. Recent advance in ultrabroadband laser oscillators in the visible has naturally resulted in the generation of high-photon-energy THz radiation, all the way down to the mid-infrared range. Mid-infrared pulses at 7 μm has been demonstrated by use of optical rectification (DFG) of sub-20-fs Ti:sapphire oscillator pulses [29]. Few-cycle or even mono-cycle pulses with a stable electric field is exactly what we desire as a seed source.

Using optical rectification (intrapulse DFG), Fuji et al. have previously demonstrated the detection and stabilization of CEP based on the interference at 1.4 μm between the fundamental spectral component and the intrapulse DFG signal [30]. This mechanism satisfies two important requirements for an OPCPA seed source. First, with an appropriate nonlinear crystal, 2-μm radiation can be acquired even more easily than 1.4 μm, because it requires less bandwidth from the fundamental radiation. Secondly, the CEP of the infrared signal generated by intrapulse DFG is automatically stabilized, regardless of the CEP stability of the fundamental pulse [14–17]. The mechanism can be easily understood by recognizing the fact that all the components in the fundamental pulse carry the same CEP, and the infrared DFG pulse has a CEP which is the difference between the long- and short-wavelength components in the same fundamental pulse. Therefore, even though the fundamental pulse's CEP may be random from shot to shot, the DFG pulse's CEP does not change, and is only defined by the envelope of the fundamental pulse. The stable CEP of the infrared seed pulse will in principle also be preserved in the OPA processes, regardless of the CEP stability of the pump pulse. The CEP stability is very important to the users of few-cycle laser pulses, when the pulses are used as a driver for HHG or other highly nonlinear processes, where CEP plays a central role in determining the electric field of the pulse.

The front end of our system consists of a commercial 1-mJ 25-fs 3-kHz Femtolasers Femtopower Pro Ti:sapphire multipass amplifier system. Five percent of the Femtopower system output is first spatially clipped by an iris, reducing the pulse energy to 12.3 μJ, and then focused into a gas cell filled with 17-bar krypton, where the pulse creates a single filament which broadens its spectrum by SPM to cover a spectral range from ~600 to ~950 nm. The pulse energy and the iris size are chosen such that the filament's output spectrum is as broad and smooth as possible without breaking into unstable multiple filaments. The white-light beam is then reflected four times on homemade bandstop mirrors, which are designed to remove the strong components between 750 and 850 nm from the reflected beam and which meanwhile have flat dispersion and uniform reflectivity in the spectral regions below 750 nm and above 850 nm (see Fig. 7.2). The pulse after the bandstop mirrors are then temporally compressed by a set of home-made chirped mirrors. An XFROG setup with the Ti:sapphire amplifier pulse as the reference pulse is used to measure the dispersion of the pulse and make sure that the long- and short-wavelength components overlap in time. The pulse is afterwards sent into a 1-mm-long 11.21-μm-period MgO-doped periodically poled $LiNbO_3$ crystal (PPMgLN), where the long- and short-wavelength components are mixed, generating a difference-frequency signal extending from 1.7 to 2.6 μm, This nJ-level infrared seed is to be used as the seed for our OPCPA system.

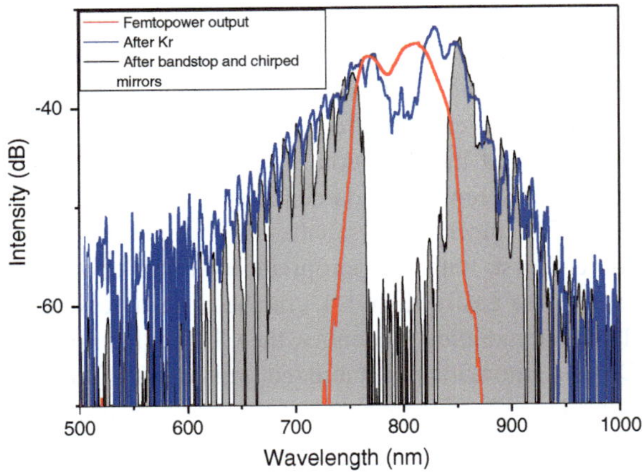

Fig. 7.2 Spectra of the Femtopower Pro laser system output pulse (*red*), the spectrally broadened pulse after the krypton cell (*blue*), and the pulse after the bandstop and chirped mirrors (*black*)

7.3.2 Nd:YLF Pump Laser

To achieve mJ-level infrared output pulse from the OPCPA system, one principal challenge is to design and build a picosecond pulse pump laser capable of delivering pulse energies of tens of mJ at kHz repetition rates. Considerable research effort has been devoted into the development of a 20-mJ 2-ps Nd:YAG thin-disk laser, which promises to be an ideal pump laser for our next-generation OPCPA sources. This laser has recently been placed into commission, and is at the time of writing being used to pump the next-generation IR OPCPA system. Meanwhile, for our prototype IR OPCPA system covered in this chapter, we have built a 49-ps 13-mJ Nd:YLF amplifier chain, consisting of a regenerative amplifier and three linear post-amplifiers, all developed in house [20].

The regenerative amplifier is seeded by the Ti:sapphire oscillator (Femtolasers Rainbow) contained in the Femtopower front end described in the previous section, and is operated at 1 kHz. The oscillator spectrum is specially tuned so that its spectrum extends just beyond ~ 1060 nm. The infrared part of the oscillator output, measuring ~ 10 pJ within the fluorescence bandwidth, is split off by a diachronic mirror, and sent as the seed into our Nd:YLF regen amplifier, which features a Northrop-Grumman laser module containing a 80-mm-long 4-mm-diameter a-cut Nd:YLF rod. This optical seeding scheme ensures that excellent synchronization is maintained between the seed and pulse pulses of the OPA, which is of great importance for the stability and reliability of the system [31]. The infrared pulse is amplified up to 3.5 mJ after ~ 40 round trips in the cavity, at which point it is ejected out of the cavity by Pockel cell switching. Two fused silica etalons of 0.7 and 1 mm are inserted in the regen cavity, in order to control the bandwidth and the pulse duration, to prevent detrimental self-focusing in the laser crystal.

Three single-pass Nd:YLF amplifier modules are used to further amplify the regen output. These amplifiers each contain a 120-mm-long a-cut Nd:YLF rod crystal, of diameter 4 or 5 mm, and are continuously pumped by laser diode arrays. The c-axes of the Nd:YLF crystals are arranged such that each crystal has an orthogonal c-axis to the one before or after, and half-waveplates are inserted between the amplifiers to rotate the beam polarization accordingly. This design is used specifically to compensate for the elliptical thermal lensing of the Nd:YLF crystal. Additionally, cylindrical lenses are also inserted at appropriate locations to maintain a good beam profile. Most importantly, delicate balance must be achieved between thermal lensing and self-focusing in these crystals, in order to extract maximum amplification without damaging the crystals. To this end, appropriate lenses are placed between the stages to optimize the beam size and divergence in these amplifiers. After the three post-amplifiers, an output energy of 13 mJ is obtained. The pulse energy stability is measured to be 1.2 % rms. A third-order autocorrelation measurement of the output pulse determines the intensity contrast ratio between the peak of the amplified pulse and the ASE floor to be $\sim 10^3$. The FWHM pulse duration, assuming a Gaussian pulse shape, is 49 ps. This pulse is used as the pump pulse for our OPCPA system.

7.3.3 Stretcher and Compressor Design

OPA being an instantaneous nonlinear process, energy transfer from the pump pulse to the seed pulse can occur only when and where both pulses overlap. Since the pump pulse is much longer than the seed pulse, in order to achieve maximum amplification efficiency, the seed pulse must be first temporally stretched to a significant fraction of the pulse pump duration, and after amplification recompressed back to its Fourier limit as closely as possible. The design of the stretcher and the compressor must allow, first of all, the required stretched pulse duration, and secondly, the cancellation of not only the GDD but also the higher-order phases in the final compressed pulse.

Traditional CPA systems typically use a positively dispersive bulk material as the stretcher, and a prism/grating pair as the compressor. For our system, we have chosen to do the opposite. We use positively dispersive AR-coated bulk silicon substrates as the compressor, which has a higher transmission throughput and requires easier alignment than a grating compressor. For the stretcher, we use a pair of 300 line/mm gratings separated by 75 mm, and a Fastlite Dazzler specially designed for the infrared spectral range. The use of the Dazzler, an acousto-optic pulse shaper, allows for fine compensation of arbitrary phase profiles within its tuning range, which is defined by the area between the group delays (GD) corresponding to the o- and e-waves in the TeO_2 crystal of the Dazzler. The design GD curves of the grating pair, silicon compressor, the phase to be compensated and the Dazzler tuning range are shown in Fig. 7.3. As one can see from the figure, the infrared seed pulse between 1.8 and 2.7 μm is stretched to 26 ps after the grating

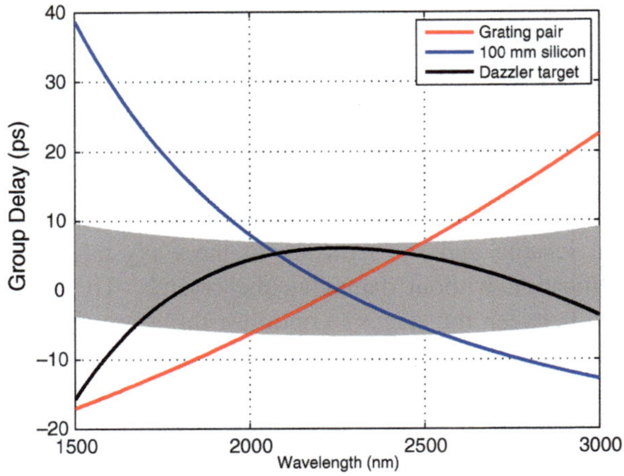

Fig. 7.3 Calculated GD curves for the grating pair (*red*), silicon compressor (*blue*) and the target GD to be realized by the Dazzler (*black*). The gray shaded area represents the tuning range of the Dazzler, delimited by the e-wave and the o-wave GDs in the TeO_2 crystal

pair and the Dazzler (which is the opposite of the silicon block's dispersion), and the phase to be compensated by the Dazzler falls well within its tuning range. Fine tuning of the dispersion can be achieved by measuring the phase of the compressed pulse, and feeding back the opposite of the measured phase to the Dazzler accordingly.

7.3.4 OPA Design

In this section, we discuss the OPA design of our system. For our system pumped by 1037-nm pulses, degenerate OPA provides a broad bandwidth around 2.1 μm. Experimentally, the signal and the pump beams are arranged to cross with a small angle (~3°) in the OPA crystals, to allow for easy separation of the beams and to avoid signal-idler interference.

Up to date, the best crystal transparent in the whole spectral range of our interest and providing the highest second-order nonlinear coefficient is lithium niobate ($LiNbO_3$). Pumped at 5 GW/cm^2, a 6.5-mm-thick MgO-doped $LiNbO_3$ crystal cut at 42.9° can support a small-signal-amplification bandwidth from 1.8 to 2.5 μm, sufficient for a sub-3-cycle pulse. Even better, periodically poled $LiNbO_3$ (PPLN) crystals can provide the same amplification with a shorter crystal length, and therefore, a broader amplification bandwidth. In a PPLN crystal, the higher d_{33} coefficient is used ($d_{33} = 25$ pm/V), and the effective second-order nonlinear coefficient for OPA is $d_{\mathrm{eff}} = 2d_{33}/\pi = 16$ pm/V. In comparison, for a type-I bulk

$LiNbO_3$ crystal, $d_{eff} = 5$ pm/V, significantly smaller than the PPLN value. The phase mismatch in a PPLN can be calculated by adding a quasi-phase-matching (QPM) term to the usual definition

$$\Delta k = k_p - k_s - k_i - \frac{2\pi}{\Lambda} \tag{7.10}$$

where Λ is the poling period.

The OPA gain calculation is the same as described in (7.2).

The calculated small-signal OPA gain curves are shown in Fig. 7.4. Clearly, PPLN supports a broader amplification bandwidth than bulk $LiNbO_3$, and therefore is the crystal of choice for our broadband IR OPCPA system. However, a practical problem with PPLN is its limited available size. Therefore, for our prototype system, we have chosen to use a 3-mm 30.2-μm poling period PPLN in the first OPA stage, where a nJ infrared seed pulse is amplified to ~ 3 μJ. In the second and the third OPA stage, we use 4-mm-thick bulk $LiNbO_3$ crystals. Large-aperture high-quality PPLN crystals are currently being developed in our collaborator Prof. Taira's group at the Institute for Molecular Science in Japan [32], and they will be key components in the next-generation IR system, which will possess an even broader bandwidth.

It should also be noted that $LiTaO_3$ crystals can also be periodically poled, and provide a similar performance as PPLN. However, periodically poled $LiTaO_3$ is less common, and the available size is smaller.

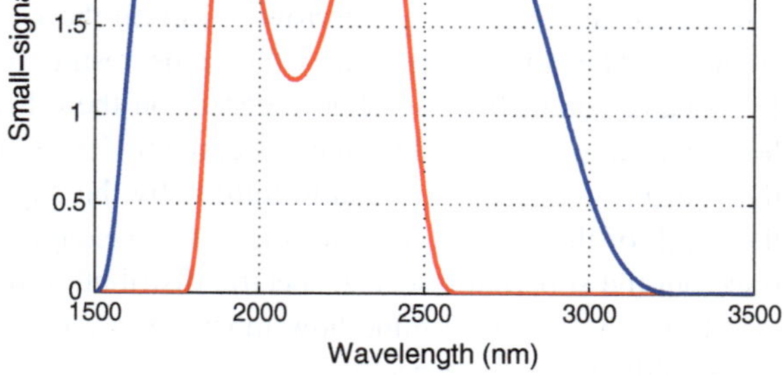

Fig. 7.4 Calculated small-signal OPA gains for a 6.5-mm-long $LiNbO_3$ cut at 42.9° (*red*), and a 1.9-mm-long 30.3-μm-period PPLN (*blue*), both pumped at 5 GW/cm^2

7.3.5 OPCPA System Output and Parametric Superfluorescence

Based on the design described in previous sections, a prototype infrared OPCPA system has been built at MPQ. The infrared seed signal generated by DFG in the PPLN is sent through a grating pair and a Dazzler, and the stretched pulse in amplified in three OPA stages, consisting of PPLN and bulk $LiNbO_3$ crystals. The first and the second OPA stages are pumped by 3 and 97 % of the 13-mJ 1053-nm pulses, and the pump beam after the second OPA stage is reused for pumping the third OPA stage. The amplified signal beam is finally sent through a 100-mm-long silicon block for compression.

The OPA system output, including possible superfluorescence, is 6 μJ after the first stage, 200 μJ after the second stage, and 920 μJ after the third stage. The energy stability of the third-stage output is 9 % rms.

In general, high-gain OPA systems have an inherent problem of background superfluorescence, and our system is no exception. Indeed, with the IR seed beam blocked, one can measure a superfluorescence output of ~780 μJ, almost equal to that of the full amplified signal with an unblocked seed. However, it is important to note that, with the pump pulse energy significantly depleted in the final OPA stage, the true superfluorescence in our system is much smaller than the superfluorescence when the seed is blocked. That is because, when the OPA system is seeded, both the seed pulse and the noise that initiates the superfluorescence experience the same saturated parametric gain, whereas when the seed beam is blocked, the noise experiences a much higher unsaturated gain. Therefore, ascertaining the true amplified signal and the superfluorescence levels in a high-gain low-seed OPA system is very difficult.

To resolve this important but difficult issue, we have employed a novel method of spectral shaping of the seed pulse. In the absence of superfluorescence, any shaping of the IR seed spectrum by the Dazzler will be directly transferred to the amplified signal. In particular, if we annihilate a narrow region in the seed spectrum (or create a zero-reaching spectral hole) using the Dazzler, the real amplified seed will preserve the spectral hole with a 100 % modulation depth. (A spectral hole can also be created by placing a narrow object on the Fourier plane within the grating stretcher. This method produces the same result as the Dazzler pulse-shaping.) Broadband parametric superfluorescence, on the other hand, is not shaped and its level unchanged if the hole is narrow enough. Therefore, in the combined spectrum, the spectral hole will be partially refilled by the superfluorescence. By measuring the depth of the spectral hole at different wavelengths across the spectrum, we can trace out the superfluorescence spectral distribution relative to the full OPA output (see Fig. 7.5), and determine how much energy is in the superfluorescence and in the amplified signal, respectively.

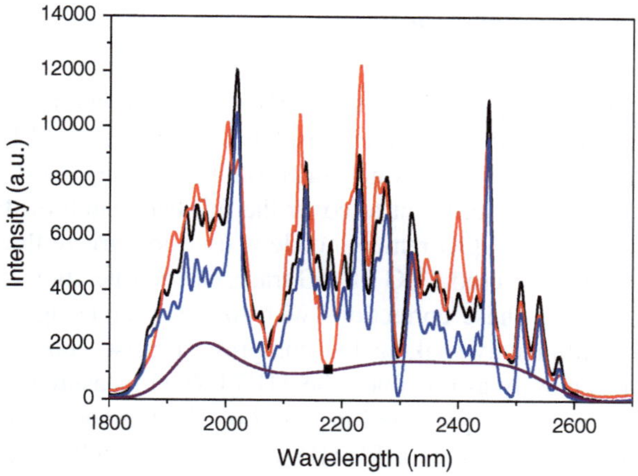

Fig. 7.5 Spectra of the full third-stage OPA output (*black*), the OPA output with a 50-nm-wide Dazzler-imposed spectral hole at 2150 nm (*red*), the superfluorescence profile traced out by hole depths at different wavelengths (*magenta*), and the true amplified signal (*blue*, the full OPA output minus the superfluorescence). The *black square* denotes the measured level of superfluorescence at 2150 nm by the spectral hole technique

Let us now have a brief discussion on the basic assumptions and limitations of this simple superfluorescence estimation method. First of all, it's easy to see that for this method to work, the seed pulse should be heavily chirped in the OPA crystals, so much so that its temporal profile faithfully reflects its spectral shape. Only then will a sharp and narrow hole in the seed spectrum be turned into a sharp and narrow gap in its temporal profile. Under this condition, the superfluorescence level outside the narrow temporal gap in the seed pulse will not be affected by the shaping of the seed pulse. In our system, the seed pulse of more than 800 nm bandwidth is stretched to ~25 ps, comfortably satisfying this condition. Secondly, the superfluorescence inside the temporal gap created in the seed pulse will become stronger at the unsaturated level, because no seed is present there to deplete the pump. Therefore, the temporal gap, or the spectral hole, in the seed pulse should be made narrow enough to avoid overestimation of the superfluorescence. In our measurement, we used a 50-nm wide spectral hole, about 6 % of the total bandwidth. We have also tried holes of varying widths, and confirmed that the choice of the hole width up to 100 nm did not change of the depth of the hole in our measurement.

We have used this simple method to measure the superfluorescence in our OPA system output, and the result is presented in Fig. 7.4. From this measurement, it is clear that out of the total 920 μJ third-stage OPA output, 740 μJ (80 %) belongs to the amplified seed and the rest (20 %) belongs to the superfluorescence. In comparison, the superfluorescence is negligible in the first-stage OPA output, and accounts for 2.5 % of the second-stage OPA output.

7.3.6 Pulse Characterization

The final compressed output pulse of the OPCPA system is characterized by a third-harmonic-generation (THG) frequency-resolved optical gating (FROG) apparatus [33]. The third-harmonic signal is generated on an air-CaF_2 interface, which possesses a large bandwidth sufficient to cover the full bandwidth of the amplified IR signal. A home-written FROG retrieval code is used to retrieve the pulse intensity and phase from the measured THG FROG trace. Several iterations have been used to make fine spectral phase corrections with the Dazzler using the result of the FROG retrieval, ultimately making the compressed pulse approach the Fourier-transform limit as closely as possible. The THG FROG measurement results of the final compressed pulse are presented in Fig. 7.6. The temporal intensity profile of the compressed pulse measures a FWHM of 15.2 fs, which is very close to the Fourier limit of 14.6 fs. At 2.1 μm, this pulse duration corresponds to merely 2 optical cycles. This is, to our knowledge, the shortest sub-mJ level pulse ever generated at this wavelength.

The compression ratio of 25 ps to 15.2 fs and the amplified signal to superfluorescence energy ratio of 2:1 means that the intensity contrast between the

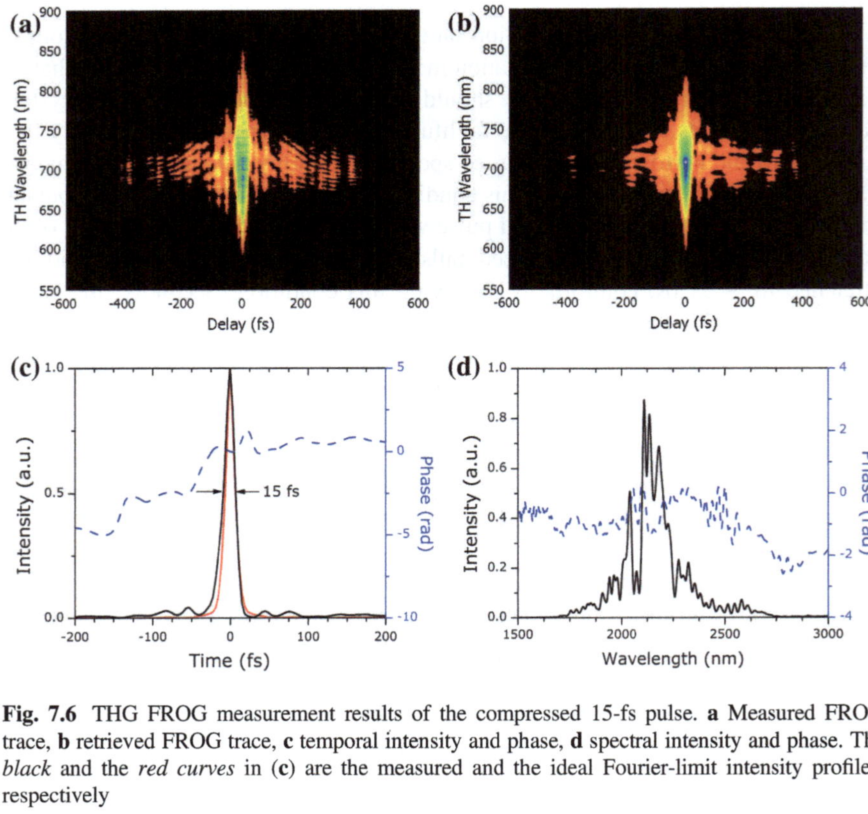

Fig. 7.6 THG FROG measurement results of the compressed 15-fs pulse. **a** Measured FROG trace, **b** retrieved FROG trace, **c** temporal intensity and phase, **d** spectral intensity and phase. The *black* and the *red curves* in (**c**) are the measured and the ideal Fourier-limit intensity profiles, respectively

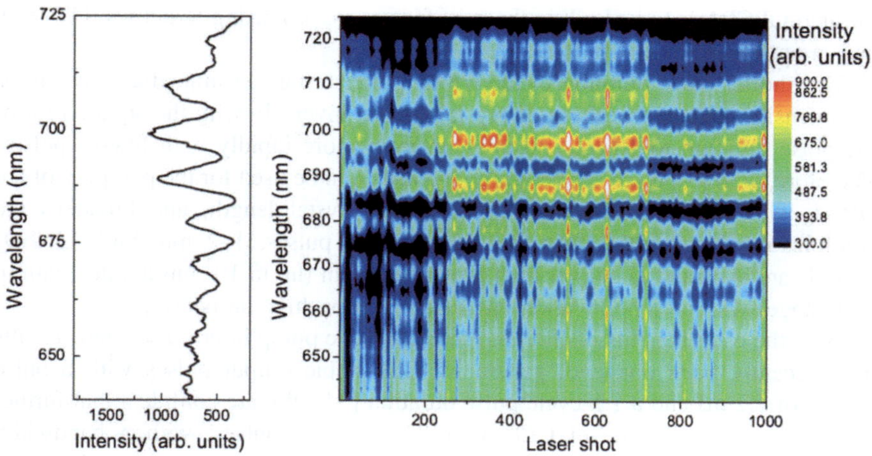

Fig. 7.7 CEP measurement of the OPCPA output pulse. *Left* A single-shot *f*-to-3*f* interferogram; *right* Interferograms measured in 1000 consecutive shots

compressed amplified signal pulse and the parametric superfluorescence is larger than 10^3, which should be sufficient for a high-quality HHG experiment.

The CEP stability of the final output of the OPCPA system has been further verified in an *f*-to-3*f* nonlinear interferometer measurement [19] (see Fig. 7.7). The spectrally broadened fundamental 2.1-μm light and its third harmonic are interfered in a spectrometer in a spectral region near 700 nm, resulting in spectral fringes whose positions indicate twice the CEP of the fundamental 2.1-μm light. This measurement experimentally confirms the preservation of the CEP stability of the infrared laser system output, which is very important for CEP-sensitive experiments. The CEP stability of the OPCPA system has also been further verified by a range of CEP-sensitive experiments, such as HHG and molecular ionization, in which the CEP is scanned by moving a thin substrate in the beam, and the CEP-dependent measurement results are recorded.

7.4 Further Upgrade of the LWS-1 System

Further upscaling of the output pulse energy of the infrared OPCPA source is limited by the fact that the Nd:YLF pump laser is already near its maximum capability. To this end, intense research in MPQ has in the past several years been devoted on the development of a Yb:YAG thin-disk-based laser amplifier. A regenerative thin-disk amplifier was recently demonstrated with an output of 1.6-ps 20-mJ pulses, with a repetition rate of 3 kHz, operating in a "pulse-doubling" mode [21]. Even higher output energy (>50 mJ) is in the plan with the addition of a Yb:YAG thin-disk power booster. Utilizing the Yb amplifier as the pump, a new infrared OPCPA is to be built in the next few years, promising to deliver TW-level near-monocycle infrared pulses.

The shorter pulse duration of the Yb amplifier brings an immediate benefit, as the intensity damage threshold of crystals scales inversely with the square root of the pulse duration to 10 ps, and goes up even more rapidly with few-ps pulses. Therefore, with ps pulses, higher pump intensity can be used for the pumping of the OPAs, which reduces the required nonlinear crystal length, and broadens the amplification bandwidth. With the 1.6-ps pump pulses, 1–2 mm bulk $LiNbO_3$ crystals are expected to deliver high parametric gain (up to 10^4 unsaturated gain in each stage) with an amplification bandwidth approaching an octave.

With the clear benefits of shorter, more energetic pump laser pulses, new results have recently been obtained, generating CEP-stable output pulses with a pulse energy of 1.2 mJ and a 1.5-cycle pulse duration [34]. We are confident that further work can deliver TW-level CEP-controlled infrared pulses, with a bandwidth reaching and exceeding an octave, thus supporting a near-monocycle pulse. This will be an exciting driver for keV HHG and a range of other experiments.

7.5 Summary

In this chapter, we have reviewed the principle, design and output characteristics of the MPQ/LMU LWS-1 mid-infrared few-cycle OPCPA system. With the output of CEP-stable 2.1-μm–700-μJ 15-fs (2-optical-cycle) pulses at 1 kHz, this laser is unique in the world, providing the shortest sub-mJ pulses in the mid-infrared. A system upgrade with a Yb:YAG thin-disk pump laser has recently produced 1.2-mJ 1.5-cycle pulses at 3 kHz, approaching the TW-level monocycle regime.

References

1. P.B. Corkum, F. Krausz, Nat. Phys. **3**, 381 (2007)
2. E. Goulielmakis, V.S. Yakovlev, A.L. Cavalieri, M. Uiberacker, V. Pervak, A. Apolonski, R. Kienberger, U. Kleineberg, F. Krausz, Science **317**, 769 (2007)
3. I.P. Christov, M.M. Murnane, H.C. Kapteyn, Phys. Rev. Lett. **78**, 1251 (1997)
4. M. Lewenstein, P. Balcou, M.Y. Ivanov, A. L'Huillier, P. Corkum, Phys. Rev. A **49**, 2117 (1994)
5. J.L. Krause, K.J. Schafer, K.C. Kulander, Phys. Rev. Lett. **68**, 3535 (1992)
6. E. Goulielmakis, M. Schultze, M. Hofstetter, V.S. Yakovlev, J. Gagnon, M. Uiberacker, A.L. Aquila, E.M. Gullikson, D.T. Attwood, R. Kienberger, F. Krausz, U. Kleineberg, Science **320**, 1614 (2008)
7. A. Dubietis, G. Jonusauskas, A. Piskarskas, Optics Commun. **88**, 437 (1992)
8. I.N. Ross, P. Matousek, M. Towrie, A.J. Langley, J.L. Collier, Optics Commun. **144**, 125 (1997)
9. R. Butkus, R. Danielius, A. Dubietis, A. Piskarskas, A. Stabinis, Appl. Phys. B **79**, 693 (2004)
10. F. Tavella, A. Marcinkevicius, F. Krausz, Opt. Express **14**, 12822 (2006)
11. D. Herrmann, L. Veisz, R. Tautz, F. Tavella, K. Schmid, V. Pervak, F. Krausz, Opt. Lett. **34** (16), 2459 (2009)

12. S. Adachi, N. Ishii, T. Kanai, A. Kosuge, J. Itatani, Y. Kobayashi, D. Yoshitomi, K. Torizuka, S. Watanabe, Opt. Express **16**, 14341 (2008)
13. S. Witte, R.T. Zinkstok, A.L. Wolf, W. Hogervorst, W. Ubachs, K.S.E. Eikema, Opt. Express **14**, 8168 (2006)
14. M. Zimmermann, C. Gohle, R. Holzwarth, T. Udem, T.W. Hansch, Opt. Lett. **29**, 310 (2004)
15. T. Fuji, A. Apolonski, F. Krausz, Opt. Lett. **29**, 632 (2004)
16. C. Manzoni, G. Cerullo, S. de Silvestri, Opt. Lett. **29**, 2668 (2004)
17. A. Baltuska, T. Fuji, T. Kobayashi, Phys. Rev. Lett. **88**, 133901 (2002)
18. X. Gu, G. Marcus, Y. Deng, T. Metzger, C.Y. Teisset, N. Ishii, T. Fuji, A. Baltuska, R. Butkus, V. Pervak, H. Ishizuki, T. Taira, T. Kobayashi, R. Kienberger, F. Krausz, Opt. Express **17**, 62 (2009)
19. T. Fuji, N. Ishii, C.Y. Teisset, X. Gu, T. Metzger, A. Baltuska, N. Forget, D. Kaplan, A. Galvanauskas, F. Krausz, Opt. Lett. **31**(8), 1103 (2006)
20. N. Ishii, C.Y. Teisset, T. Fuji, S. Kohler, K. Schmid, L. Veisz, A. Baltuska, F. Krausz, IEEE J. Sel. Top. Quantum Electron. **12**, 173 (2006)
21. T. Metzger, A. Schwarz, C.Y. Teisset, D. Sutter, A. Killi, R. Kienberger, F. Krausz, Opt. Lett. **34**(14), 2123 (2009)
22. E. Sorokin, S. Naumov, I.T. Sorokina, IEEE J. Sel. Top. Quantum Electron. **11**, 690 (2005)
23. Q. Fu, G. Mak, H.M. van Driel, Opt. Lett. **17**, 1006 (1992)
24. W.S. Pelouch, P.E. Powers, C.L. Tang, Opt. Lett. **17**, 1070 (1992)
25. M.H. Dunn, M. Ebrahimzadeh, Science **286**, 1513 (1999)
26. D.H. Auston, K.P. Cheung, P.R. Smith, Appl. Phys. Lett. **45**, 284 (1984)
27. B.B. Hu, X.C. Zhang, D.H. Auston, Appl. Phys. Lett. **56**, 506 (1990)
28. A. Rice, J. Jin, X.F. Ma, X.C. Zhang, Appl. Phys. Lett. **64**, 1324 (1993)
29. A. Bonvalet, M. Joffre, J.L. Martin, A. Migus, Appl. Phys. Lett. **67**, 2907 (1995)
30. T. Fuji, J. Rauschenberger, A. Apolonski, Y.V. S., G. Tempea, T. Udem, C. Gohle, T. Hansch, W. Lehnert, M. Scherer, F. Krausz, Optics Letters **30**, 332 (2005)
31. C. Teisset, N. Ishii, T. Fuji, T. Metzger, S. Köhler, R. Holzwarth, A. Baltuška, A. Zheltikov, F. Krausz, Opt. Express **13**(17), 6550 (2005)
32. H. Ishizuki, T. Taira, Opt. Lett. **21**, 2918 (2005)
33. T. Tsang, M.A. Krumbugel, K.W. Delong, D.N. Fittinghoff, R. Trebino, Opt. Lett. **21**, 1381 (1996)
34. Y. Deng, A. Schwarz, H. Fattahi, M. Ueffing, X. Gu, M. Ossiander, T. Metzger, V. Pervak, H. Ishizuki, T. Taira, T. Kobayashi, G. Marcus, F. Krausz, R. Kienberger, N. Karpowicz, Opt. Lett. **37**, 4973 (2012)

Part II
Materials Processing

Chapter 8
Micromachining

Alexander Horn, Ulrich Klug, Jan Düsing, Javier Gonzalez Moreno, Viktor Schütz, Oliver Suttmann, Ludger Overmeyer, Andreas Lenk and Bodo Wojakowski

Abstract Micromachining using ultra-short pulsed laser radiation can be used to manipulate matter without interacting with the surrounding matter. This ideal processing, called “cold ablation”, is attributed to picosecond and femtosecond laser radiation, and nowadays there are applications in some special industrial processes. But ultrafast laser radiation can also be used to heat matter very locally, allowing new processing strategies for welding or annealing. Starting from typical micromachining conditions, the limitations and scaling up techniques for industrial ultra-short material processing are presented. Significant examples of possible industrial applications using ultrafast laser micromachining are presented, elucidating the applicability of this unique radiation source.

8.1 Ultra-Short Laser Pulses and Their Way into Industrial Applications

The first experiments on the interaction of ultra-short laser pulses with matter revealed a new kind of laser ablation mechanism characterized by the almost complete absence of recast and heat affected zones. “Cold ablation” immediately sparked the idea of pushing the accuracy of laser material processing to a new dimension. Simultaneously with the invention of the post-amplified solid state Ti: Sapphire femtosecond laser, the new ablation mechanism quickly found its way

A. Horn · U. Klug (✉) · J. Düsing · J.G. Moreno · V. Schütz · O. Suttmann · L. Overmeyer
Laser Zentrum Hannover e.V, Hollerithallee 8, 30419 Hannover, Germany
e-mail: u.klug@lzh.de

L. Overmeyer
ITA, Leibniz Universität Hannover, An der Universität 2, 30823 Garbsen, Germany

A. Lenk
Continental Automotive GmbH, Ostring 7, 09212 Limbach-Oberfrohna, Germany

B. Wojakowski
ALLTEC GmbH, An der Trave 27-31, 23923 Selmsdorf, Germany

S. Nolte et al. (eds.), *Ultrashort Pulse Laser Technology*,
Springer Series in Optical Sciences 195, DOI 10.1007/978-3-319-17659-8_8

into other research groups that focused on new laser machining applications. Since then, the number of publications regarding the effect of ultra-short laser pulses on structure resolution and the machining of delicate and heat-sensitive materials has progressively increased [1, 3, 9]. Negative effects known from conventional laser processing with continuous or nanosecond-pulsed lasers, such as melt spilling, could be avoided by using ultra-short laser radiation. Material is mostly vaporized, and is deposited only as re-solidified, removable debris. From the beginning, microstructures generated by laser ablation using Ti:Sapphire systems have been of superior quality, compared to longer pulsed systems, and is thus technically highly interesting for the micro- and nanotechnology community (Fig. 8.1). However, only a very small number of micromachining applications could successfully meet the requirements for commercial exploitation. Particularly, the technical complexity and the achievable system parameters, such as average power and pulse repetition rates of the Ti:Sapphire systems in the 1–10 kHz–regime were a limiting factor for economically interesting medium- and large-scale production. This situation dramatically changed in the beginning of this century, when ultra-short pulse generation and amplification concepts based on new laser media such as Nd:YVO_4 and Yb:YAG emerged. Different laser concepts were applied, starting from the known rod to the fiber, disk or slab designs. The new developments quickly provided reliable laser sources with ultra-short pulse durations less than 10 ps at high average power levels of up to multiple 100 W and repetition rates up to the MHz regime. This new generation of high-repetitive, ultra-short pulsed lasers offers a good tradeoff between process quality and process speed, that in return will provide profitable, high-quality laser micromachining processes, including the non-academic end user market.

A new microstructuring quality resulting from laser ablation using ultra-short pulsed laser radiation is, apart from production aspects, important for industrial applications. These applications show a possible technological advance:

- Burr-free drilling, cutting and texturing, without post-processing
- Processing of heat-sensitive materials such as biopolymers (poly lactate), or NiFe-alloys with magnetic properties
- Real layer-by-layer ablation (2.5-D structures) for shaped holes and texturing
- Post-processing of prefabricated microstructures, such as trimming and repairing of sensors
- Use of sacrificial layers without bonding to the substrate
- Selective thin-film structuring for sensor applications
- Multi-photon processes like 2-photon polymerization, in-glass scribing, and in-glass structuring

In fact, combining the processes mentioned with high-repetition, ultra-short lasers is still a challenge for the system technology, for example high-velocity, high-precision focused beam positioning. In order to maintain high ablation quality with the beam sources in the range of several hundred watts, very fast scanning techniques in the range of several 100 m/s are necessary. Already, industrial applications demonstrate the power of ultra-short laser microstructuring.

Fig. 8.1 **a** Laser drilled match tip, **b** laser structured chip breaker in polycrystalline diamond, **c** laser patterning of metallic cylinders

Fig. 8.2 Ablation depth per pulse for X20Cr13 versus incident laser fluence for 12-ps laser radiation. *Inlet* SEM of ablated grooves within the two regimes [15]

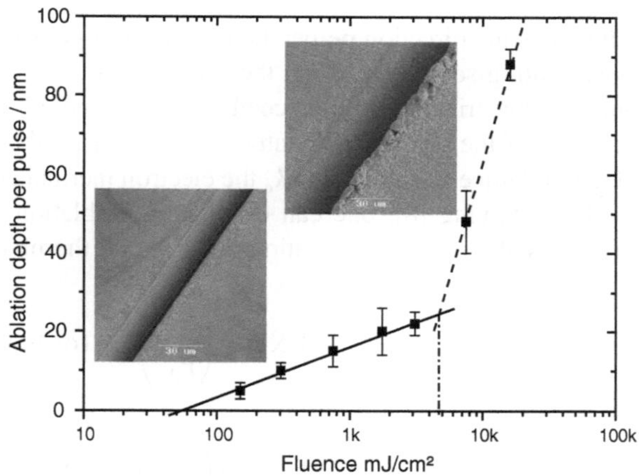

8.2 Residual Heat Due to Excessive Fluence

The characteristic ablation behavior with two ablation regimes [8] (Figs. 8.2 and 8.3) is typical for ultra-short laser processing. The ablation rate per pulse in dependence of the logarithmic fluence features a linear dependency. In this regime, also called the optical regime, single pulse ablation results in smooth surfaces, and for overlapping multi-pulses a periodic to quasi-periodic topology [10]. Above a transition fluence $F_{trans} > F_{thr}$, the linear dependency of the ablation rate continues, but with an increased slope (Fig. 8.2). Single pulse ablation is accompanied by melting, and multi-pulse ablation results in strong heating and melting of the substrates. Therefore, this regime is called the thermal regime (Fig. 8.2 and inlets).

The ablation depth for metals can be calculated by applying the two-temperature diffusion model, [8] and citations therein. One dimensionally, the temperature evolution of the electron and the lattice can be described by

$$C_e \frac{\partial T_e}{\partial t} = \frac{\partial Q(z)}{\partial z} - \gamma(T_e - T_i) + S \tag{8.1}$$

$$C_i \frac{\partial T_i}{\partial t} = \gamma(T_e - T_i) \tag{8.2}$$

$$Q(z) = -k_e \frac{\partial T_e}{\partial z} \tag{8.3}$$

$$S = I(t) A \alpha e^{-\alpha z} \tag{8.4}$$

where z is the direction perpendicular to the target surface, $Q(z)$ the heat flux, S the laser heating-source term, $I(t)$ the laser intensity, A and α the surface absorptivity and the material absorption coefficient, C_e and C_i the heat capacities (per unit volume) of the electron and lattice subsystems, γ the parameter characterizing the electron–lattice coupling, and k_e the electron thermal conductivity. In this simplified model following [8], one can calculate the ablation depth per pulse l being also driven by the optical penetration depth λ and thermal penetration depth δ:

$$1 \propto \delta \ln\left(\frac{F_a}{F_{thr}^{\delta}}\right), \quad (\delta \gg \lambda) \tag{8.5}$$

$$1 \propto \lambda \ln\left(\frac{F_a}{F_{thr}^{\lambda}}\right), \quad (\delta \ll \lambda) \tag{8.6}$$

The proportional factors detected for the ablation depth per pulse (Figs. 8.2 and 8.3) are represented by the optical or respectively thermal penetration depth. Precise microstructuring with smooth ablation features with little melt and small heat affect zones at fluencies below the transition fluence (inlet Fig. 8.3) are represented by an optical absorption behavior (8.5). Above the transition threshold, microstructuring is accompanied by re-solidified melt and large heat-affected zones.

8.3 Accumulated Heat Due to High Repetition Rates

The two-temperature model depicts ablation only in a semi-qualitative manner, and does not consider thermo- or hydrodynamics of the solid, the melt and the plasma. Also, a multi-dimensional approach will not sufficiently describe the observed heating due to high-repetitive, ultra-short laser radiation. In fact, apart from the vapor and the plasma plume interacting with the surface, excessive pulse energy not

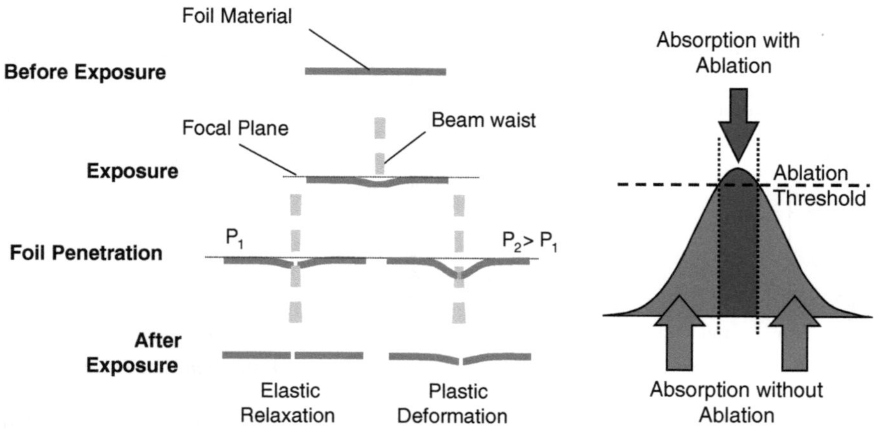

Fig. 8.3 Schematics on accumulated heat in percussion drilled foil material (*left*); description of the threshold definition and of accumulated heat (*right*)

contributing to ablation is partly converted to heat. At large repetition rates, the heat accumulates and the temperature rises within the interaction volume to values up to melting temperatures. This effect is disadvantageous for machining heat sensitive, microstructured materials (e.g. special alloys, ion-doped semiconductors, organics), but can also be used as a heat source, for example to weld glass [7].

Exemplary, drilling and cutting of thin sheets of metal, nylon or silicon using picosecond pulsed laser radiation [5] have been investigated. The thermally induced vertical displacement has been detected while drilling thin sheets irradiated at intensities above ablation threshold (Fig. 8.3 left).

By irradiating a metallic surface with focused Gaussian distributed laser radiation at fluencies above ablation threshold, two regions are characterized, see Fig. 8.3 right: the region within the dashed lines represents the ablated region, and the region outside the dashed lines represents the heated region. Multi-pulse irradiation of the metal surface results in an accumulation of heat followed by mechanical expansion of the material (Fig. 8.3 left). Depending on the thermo-mechanical material properties, a linear or nonlinear displacement is observed: For example, when irradiating thin foils of silicon with 50 μm thickness at a constant fluence, a linear dependence of the vertical displacement on increasing repetition rate is observed (Fig. 8.4 left), whereas steal and nylon depict a nonlinear one. Also, exposing the investigated material to laser radiation at a constant average power, but altering the repetition rate shows an increasing foil bending for steel foils, due to an increasing heat accumulation, though the fluence is decreasing. Applying high-repetitive, ultra-short laser radiation e.g. for high-density drilling thin sheets, need a drilling strategy, like ablating with a chaotic strategy reducing the local heat load. Or, the repetition rate must be reduced below 100 kHz, until the displacement is smaller than the precision needed (Fig. 8.4 right).

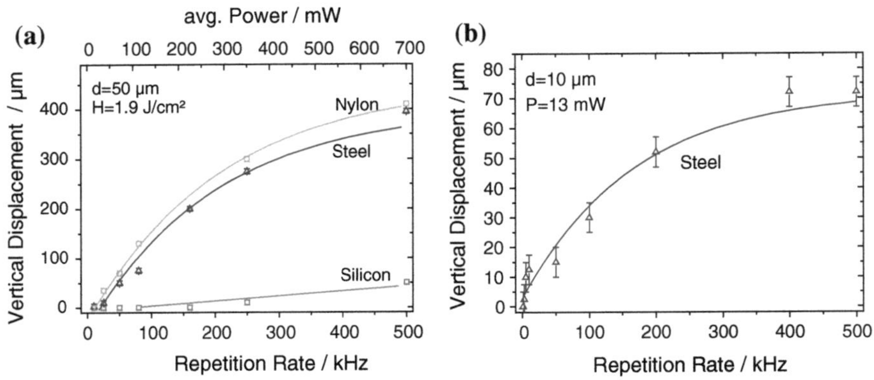

Fig. 8.4 **a** Displacement of 50 μm foil substrates versus average power. **b** Displacement of 10 μm thick steel versus repetition rate at constant average power

8.4 Typical Micromachining Conditions and Methods for Up-scaling

8.4.1 Small Aspect Ratio Processes

Microstructuring with ultra-short pulsed laser radiation can be subdivided in small aspect ratio[1] processes and high aspect ratio processes. Small aspect ratio processes like cutting substrate with an aspect ratio 3, thin-films, or free-form texturing need ultra-short pulsed laser radiation with low pulse energies and high repetition rates. Cutting depends on the material properties, the thickness, and the pulse overlap (Fig. 8.5). Thin-films can be removed with ultra-short pulsed laser radiation with little or no pulse overlap, either by ablation (melting and vaporization), or for a substrate transparent for the laser radiation, by chipping the metal from the backside of the substrate, see Fig. 8.6.

An advantage of microstructuring thin-films close to the ablation threshold is precise ablation with little plasma plumes at small electron and ion densities. Therefore, the subsequent laser radiation does not interact significantly with the plasma, and does not induce plasma heating of the surface. This allows an increase of the repetition rate into the multi-100 kHz regime without altering the ablation quality. Nevertheless, care must be taken not to alter the material, because heat induced by the tail of Gaussian shaped radiation is localized in a thin layer.

[1]The aspect ratio is defined here as the ratio of the structure lateral size and the structure depth.

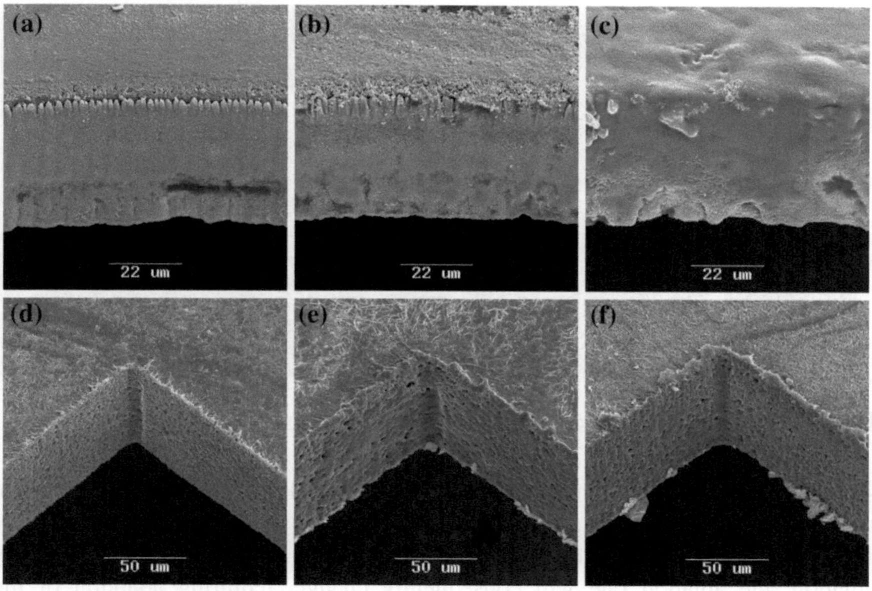

Fig. 8.5 SEM images of cutting edge in silicon (**a–c**) and Nylon (**d–f**) produced with a variable pulse overlap (**a/d**: 55 %, **b/e**: 75 %, **c/f**: 95 %) [5]

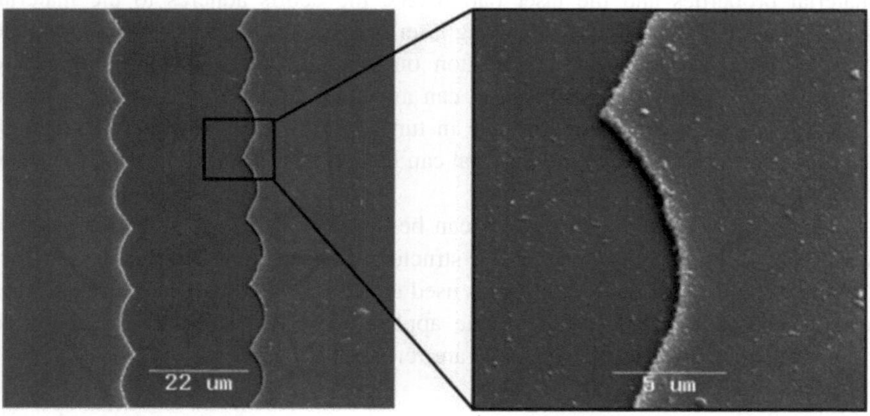

Fig. 8.6 Chipping of thin-film molybdenum (thickness 500 nm) from a glass substrate with little pulse overlap using laser radiation with 600 ps pulse duration [4]

8.4.2 High Aspect Ratio Processes

Contrary to small aspect ratio processing, hole drilling, excavation, or cutting on substrate thicknesses >250 μm (aspect ratios >> 3) requires high fluencies at moderate repetition rates (<150 kHz). The repetition rate must be decreased

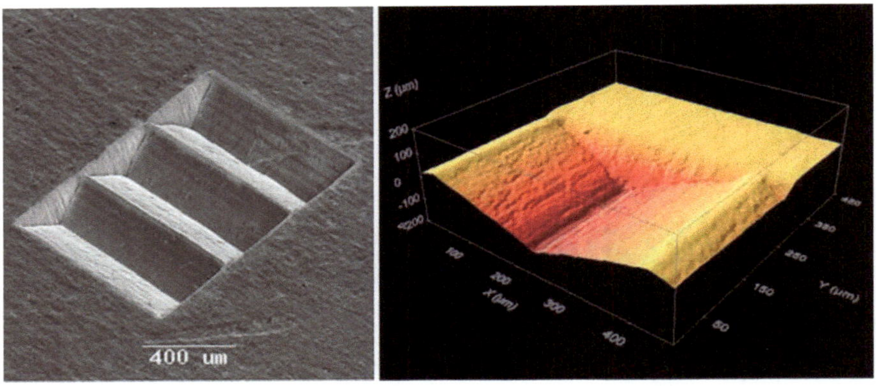

Fig. 8.7 Prism structure in tungsten: SEM (*left*), Laser scanning micrograph (*right*)

significantly to increase fluence, because plasma shielding inhibits laser ablation, and the high energy plasma plume acts as a secondary heat source, negatively altering ablation. Because the optical properties of the material are also altered by ablation, the ablation rate will consequently change, requiring adaption of the fluence during processing. Ablation of multiple layers requires a dynamic adaption of the focal plane.

Removing large amounts of material generates a lot of debris. Depending on the material properties and the laser parameters, the debris adheres to the material surfaces and cannot be removed during laser processing, which further inhibits the ablation process. High-power aspiration on the one hand, and a high-pressure process gas stream on the other hand can avoid the re-deposition of most debris. Complex 3-D structures, for example in tungsten, can then be generated at high aspect ratios (Fig. 8.7), and the material can be removed layer by layer in real-time by adapting the process parameters.

Alternatively a combined process can be developed, first ablating the material with large fluencies to achieve rough structures. Then, high-precision laser structuring at small to moderate fluencies is used to generate acceptable surface qualities with a roughness in the range of the applied wavelength. Thus, the excessive roughness and the heat affected layer are removed.

8.4.3 Limitations

Ultrafast laser processing can be used to very precisely ablate material. However, materials emit surface plasmons during irradiation. The surface plasmons interact coherently with the incoming radiation and change the absorption properties of the surface drastically, e.g. imposing a spatial, periodically varying absorptivity which limits the resolution of micromachining. As a consequence of the changed absorptivity, also the ablation changes by locally generating ripples [10] (Fig. 8.8).

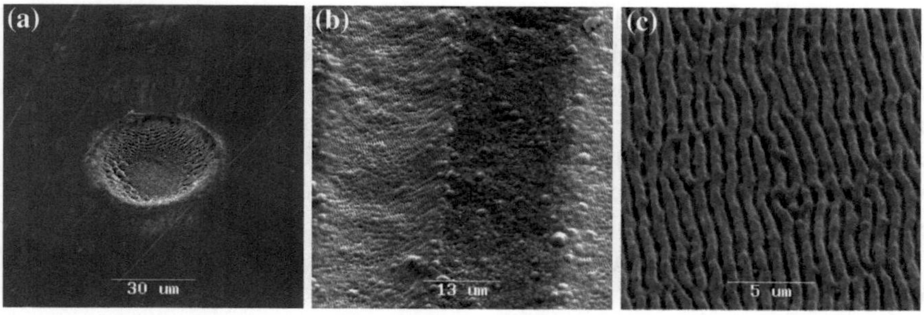

Fig. 8.8 Laser-induced periodic structures (LIPS) **a** in an ablated crater of NiCr film, **b** in linear scribes on polycrystalline diamond, **c** on steel surface

Ripples, also called LIPS (Laser-Induced Periodic Structures) are periodic structures with about the laser wavelength in periodicity, and with an orientation perpendicular to the laser polarization direction. Looking closer at LIPS, two types of ripples are observed, the HF-LIPS and the LF-LIPS [11, 16, 19]. Avoiding ripple formation is still a topic of research. Micro structuring close to the ablation threshold using circular polarized laser radiation might reduce ripple formation. But, because the manipulation of the laser radiation to ideal circularity is a nearly unreachable effort, real circular polarized radiation will even induce ripples itself. Processing with long pulsed laser radiation in the nanosecond regime can smooth the surface by melting the ripples, or chemical etching can be used.

Even processing metals and semiconductors at moderate to high fluencies, 2–5 F_{thr} will induce quasi-periodic features, so called cones [10, 17]. The periodicity depends not only on the radiation wavelength, but also on the processing parameters (such as fluence, overlap, focusing diameter, repetition rate, processing gas). Cones are applied for reflectivity reduction in electronics, especially for semiconductor detectors and photovoltaic elements.

8.4.4 Frequency Conversion

Picosecond and femtosecond laser radiation can efficiently be frequency-converted into lower or higher frequency harmonics of the fundamental wavelength with very high power densities. The fundamental wavelength of the laser radiation is typically in the near-infrared regime. Optical equipment such as dielectric mirrors or lenses, and optoelectronic components such as optical switches, nowadays feature industrial standards with high quality, and offer a nearly loss-free transmission or deflection of radiation. Generally, converting laser radiation to higher harmonics can better focus the radiation, e.g. the focal diameter scales with about half the wavelength, and allows micro structuring close and even below diffraction limit.

Diffraction limited infrared ultrafast laser radiation converted by second harmonic generation (SHG) into the visible spectrum features radiation with excellent beam quality $M^2 \approx 1$, and conversion efficiency as high as 80 %. All available conventional optics for visible radiation can be adopted for micromachining, also allowing the use of microscope lenses for focusing and microstructuring in the micrometer range. Because conventional optics is designed for low intensity radiation, care must be taken not to damage the optics due to overloading. Specially designed high-power optics for ultra-short pulsed laser radiation allow microstructuring at high fluencies.

Microstructuring with ultraviolet laser radiation can be used to "ablate cold" organic materials such as polymers, by photo-chemically breaking the chemical bonds. Converted IR laser radiation into the ultraviolet regime by third harmonics (THG) or forth harmonics generation (FHG) is applicable today, especially the THG. But due to aging, the lifetime of the converting crystals is limited, and continuous operation on an industrial level (24/7) is not achievable. Conversion efficiency up to 30 % has been reported for THG. Also, the THG beam quality has negative effects on suffers little on the crystal quality, resulting in a non-diffraction limited radiation $M^2 \approx 1.5$. High-power optics for UV converted YAG radiation are available today.

8.4.5 Scaling Up

Scaling up of micromachining using ultra-short laser sources is predominantly done by increasing the repetition rate, or by applying multiple beams. However, fast micromachining with all the features of cold ablation limits the maximum repetition rate, since:

(a) residual heat due to incomplete transfer of absorbed pulse energy into ablation, and
(b) temporal and spatial overlap of pulse sequences

results in accumulated and residual heat. Apart from scaling up the microstructuring process by increasing fluence (see Sect. 8.2) and repetition rate (see Sect. 8.3) of the laser radiation, a process can be parallelized by multiplying the number of laser beams using conventional beam splitting optics, or diffractive optical elements (DOE). Depending on the application, the multi-beams are either separated, and each of the beams delivered to different machining stations, or the beams are focused on one micromachining station (Fig. 8.9).

A single laser beam can be multiplied using DOEs, increasing the number of beams linearly up to about 100 spots, or distributed on a matrix up to 10 × 10 beams (Fig. 8.10). The diffraction efficiency is now about 80 % of the incoming radiation. Zero-order diffraction is often implemented in the diffraction scheme.

Speeding-up is achieved by increasing the scanning velocity using high-speed galvano scanners (limited to about 10 m/s at 100 mm focus length), or polygon scanners (limited to 360 m/s at 100 mm focus length). Volumes up to 2 mm^3/s are removed using high-power, high-repetition rate ultra-short laser machining systems

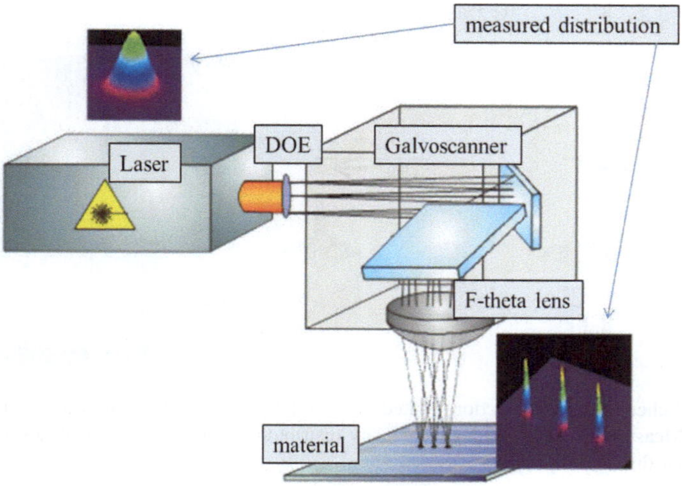

Fig. 8.9 Schematics of multi-beam processing using diffractive optical elements (DOE), scanning technology and focusing optics

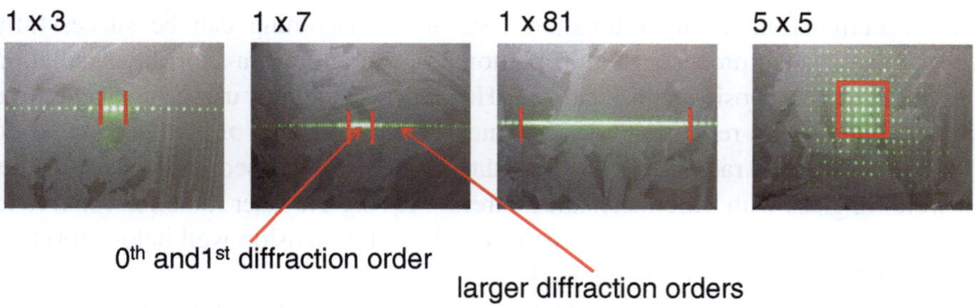

Fig. 8.10 Multi-spot generation using DOE: 3, 7, 81 spots in a line; 5 × 5 matrix, larger diffraction orders outside the red lines contain <5 % of the pulse energy

combining high-speed scanning for one axis, and an electro-optic deflection with low deflection in the other axis. Also, combing a high-speed rotational stage with a slow co-axial axis allows micromachining on the shell of cylindrical samples with high processing velocities.

Spatially shaping the intensity distribution of the focused laser radiation reduces the amount of wasted energy, i.e. energy not used for ablation. For example, micromachining with diffraction-limited laser radiation suffers of the heat load given by the tails of the Gaussian distribution and the ablation threshold (Figs. 8.3 and 8.11). Spatially shaping the laser radiation to achieve a top-hat shaped focused radiation can be used for ablating materials with a reduced heat-affected zone (Fig. 8.11). Exemplary, patterning of molybdenum layers using shaped laser radiation reduces the heat load, and also the pulse overlap.

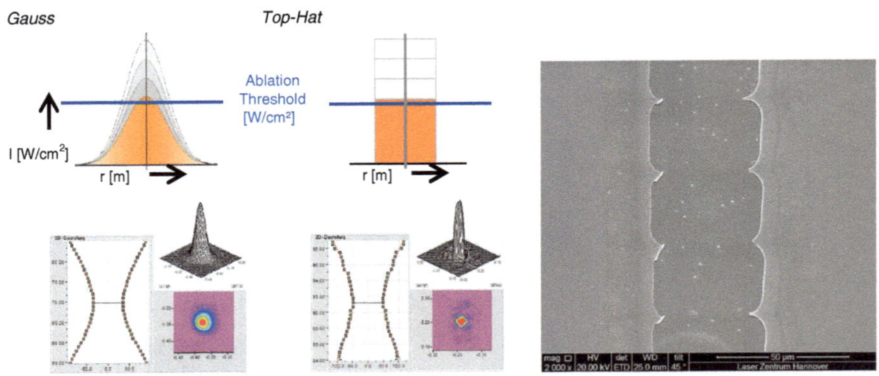

Fig. 8.11 Schematics of diffraction-limited and top-hat distributions (*top*) with defined ablation threshold. Measured caustic and focal intensity distributions (*bottom*). SEM of ablated Mo-layer using top-hat distribution (*right*)

8.5 Exploitation of Residual Heat

Heat accumulation during ultra-short laser micromachining can be successfully applied in different processes: Multi-photon absorption of ultra-short laser radiation can be used to deposit much localized. Heating and melting using high-repetitive laser radiation with repetition rates beyond 100 kHz benefits of heat accumulation [6, 7]. Focusing the radiation at the boundary of two compressed glass plates results in welding glass with little mechanical stress. Moving the laser radiation relative to the substrates generates a welding seam with lateral dimension well below 100 μm, with excellent bonding strengths [12].

Thin-film metals and semiconductors such as ITO (Indium tin oxide) and zinc oxide are adopted for electronics, photovoltaic, and lighting. Their optical and electrical properties often need to be altered after deposition on the substrate. For example, ZnO_2 thin-films exhibit poor optical properties and low electrical conductivity after deposition. Conventional thermal treatment, so called annealing, is achieved by furnace treatment, a slow and expensive process. Using laser radiation optical energy can be localized within the layer, not stressing the substrate. Apart from continuous wave laser treatment, also ultra-short high-repetition laser treatment is applicable. A very precise and spatially localized heat input is used to thermally treat thin layers, preserving thermally sensitive substrates [4].

Conventional laser microstructuring of ceramics induces microcracking because of the excessive heat load. By using ultra-short laser radiation crack-free ablation is possible. The re-depositions generated during multi-pass irradiations of up-scaled ultra-short laser processes, however, firmly adhere to the inner surface of the ablation trench which cannot be removed by compressed air or by ultrasonic cleaning any more. In case of an electromechanic component such as a piezo stack, where a metallic inter layer has to be recessed within the ceramic bulk material

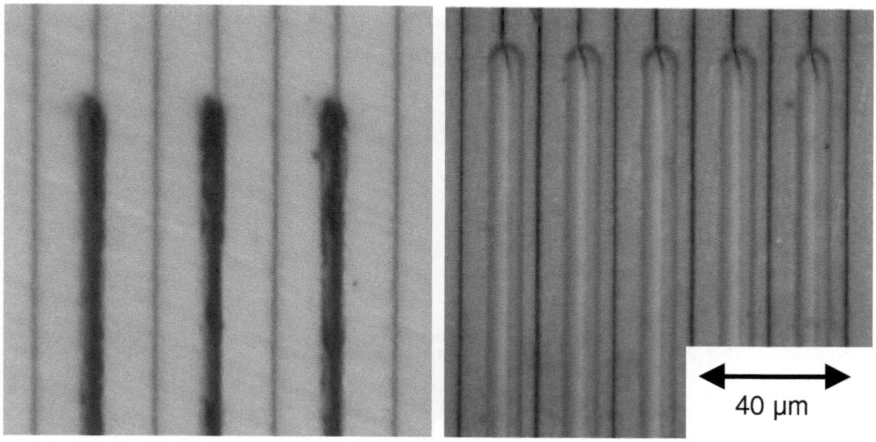

Fig. 8.12 Laser-structured piezo stacks emphasizing "cold" ablation: metallic recast blocks functionality (*left*). Well-balanced laser processing applying residual heat results in recast-free structures on piezo stacks (*right*)

(green part), the metallic re-depositions alter the electric properties of the post-processed (debindered) ceramic (Fig. 8.12 left). In the worst case the piezo stack is then non-functional. A novel process has been developed, applying ultra-short laser radiation with well-balanced laser process parameters, especially pulse energy and repetition rate. Localized heating of the immediate vicinity of the ablation trench results in delayed flaking of the r-depositions by heat induced exudation of the surrounding binder material of the green body. A simple ultrasonic post-treatment completely removes the metallic debris (Fig. 8.12 right).

8.6 Scenarios for the Transfer of Ultra-fast Pulsed Processes to Large Scale Industrial Applications

8.6.1 Electrical Deactivation of Piezo Stacks Using Laser Ablation

Laser microstructuring of piezo stacks for automotive injection modules using ultra-short pulsed laser radiation is investigated here. The aim was to find a way to save costs with a new process, and at the same time to increase the performance and the durability of the product. The piezo stacks consist of a series of piezo ceramics layers (thickness 70 μm) divided by metallic, electric conducting layers (thickness 2 μm). In order to connect two piezo elements in series, the electrical conductor of every second piezo layer has to be removed locally using laser ablation, as shown in Fig. 8.13. Using ultra-short laser radiation, full active stacks were generated with

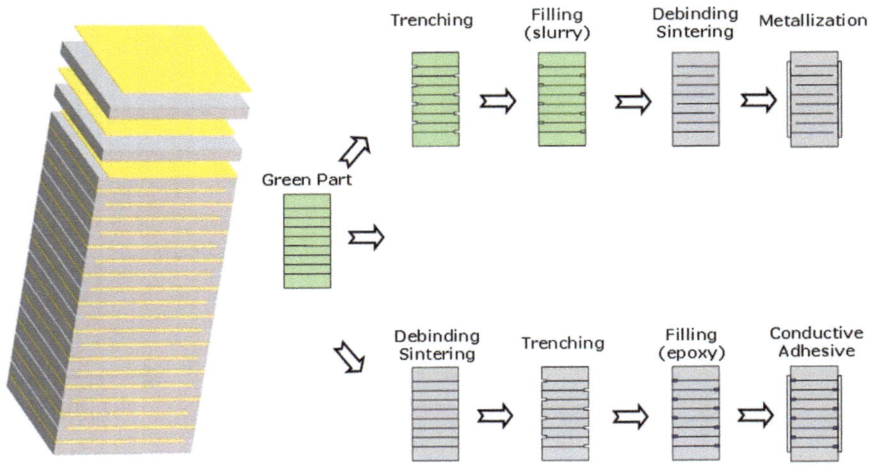

Fig. 8.13 Principle of electrode deactivation (*left*), technological variants for the fabrication of piezoelectric stacks with alternating electrodes (*right*): structuring of *green* part first (*upper scheme*) and first sintering then structuring (*lower scheme*)

reduced micro-cracking. Two strategies for accomplishing electrical isolation have been investigated (Fig. 8.13 right):

1. After sintering the green part (see Fig. 8.13), the electrical isolations of the piezo stack electrodes is achieved by ablating gaps between two piezo cells using "cold" ablation of the metallic layer
2. Using laser structuring, the metallic layer is removed from the green part. Afterwards, the gap is closed with slurry, and the complete piezo stack is sintered.

Non-thermally loaded ceramic results from using "cold" ablation with ultra-short laser radiation at reduced repetition rates smaller than 100 kHz. However, the re-deposited metal on the walls of the gap influences the electrical conductivity of the filling (Fig. 8.14 left) in post-processing. The re-deposited metal could not be removed by post-processing, such as ultrasound cleaning. Electrical isolation could not be achieved.

A functional piezo stack could be achieved by first structuring the green part and locally removing the metallic layer, and afterwards using ultrasound cleaning, filling and sintering. In order to completely and easily remove the re-deposited metals, the heat load during laser processing was balanced, resulting in a cloudy recast with in clean structures (Fig. 8.14 right).

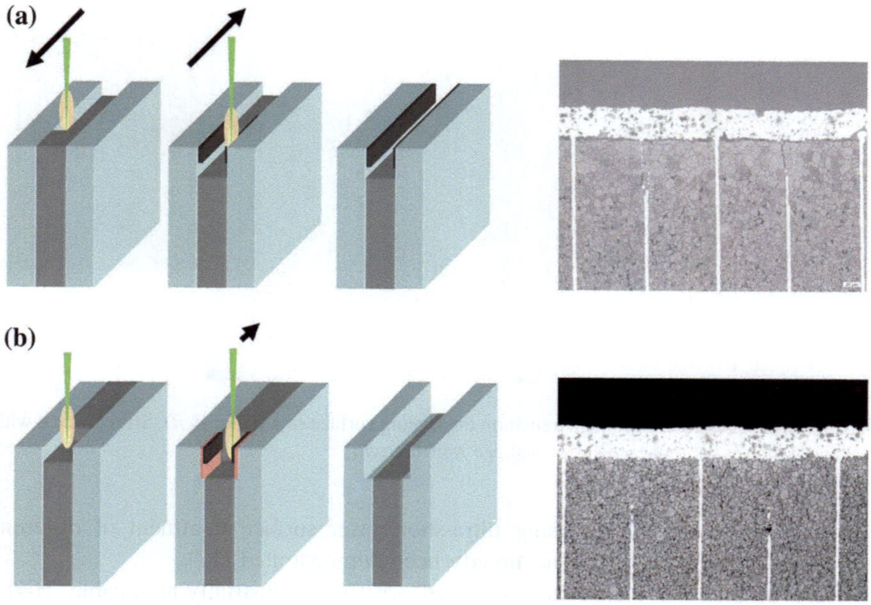

Fig. 8.14 Strategies for gap generation: **a** "Cold" ablation of metallic layer resulting in metallic recast on the walls. **b** Heat accumulated ablation removing metallic layer, and by post-processing the re-deposited material. *Right* SEM of cross-section with piezo-ceramic layers, metallic layer and fillings

8.6.2 Micromachining of Seal Faces Without Post-processing

A leakage trench on a surface of a seal face (Fig. 8.15 left) was accomplished, ablating grooves with a width of about 40 μm and a comparable depth using ultra-short pulsed laser radiation (Fig. 8.15 right). The processing parameters were adapted to form grooves with negligible bulging and recast on the surface. Post-processing of the component was not necessary, reducing production costs.

8.6.3 Efficiency Enhancement by Patterning Si-Solar Cells

The efficiency of multi-crystalline silicon (mc-Si) solar cells can be improved by reducing the amount of reflected solar radiation. Today common techniques are

1. Iso-texture etching for mc-Si,
2. Applying antireflective coatings, and
3. Anisotropic etching for mono-crystalline silicon solar cells for pyramid structures.

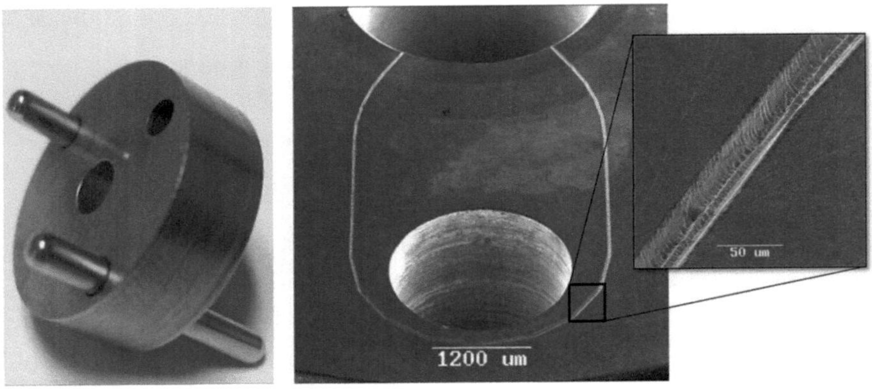

Fig. 8.15 Laser engraved leakage trench on the sealing surface of a buffer (*left*). SEM surface with leakage trench, *inlet* enlargement of leakage trench (*right*)

A reduction of reflectivity using ultra-short laser surface treatment of different materials, including silicon, has already been demonstrated [17].

Here, an alternative approach is shown, which is industrially applicable. High-repetition rate ablation is achieved, by scanning meandrian trajectories with focused laser radiation. By using high-repetition rate laser radiation from an industrial laser in the multi-100 kHz regime, productivity, expressed in m^2/scan, can increase into the productivity range of one 5″ wafer per second, at an appropriate mean laser power. Processing is parallelized by using diffractive optical elements to generate multiple laser spots (Sect. 8.4.5) [14].

Depending on the applied fluence, two regimes for ablation are detected (Sect. 8.2). In the optical regime at fluencies below about 4 J/cm^2, a cone-like topology can be generated, whereas when fluencies are above 4 J/cm^2, smooth structures attributed to the thermal regime are generated [10, 13]. The surface is modified by ablation, depending on the laser parameters in the optical regime of silicon, by generating cones with sizes of up to 10 μm with dimensions in the range of a few μm. The reflectivity is reduced absolutely to about 11 % over the spectral distribution of solar light, in comparison to the standard iso-textured surface on mc-Si solar cells (Fig. 8.16).

8.6.4 3-D Laser Patterning of Thin-Film Strain Sensors

Thin-film strain sensors which are directly deposited onto the surface of mechanical components can be used for measurement of forces, pressure or strain in harsh environments and at high temperatures [18]. Femtosecond laser patterning is used for thin-film sensors on component surfaces. Ultra-short pulsed laser radiation allows high selectivity and quality during ablation of the different thin-film layers

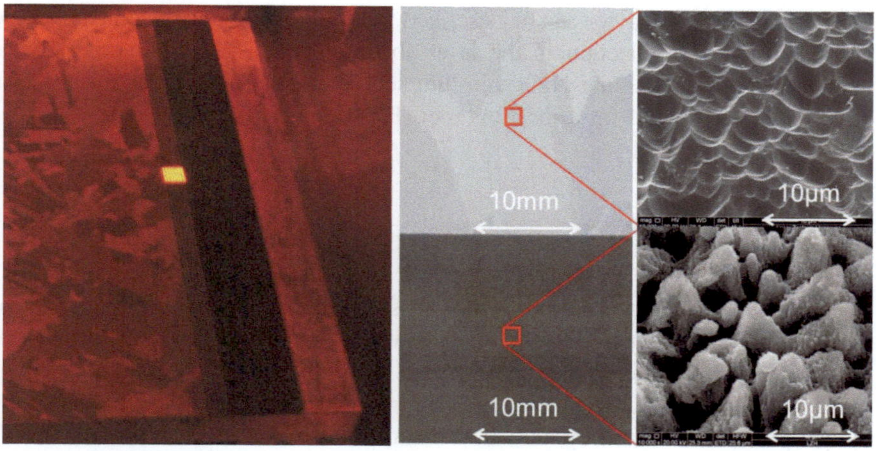

Fig. 8.16 Multi-spot processing of silicon using ultrafast laser radiation (*left*), non-processed, iso-textured silicon surface (*top*) and laser-processed surface (*bottom*)

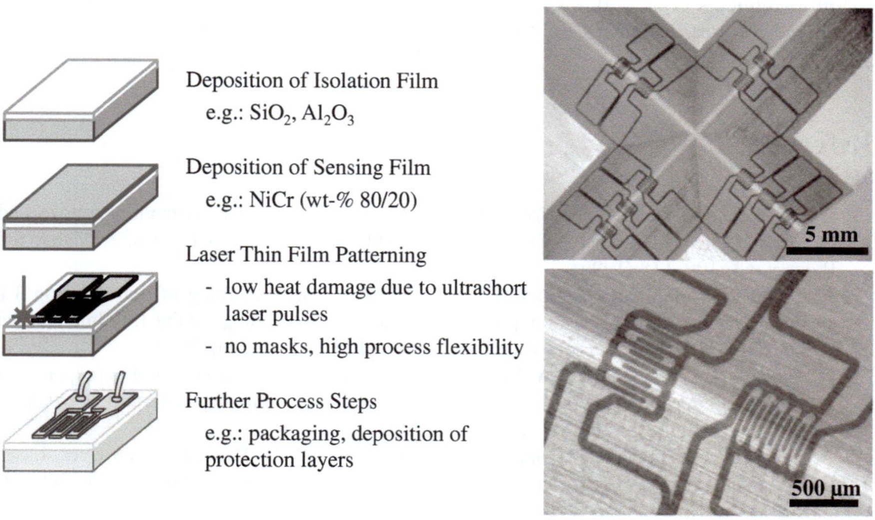

Fig. 8.17 Schematic procedure for laser-processing of thin-film sensors (*left*); laser patterned thin-film strain sensors processed into a *v-shaped* groove of an aluminum component. The top metallic layer is selectively removed using ultra-short laser pulses (*dark area*, *right*)

(Fig. 8.17 left). This method is particularly useful for patterning strain sensors which are sputter deposited directly onto curved surfaces.

The ablation behavior of NiCr film irradiated by femtosecond laser pulses at non-normal angles of incidence has been modeled and experimentally verified with linear and circular beam polarization for incidence angles up to 80° [2, 9]. The ablation threshold behavior can be described when including polarization and angle dependent Fresnel reflection in the laser ablation model. The laser process was demonstrated by patterning NiCr thin-film sensors on the non-planar surface of a mechanical component for a machine tool (Fig. 8.17 right).

8.7 Conclusion

Micromachining using ultra-short pulsed laser ablation is nowadays part of production chains in industry. Today, high-power laser systems in the kW-range are available, with very high repetition rates up to the MHz range, which allow high-quality, highly productive microstructuring of all kind of materials. Due to the unique processes induced by ultra-short laser radiation, on one hand nearly "cold" ablation is achievable, and on the other hand a very precise heat load is applicable for sensitive thermal treatments of materials such as welding of glass and annealing of thin-films.

Acknowledgments This work was supported by German Research Foundation (DFG) within the Collaborative Research Centre CRC 653, "Gentelligent Components in their Lifecycle".

References

1. B.N. Chichkov Momma, C. Nolte, S. von Alvensleben, A. Tünnermann, Femtosecond, picosecond and nanosecond laser ablation of solids [Article] //. Appl. Phys. A Mater. Sci. Process. **63**, 109–115 (1996)
2. J.F. Düsing, O. Suttmann, J. Koch, U. Stute, Ultrafast laser patterning of thin films on 3-D shaped surfaces for strain sensor applications [Article] //. Proceedings of the 13th International Symposium on Laser Precision Microfabrication (LPM), 12–15 June 2012
3. E. Fadeeva, S. Schlie, J. Koch, A. Ngezahayo, B.N. Chichkov, The hydrophobic properties of femtosecond laser fabricated spike structures and their effects on cell proliferation [Article] //. physica status solidi (a) **206**(6), 1348–1351 (2009)
4. A. Horn C.-C. Kalmbach, J. González Moreno, V. Schütz, U. Stute, L. Overmeyer, Laser-surface-treatment for photovoltaic applications [Article] //. Physics Procedia **39**, 709–716 (2012)
5. U. Klug, U.B. Kamlage-Rahn, J. Koch, R. Knappe, U. Stute, B. Chichkov, Picosecond laser material processing—prospects and limitations [Article] (2006)
6. I. Miyamoto, K. Cvecek, Y. Okamoto, M. Schmidt, Novel fusion welding technology of glass using ultrashort pulse lasers [Article] //. Physics Procedia **5A**, 483–493 (2010)
7. I. Miyamoto, A. Horn, J. Gottmann, D. Wortmann, F. Yoshino, Fusion welding of glass using femtosecond laser pulses with high-repetition rates [Article] //. JLM-N J. Laser Micro/Nanoeng. **2**(1), 57–63 (2007)
8. S. Nolte, C. Momma, H. Jacobs, A. Tünnermann, B.N. Chichkov, B. Wellegehausen, H. Welling, Ablation of metals by ultrashort laser pulses [Article] //. JOSA B. **14**, 2716–2722 (1997)
9. L. Overmeyer, J.F. Duesing, O. Suttmann, U. Stute, Laser patterning of thin film sensors on 3-D surfaces [Article] //. CIRP Annals Manufact. Technol. **61**(1), 215–218 (2012)

10. L. Overmeyer, V. Schütz, A. Horn, U. Stute, Laser induced quasi-periodical micro-structures with external field modulation for efficient gain in photovoltaics [Article] //. CIRP Annals Manufact. Technol. ed. Elsevier **62**(1), 207–210 (2013)
11. H. Raether, Surface plasmons on smooth and rough surfaces and on gratings [Book], vol. 111 (Springer, Berlin, 1988)
12. S. Richter, S. Nolte, A. Tünnermann, Ultrashort pulse laser welding—a new approach for high-stability bonding of different glasses [Article] //. Physics Procedia. **39**, 556–562 (2012)
13. V. Schütz, A. Horn, U. Stute, Investigations into laser edge isolation (LEI) of mc-Si solar cells using ns- and ps-laser radiation [Article] //. Photovoltaics International 14th Ed. vol. 11, pp. 69–76 (2011)
14. V. Schütz, A. Horn, U. Stute, High-throughput process parallelization for laser surface modification on Si-Solar cells: determination of the process window [Article] //. Proc. SPIE. **8244**, 33 (2012)
15. F. Siegel, U. Klug, R. Kling, Extensive micro-structuring of metals using picosecond pulses—ablation behavior and industrial relevance [Article] //. JLMN-J. Laser Micro/Nanoeng. **4**, 104–110 (2009)
16. M. Straub, M. Afshar, D. Feili, H. Seidel, K. König, Surface plasmon polariton model of high-spatial frequency laser-induced periodic surface structure generation in silicon [Article] //. J. Appl. Phys. **111**, 124315 (2012)
17. B.R. Tull, J.E. Carey, E. Mazur, Silicon surface morphologies after femtosecond laser irradiation [Article] //. MRS Bull. **31**, 626–633 (2006)
18. G.R. Witt, The electromechanical properties of thin films an the thin film strain gauge [Article] //. Thin Solid Films **22**, 133–156 (1974)
19. A.I. Zayats, I. Smolyaninov, Near-field photonics: surface plasmon polaritons and localized surface plasmons [Article] //. J. Opt. A: Pure Appl. Opt. **5**, 16 (2003)

Chapter 9
Drilling with Ultrashort Laser Pulses at High Repetition Rates

Felix Dreisow, Sven Döring, Antonio Ancona, Jens König and Stefan Nolte

Abstract Ultrashort laser pulses offer extraordinary precision in microprocessing a variety of materials, especially metals. Thermal and mechanical damage can be minimized by working at fluences not too far above the ablation threshold. However, this comes at the expense of low ablation rates and thus high processing times. A scaling of processing speed by increasing the fluence results in degradation in quality. Therefore, in this chapter we investigate the potential for scaling the processing speed by increasing the pulse repetition rate to several 100 kHz up to the MHz regime with average laser powers of up to 100 W exemplary for percussion drilling of metals. Limiting factors like particle shielding and heat accumulation are identified, their dependence on laser parameters as well as material properties are discussed and options for drilling at significantly improved speeds are highlighted.

9.1 Introduction

The advent of reliable ultrashort pulse lasers at the end of the last century has led to their widespread use in research laboratories, with various applications such as chemistry [1], microscopy [2], high field physics [3, 4], filamentation and

F. Dreisow · S. Döring · S. Nolte
Institute of Applied Physics, Abbe Center of Photonics, Friedrich Schiller University of Jena, Max-Wien-Platz 1, 07743 Jena, Germany

A. Ancona
Institute for Photonics and Nanotechnologies, (IFN)-CNR U.O.S. Bari, Physics Department "M. Merlin", via Amendola 173, 70126 Bari, Italy

J. König
Robert Bosch GmbH, Corporate Sector Research and Advance Engineering Laser Material Processing and Joining Technology (CR/APJ2), Postfach 30 02 40, 70442 Stuttgart, Germany

S. Nolte (✉)
Fraunhofer Institute for Applied Optics and Precision Engineering IOF, Albert-Einstein-Straße 7, 07745 Jena, Germany
e-mail: stefan.nolte@uni-jena.de

S. Nolte et al. (eds.), *Ultrashort Pulse Laser Technology*,
Springer Series in Optical Sciences 195, DOI 10.1007/978-3-319-17659-8_9

supercontinuum generation [5, 6] as well as metrology [7, 8]. A very important application, especially from an industrial point of view, is ultrafast laser materials processing. The use of ultrashort laser pulses promised a significant increase in quality. With the right choice of processing parameters well-defined patterns were ablated in metals [9] and other solid targets [10, 11]. The minimized thermal and mechanical damage to the surrounding material led to the term "cold" ablation. However, despite of these advantages a use in serial production in a harsh industrial environment was not possible at that time. This was mainly due to the fact that, although commercial amplified ultrashort pulsed laser sources became available, their output power was limited to the Watt level and the repetition rate to the kHz range. Therefore, production times were substantially too high.

Within the past 10 years tremendous advancements have been made in ultrashort pulse laser development. Today, average powers around 1 kW have been demonstrated [12–14] and >100 W are commercially available (see Chaps. 4, 5 and 6). These systems offer high repetition rates up to the MHz range, at pulse energies appropriate for various processing applications. However, to make full use of these high-power, high repetition-rate laser sources for scaling the processing speed, a detailed understanding of the processes and time-scales involved under these conditions is required.

This chapter provides the necessary fundamentals of ultrashort pulse laser micromachining and describes possible effects and relevant time-scales for processing at high repetition-rates and high average powers. The first section gives an introduction to laser materials processing, defines the required quantities and presents basic concepts of laser ablation. Subsequently, the physical background of the ablation process is discussed as well as the light absorption by opaque materials, energy relaxation processes and in particular the material removal itself. We focus on the different relevant time scales that cover the range from sub-picoseconds to a few microseconds, even for ultrashort laser pulses. The third section finally analyses the drilling of holes in metal sheets at various repetition rates and pulse energies. The relevant aspects are linked to the time scales of the particle removal process.

9.2 Ultrashort[1] Laser Ablation of Opaque Materials

The laser is an indispensable tool in metalworking and materials processing. It is used in various fields for cutting, welding, hardening, joining, bending, drilling, milling, surface structuring and texturing. Most of them are large-scale applications driven by the metal and automotive industry. The use of lasers leads to benefits in the achievable precision exceeding most mechanical manufacturing methods. Moreover, it offers the advantage of contactless processing most common metals

[1]In this chapter we use the term "ultrashort" for sub-picosecond and picosecond pulse durations. With "long" we refer to pulse lengths of nanoseconds and longer.

like aluminum, copper, tungsten, nickel, alloys like stainless steel and brass, and even composites and ceramics.

Even though conventional laser processing methods often provide high precision and flexibility, one typically cannot achieve the theoretical resolution limit as defined by Abbe's diffraction law. According to this law the laser can be focused down to minimal dimensions in the order of the laser wavelength, i.e. from approximately ten microns for CO_2 lasers to one micrometer for neodymium or ytterbium based systems and several 100 nm for UV laser systems. However, such structural sizes typically cannot be attained with standard laser processing methods. Instead, the structure size is determined by the light-matter interaction process and heat diffusion into the surrounding material. As a result, typical structural dimensions are several tens of microns, i.e. much larger than the wavelength of the light used. One reason is the occurrence of a molten phase during the ablation process. The residues thereof can often be seen by the rounded shapes at the edges of the ablated structure. In addition, sharp peaks of rapidly cooled material occur as burr. However, even larger regions surrounding the laser ablation spot may be affected by heat diffusion. In these so-called heat affected zones (HAZ) the material morphology and physical properties may have changed. Also, in certain cases phase changes or coloring of the material can occur. Particularly the latter one is a very famous effect for stainless steel, which shows tempering colors, which decode the temperature profile [15].

Such severe thermal effects are typically occurring during laser processing using continuous wave light or long laser pulses (~ 1 ns). For a deeper understanding let us consider a hole drilling process in a metallic workpiece using a single laser pulse with long pulse duration and an energy high enough for ablating a certain amount of material. The laser beam is focused by a lens onto the sample surface with the aim of achieving a small structure size and a high laser fluence. When the laser pulse hits the sample, a certain fraction of the incident energy is absorbed in a shallow region. For most metals the absorption coefficients are typically on the order of $\alpha \approx 10^6$ cm^{-1}, which corresponds to an optical penetration depth of 10 nm [16]. Compared to the wavelength of the laser light, this is two orders of magnitude smaller and therefore one can assume in good approximation that light is absorbed at the surface of the sample, where the temperature increases rapidly. The material melts and heat begins to dissipate into the surrounding. With continuing laser irradiation, the temperature increases further and the surface will shortly reach the vaporization temperature leading to ablation and the formation of a plasma. The plasma expansion and the vaporization process create a strong recoil pressure that expels molten material. This results in the formation of droplets and debris on the target surface. Expelled liquid material, which resolidifies at the edges of the hole, results in the formation of burr, which can attain heights in the order of hundred micrometers. The plasma and the ejected particles can cause shielding and scattering of the laser beam. Consequently, the laser power at the target and thus the ablation rate is reduced. Plasma and particle shielding become more prominent the longer the pulses are or, as we will see later on, when high pulse repetition rates are used.

On the other hand, the recoil pressure of the ablated material drives a shock pulse into the target. This shock wave and the large temperature gradients impose stress, which in most cases causes micro-cracks ranging far into the region which is not directly affected by the laser radiation. Heat diffusion and crack formation continue even after the laser pulse. The thermal evolution, which typically takes place at a timescale of several tens of nanoseconds to microseconds, mainly drives the modification after the laser pulse. The diffusion yields a broadening of the high temperature region and increases the heat affected zone. Therefore, stress is applied continually to the surrounding material and drives micro-crack formation. Also after the laser pulse the melt film at the sidewalls of the hole cools down and solidifies, resulting in a changed material morphology compared to the bulk material.

Finally, a view on the surface of the specimen shows a rather crater like drilling, which is surrounded by burr or molten material. Around the hole one can find a HAZ, which can extend overall several times the hole size. Here, permanent changes of the material can be recognized, which originate from the effects described above. The HAZ can be distinguished clearly from the untreated material. Particularly, the typical surface patterns resulting from a previous mechanical treatment, e.g., grinding or polishing, fade out. Additionally, micro cracks may be visible on the surface, which are evidence of the stress induced by the thermal load applied to the material. Furthermore, the surface is polluted by debris of ablated particles. The debris consists of particles in the size of a few to several tens of microns and can be clearly seen in the regions around the hole. However, it can be removed often by chemical or mechanical cleaning, but in certain cases an additional surface treatment is not allowed, particularly when the surface is finished with delicate structures or coatings.

All the effects discussed above are reasons for the modifications of the surrounding region of the ablated zone, which are in most cases unwanted, uncontrolled, and reduce precision and quality. Even more, it may not be possible at all to process certain structures, such as thin films with thicknesses of a few micrometers, which would melt and could be destroyed when treated with nanosecond pulses.

Ultrashort laser pulses (≤10 ps) offer tremendous potential to overcome these limitations. Figure 9.1 depicts a hole drilled with 10,000 laser pulses of 200 fs pulse duration and a fluence of 0.5 J/cm^2 in a 100 μm stainless steel foil [9]. The image shows a clean shape without any burr and almost no debris and HAZ. Additionally, it turned out that for ultrashort laser pulses the drilling quality is dependent on the laser fluence. A processing without HAZ and melting can be only achieved with low fluences, where the amount of ablated material per pulse corresponds to the optical penetration depth [17–19]. While the thermal influence at slightly elevated fluences is still negligible, melting increases dramatically for very high fluences even for ultrashort pulses and a hole similar to that produced with much longer pulses can form (Fig. 9.2) [20].

The ablation with ultrashort laser pulses cannot be described using a simple heat diffusion model, like for long laser pulses, where the material is strongly heated and consequently melted and finally vaporized. For ultrashort pulses different processes happen on distinct timescales. The overall process begins with absorption of the laser energy by electrons (time duration determined by the pulse length) and continues with fast relaxation amongst them (shortly after the laser pulse, $\approx$1 ps) [21]. The energy is transferred to the lattice by electron-phonon coupling

Fig. 9.1 Hole drilled in a 100 μm-thick stainless steel foil with 10,000 pulses of 200 fs pulse duration and a fluence of 0.5 J/cm^2

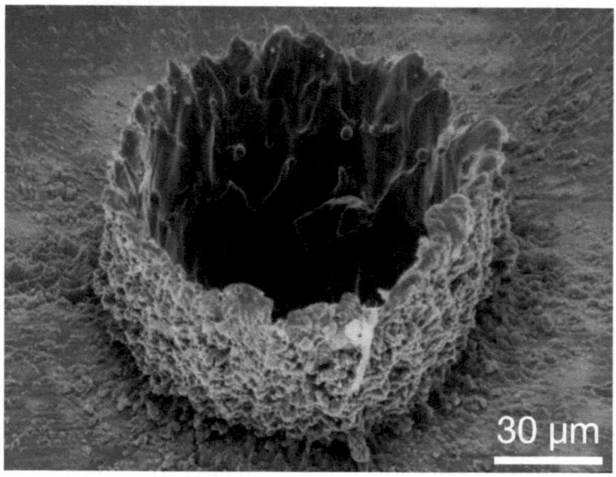

Fig. 9.2 Blind hole drilled in stainless steel with 200 fs pulses at a fluence of 200 J/cm^2. With just 100 pulses a depth of 140 μm was achieved

(10–100 ps) and a particle plasma is ignited (1–10 ns), which expands with high velocity [17, 22]. Particles are emitted approximately up to 100 ns after the laser pulse and the whole ablation process may last even microseconds. The following chapter discusses details of these processes and gives an overview on the temporal scales for ultrashort laser ablation.

9.3 Metal Ablation Processes Using Ultrashort Laser Processes

The laser ablation of metals with ultrashort laser pulses covers a long and complex sequence of linked physical processes, which occur on a time scale from 100 fs to a few microseconds. The first step is the absorption of light by free electrons. As discussed above the absorption length is given by the optical penetration depth, which is typically in the order of 10 nm for most metals. This absorption is of fundamental nature and as it is no special property of ultrashort pulses we refer here to other sources, see e.g. [23]. It is obvious that the absorption and the energy transfer from light to material can only take place as long as the pulse is present.

The next process is a fast thermalization of the high energy electrons to the Fermi energy level [21], which can be understood as a relaxation amongst the electrons. Spectral measurements of the energy density of states have shown a deviation from the Fermi-Dirac distribution for timescales shorter than 500 fs while after that the Fermi-Dirac fit describes the spectrum very well. This means the relaxation time is on a sub-picosecond timescale [21] and can be treated in good approximation as instantaneous for laser ablation. Following this initial relaxation process an energy transfer to the lattice takes place. This process can be described using a two-temperature model [24, 25]. It introduces dissimilar temperatures for the electrons and lattice, respectively and treats both systems independently. Due to the electron-phonon coupling, energy is transferred from the electrons to the lattice, finally resulting in an equilibrium state with a common temperature. The maximum achievable lattice temperature is mainly linear proportional to the absorbed laser fluence with material dependent proportionality factors [9].

In parallel to this thermal evolution photoelectric and thermionic emission from the surface is responsible for plasma formation. This plasma formation and in particular its evolution dynamics is discussed in the following.

9.3.1 Plasma Luminescence

The white light emitted from a hot particle plasma can be analyzed and studied to obtain information about the process of its generation. The plasma light intensity can be related to the amount of emitted particles. A spectral analysis can be performed to measure the energy distribution and the temperature of the plasma. We will use here the temporally resolved plasma luminescence to reveal the ablation dynamics of ultrashort laser pulse ablation.

For this purpose stainless steel has been ablated with ultrashort pulses. The laser source was a Ti:Sapphire amplifier (Spectra-Physics, Spitfire) delivering 800 nm pulses with a duration of 200 fs, which were focused with an achromatic lens with a focal length of 100 mm and the pulse energy was adjusted to obtain a fluence of 20 J/cm^2, which is a typical value for obtaining high ablation rates and still acceptable thermal influence. The plasma expansion dynamics is expected to occur on a timescale of nanoseconds. Therefore, the plasma is monitored with a gated intensified camera (ICCD; LaVision, "PicoStar TH7863") with an exposure time of 4 ns. The plasma is imaged perpendicularly to the laser beam onto the camera. Images were taken each 10 ns after the pulse with an uncertainty of 4 ns given by the gating time of the CCD camera. The plasma plume is of Gaussian shape and its intensity varies strongly in time. The plasma can be observed in a timespan until

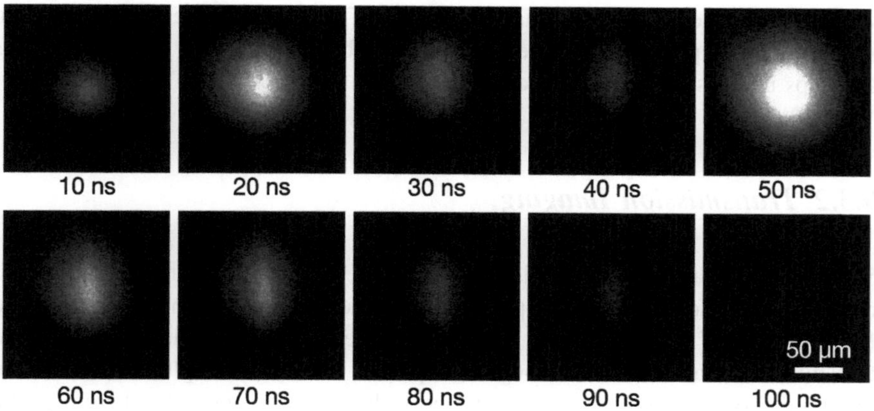

Fig. 9.3 Plasma luminescence for different delays after ablation of stainless steel with a 200 fs pulse

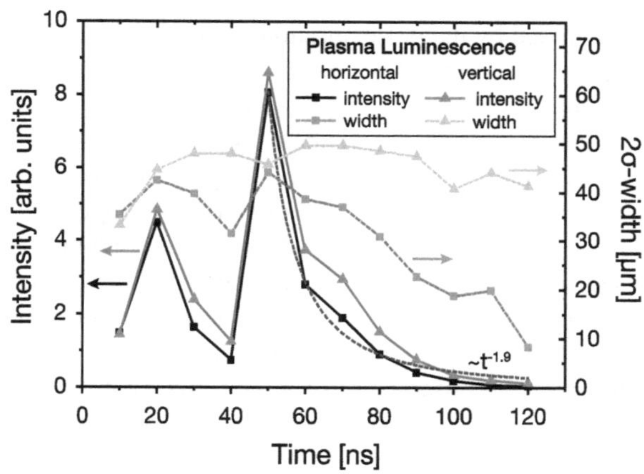

Fig. 9.4 Temporal evolution of the plasma luminescence in horizontal and vertical direction for the ablation of stainless steel with a pulse duration of 200 fs and a fluence of 20 J/cm^2

100 ns after the laser pulse (Fig. 9.3) [26]. Thereafter, the luminescence vanishes completely. Within this timespan, two strongly pronounced intensity maxima can be found at 20 and 50 ns after the laser pulse hits the surface, which indicates that the ablation process takes place in two steps. Additionally, the plasma has a circular shape at the beginning of the observation and at the second maximum. For all other data points one can observe an elliptical shape with the short axis parallel to the sample surface. This is an additional indication of the two step process, which starts at the surface. The following expansion is faster in the vertical direction resulting in and elliptical shape. The results are summarized in Fig. 9.4 [22]. The data points correspond to Gaussian fits of the imaged plasmas, where the arbitrary intensity corresponds to the maximum of the Gaussian and the width is given as twice the standard deviation 2σ. The grey and black curves in Fig. 9.4 clearly show the two maxima at 20 and 50 ns. After the second maximum the intensity decreases rapidly,

which can be fitted with a potential law $\sim t^{-1.9}$. This decay can be also observed in spectroscopic experiments, where aluminum is ablated with laser pulses of 500 fs and 5 ps under similar fluence conditions [27].

9.3.2 Transmission Imaging

Since the plasma luminescence is a consequence of the recombination process, it cannot account for particles still present above the ablation region also responsible for shielding of the laser light. Therefore, instead of the self-luminescence the transmission characteristics through the plasma and particle plume have to be observed. For this purpose a pump-probe technique was applied, where a probe beam with an adjustable delay was aligned perpendicular to the high power beam used for ablation (pump beam). The probe beam was a fraction separated from the main beam, shifted in time by a mechanical delay stage, transmitted above the target surface and imaged onto the ICCD camera. The temporal resolution is determined by the pulse duration of the probe beam and the absolute temporal delay has an uncertainty of 0.2 ns. To further increase the image contrast the probe beam is frequency doubled using a second harmonic crystal. A spectral filter in front of the camera blocks the fundamental frequency and thus scattered light from the intense pump beam. Therefore, only the transmitted probe light is analyzed with low noise and vanishing background. For longer delays an additional femtosecond laser oscillator running at 76 MHz corresponding to a pulse spacing of 13 ns was used for illumination, since the length of the mechanical delay stage limited the maximum observable timeframe. The concept of using a second laser source for the probe beam enabled a high temporal sampling resolution of a few nanoseconds but allows for an observation of a long time span, i.e. several microseconds.

Figure 9.5 shows the recorded transmission images in steps of 10 ns in the interval 4–180 ns. For achieving better image quality and increased contrast the plasma illumination has been subtracted and background light measured without an ablating pulse was removed. The image series clearly reveals the structure and the speed of the ablation products. The laser parameters used here for the ablation of stainless steel are the same as in Sect. 9.3.1. The image at 4 ns is magnified to highlight the emerging plasma. It shows a flat and horizontally oriented ablation front with a width corresponding to the focus diameter of the laser beam. This front evolves to a more spherical shape. The image recorded 50 ns after the laser pulse depicts a second ablation process. It can be seen as a dark spot in the center of the sphere. This second component can be clearly distinguished from the surrounding. These dissimilar components are a shock wave followed by vaporized ablated material [23]. The shock wave expands with a high speed of 2.8 km/s in vertical direction and slows down after 30 ns. The behavior of the vapor shows a very similar evolution but the expansion speed is with 3 km/s even slightly higher. After 60–70 ns also this evolution slows down.

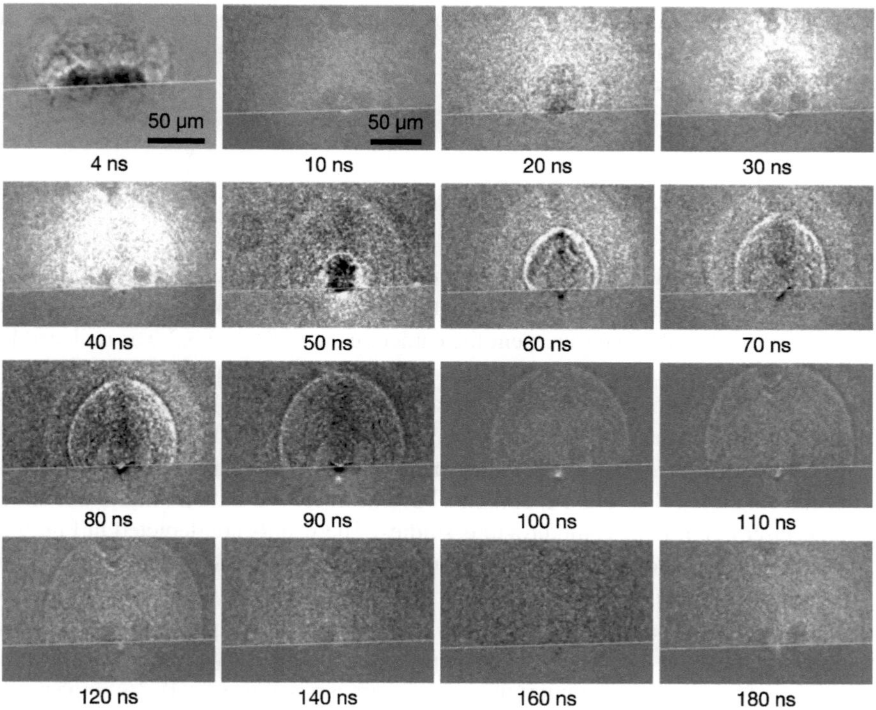

Fig. 9.5 Transmission images of a plasma generated by ablating stainless steel with 200 fs pulses. Images are obtained by subtracting the luminescence and a reference image without ablation pulse [26]

The measurements of plasma luminescence and the transmission imaging show independently that the material removal occurs in two steps. Both measurements show that a first ablation phase takes place immediately after the irradiation. This ablation phase is due to plasma formation by direct ionization, sublimation, photoelectric and/or thermionic electron emission [22, 25, 28–30].

The second ablation phase can be seen in both experiments (Figs. 9.4 and 9.5) either as a strong increase of the plasma luminescence intensity or a sudden formation of a particle wave. This process is delayed by 30–50 ns with respect to the incident laser pulse, which is a clear indication of involvement of thermal processes. A comparison of the timescales with thermal ablation processes occurring during processing with nanosecond laser pulses gives the clue that boiling and vaporization cause the formation of the expanding vapor plume [31]. The maximum in the luminescence can be explained due to recombination- and thermal radiation after emission of hot and partially ionized metal vapor. The short duration and the high temperatures indicate an ablation due to a phase explosion after heterogeneous or homogeneous nucleation. The subsequent decrease of the luminescence means that after the phase explosion a rapid cooling sets in. A similar explanation can be found by analyzing the transmission images. A fast developing process is depicted in Fig. 9.5, where a sudden change can be noticed between 40 and 50 ns also indicating the ablation via phase explosion [32, 33].

9.3.3 Quantification of the Plume Transmission

For an evaluation of the shielding of the following laser pulses by the particles present from the ablation process a quantitative analysis of the transmission has to be performed. For this measurement the camera used in Sect. 9.3.2 was replaced by a photodiode to measure the transmitted light through the particle plume. A comparison to the transmission without the ablation laser (pump) pulse gives normalized transmission data, which can be later on interpreted and evaluated together with a numerical analysis. The measurements cover a timespan from 400 ps to 5 μs after the ablation pulse. The experiments have been carried out on Aluminum under similar laser conditions as the previous studies. The results are depicted in Fig. 9.6. The transmission shows two prominent minima. The first one occurs at 4 ns after the pump pulse, mainly due to the fact that the particles have to travel a certain distance to block the probing beam. The transmission of the probe pulse is almost completely blocked but recovers to approximately 80 % within a few tens of nanoseconds. A second minimum with a transmission of 60 % appears at approximately 200 ns and the total recovery of the transmission is obtained at 2 μs. Even though a different material is used in these experiments a two-step process can be clearly recognized. The first minimum can be attributed to the blocking of the probe beam by a dense plasma created at the surface of the sample. Assuming an expansion velocity of 2.8 μm/ns, as measured in Sect. 9.3.2, the extension of the plasma can be estimated to be 11 μm after the 4 ns measured, which corresponds to

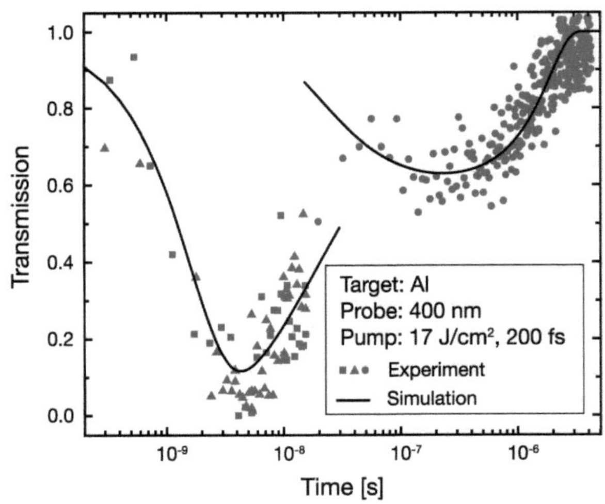

Fig. 9.6 Temporal evolution of the transmission of probe pulses with a wavelength of 400 nm for the ablation of Aluminum with 200 fs pulses at a fluence of 17 J/cm^2

the size of the probe beam. Additionally, due to the almost complete inhibition of the light transmission one can assume that the electron density n_e is at least as high as the critical electron density $n_e > n_c \approx 6.8 \cdot 10^{21}\ \mathrm{cm}^{-3}$ for the probe wavelength of 400 nm, and therefore absorbs and reflects the probe beam. The transmission increases after 6 ns, corresponding to a decreasing electron density through a further expansion of the plasma.

A careful analysis of the second transmission minimum by performing the measurements for different wavelengths revealed that here the transmission does not depend on the wavelength of the probe laser beam. Thus, this second transmission minimum cannot be originating from the absorption or reflection of free electrons and particles smaller than the wavelength. The independency on the wavelength suggests Mie scattering from particles larger than the wavelength [34] as the main reason for this transmission decrease, which gives rise to the assumption that the second process is ablation of clusters and droplets of a few microns size [28]. These comparatively large particle sizes for femtosecond laser ablation can be attributed to the relatively high laser fluences around 20 J/cm^2 used for ablation.

9.3.4 Timescales

Finally we conclude this section with an overview of the relevant timescales. The absorption of light by free electrons and/or by multi photon absorption can take place only during irradiation. Thermalization of the electrons occurs on a femtosecond time-scale, while the energy exchange with the lattice typically takes several picoseconds. Shortly thereafter an electron-ion plasma is formed by photoelectric and thermionic emission. This plasma expands with high velocities in the range of several km/s, as obtained from plasma luminescence and transmission imaging. As a consequence, the probe beam transmission in the experiment presented in Sec. 9.3.3 was minimal after about 4 ns. At this time the highest electron density across the probe beam diameter was reached. In addition, a second ablation phase occurs approximately 50 ns after the laser pulse. This second ablation phase is due to a phase explosion after heterogeneous or homogeneous nucleation. As such it takes a significant amount of time for the nucleation bubbles to develop and grow, explaining the temporal delay. The particles emitted in this phase have sizes in order of the wavelength or above, leading to Mie scattering of the probe beam. Thus, the transmission of the probe beam is reduced again approximately 200 ns after the laser pulse when the particles block the probe beam and full transmission is recovered only after 2 μs. Thus, particle shielding will occur if the next laser pulse arrives earlier than these 2 μs, which corresponds to a maximum repetition rate in the order of 500 kHz if interaction with the emitted particles shall be avoided. Figure 9.7 gives a simplified overview on the relevant timescales involved in ultrashort pulse laser ablation.

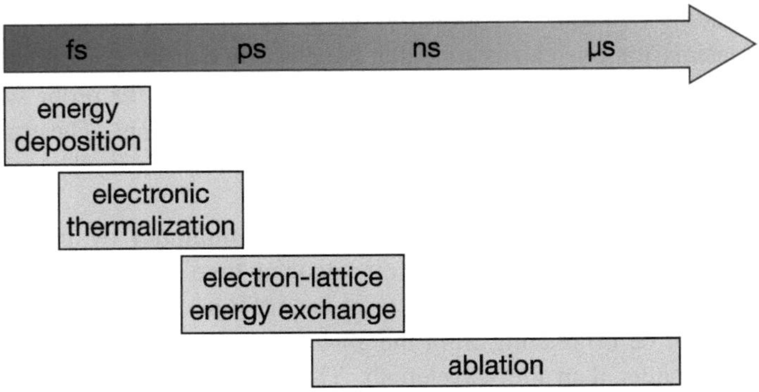

Fig. 9.7 Schematic illustration of the relevant timescales in ultrashort pulse laser ablation

9.4 High Repetition Rate Drilling

So far, laser micro processing was discussed for single pulses. Such a description is applicable for laser systems with low repetition rates, i.e., typically in the order of 1 kHz, where one can assume vanishing influence of consecutive pulses. In this case the cumulative action of the whole pulse series can be approximated by summing up the individual action of the single pulses. However, this assumption does not hold for higher repetition rates of 100 kHz or above. Laser systems with such repetition rates and sufficient pulse energies for ablation became available recently (see Chaps. 4, 5and 6) [12–14]. For investigating this regime, we used a fiber laser delivering 68 W average power at almost 1 MHz repetition rate, a maximum pulse energy of 70 μJ and a pulse duration adjustable from 800 fs to 19 ps [35]. The following sections present the results on percussion drilling and laser trepanning.

9.4.1 *Percussion Drilling*

If a single pulse is not sufficient to drill through the sample, the simplest case for hole drilling with a pulsed laser is percussion drilling, where consecutive laser pulses are focused on the same spot and every pulse ablates a certain but thin layer of the material. To drill through a sample of given thickness d requires NP pulses each ablating an average ablation depth per pulse of l_{th}:

$$d = NP \cdot l_{th}$$

The time required to drill the hole is then given by $T = NP/v$, where v is the laser repetition rate.

For the experiments we used high purity copper (99.9 %) sheets with 0.5 mm thickness, as well as 0.5 and 1.0 mm thick stainless steel sheets (Fe/Cr18Ni10). The

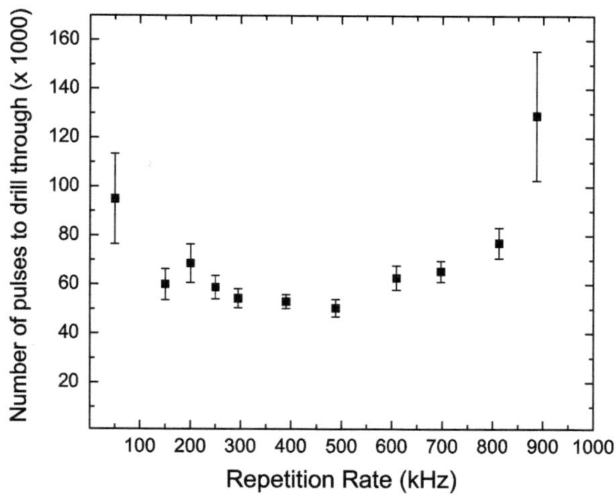

Fig. 9.8 Average number of pulses to drill through a 0.5 mm thick stainless steel foil at various repetition rates with 20 μJ pulses of 800 fs duration

drilling time was measured with a photodiode placed perpendicular to and at the side of the sample collecting light from both the top and bottom of the sample simultaneously. After breakthrough of the hole, additional light is reflected to the detector by an inclined plate below the sample yielding an increased signal. The detection routine starts with a rising slope of the photodiode's signal due to plasma formation on the top of the sample. The end is determined by an additional increase of the signal originating from the reflection from the plate below the sample.

In a first experiment comparably low laser pulse energies of 20 μJ with a duration of 800 fs were used. To obtain similar fluence conditions as before ($\sim$20 J/cm^2) the focal length was chosen to be 11 mm. The pulse repetition rate was varied from 50 to 975 kHz. The number of pulses to drill through the thinner (0.5 mm) steel plate was measured to be between 50.000 and 70.000 for all repetition rates below 500 kHz (see Fig. 9.8) [36]. The deviations and the larger measurement errors below 150 kHz can be explained by partial or complete closure of the hole due to material redeposition inside the capillary [37]. For all repetition rates below 500 kHz the average ablation depth per pulse is practically independent of the repetition rate, indicating that the ablation process is not depending on the previous pulse and its residues, e.g. particle vapor or residual heat.

This behavior changes for repetition rates above 500 kHz corresponding to pulse intervals shorter than 2 μs. Here, the number of pulses required to drill through increases with increasing repetition rate. At 975 kHz the amount of pulses to drill through is doubled approximately. This result is in agreement with those presented in the previous chapter, where particle shielding takes place approximately until 2 μs. Therefore, one can assign the increasing number of pulses necessary to drill through the sample to particle shielding. Particles and clusters can be found above the target and they consequently hinder the subsequent pulse to reach the target. Absorption, scattering and reflection by that particle plume reduce the laser power at the target and hence lead to lower ablation efficiency. This result is independent of the material thickness. The same experiments have been carried out for 1 mm thick stainless steel, which of course leads to an overall increase of the number of pulses required, but the general tendency remains the same.

If higher laser pulse energies are used this behavior changes. To reveal this, three different pulse energies (30, 50 and 70 μJ) have been used for ablation. In this case we used a lens with a longer focal length of 25 mm for focusing in order to obtain fluences in the same order of magnitude as before. Figure 9.9 depicts the different results of drilling through 0.5 mm thick stainless steel plates at the three pulse energies. The increased pulse energy leads to the occurrence of particle shielding already at lower repetition rates. However, in the particle shielding regime the number of pulses required for drilling through the sample does not increase monotonously with the repetition rate. Instead, a maximum is reached at 400, 300 and 200 kHz repetition rate for pulse energies of 30, 50 and 70 μJ, respectively.

Additionally, it is important to note that the number of pulses to drill through decreases with higher repetition rates beyond this maximum, even below the values for low repetition rates. This can be attributed to heat accumulation of the successive pulses. The energy of each laser pulse deposited in the material is partially used for plasma formation, bond breaking and particle removal. However, a certain fraction of the deposited pulse energy remains in the bulk material in the form of heat. For a single ultrashort pulse the thermal energy deposited in the material and therefore the temperature change is low and only marginal permanent changes occur in the surrounding material. However, if many pulses hit the sample with a temporal spacing shorter than the time required for the heat to diffuse out of the focal region, the temperature gradually increases and finally can lead to melting [38] around the focal volume. Since material removal from a sample at elevated temperature or even from molten material requires less energy than for a sample at room temperature, the ablation efficiency is expected to increase. This effect can be observed in the experiments shown in Fig. 9.9. The number of pulses necessary to drill through falls below the low repetition values at 975, 600 and 400 kHz for pulse energies of 30, 50 and 70 μJ, respectively. This means that the effect of heat accumulation is dependent on the pulse energy and cannot be seen for low energies (Fig. 9.8) at the repetition rates investigated here. For higher pulse energies the effect becomes more prominent and is observed at lower repetition rates, since more energy of every pulse remains in the sample. The quantitative analysis of the measured data shows that heat accumulation can enhance the ablation efficiency by a factor of 10, if one compares the values for 50 μJ: 2.5×10^3 and 2.5×10^4 pulses at 1 MHz and 100 kHz and for 70 μJ: 2.0×10^3 and 2.0×10^4 pulses at 700 and 100 kHz, respectively.

A simple theoretical model can approximate the effect of heat accumulation. Therefore, let us assume a laser with a Gaussian beam distribution irradiating a semi-infinite substrate with finite absorption A and temperature independent material properties. If we neglect any phase changes, the temperature evolution after the laser pulse is only determined by the heat diffusion and can be calculated analytically. For times t much longer than the pulse duration τ_l one can express the temperature change ΔT at the focal point as [23]:

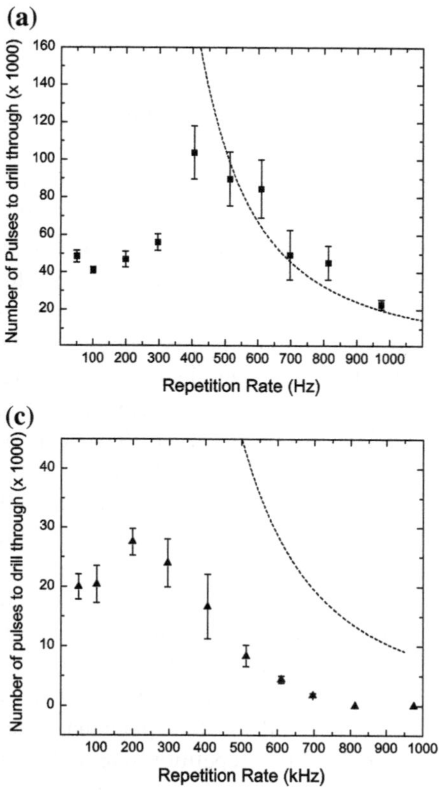

Fig. 9.9 Average number of pulses to drill through 0.5 mm thick stainless steel foils at various repetition rates with 800 fs pulses and for **a** 30 μJ, **b** 50 μJ and **c** 70 μJ pulse energy. In all graphs experimental data are plotted in *dots*, while the *lines* represent the estimated melting threshold due to the heat accumulation effect

$$\Delta T(t) = \frac{I_a w_0^2 \tau_l}{4\sqrt{\pi}\kappa t (Dt)^{1/2}}. \tag{9.1}$$

Here, $I_a = A \cdot I$ is the average absorbed laser-light intensity, w_0 is the radius of the focal spot and κ and D are the material properties thermal conductivity and heat diffusivity.

A significant change in the energy required for ablation occurs, once the sample is melted. It is thus important to calculate, when the melting temperature is reached. As the temperature change in the approximation made here is independent on its initial value it increases from pulse to pulse by the same amount. Thus, the number of pulses NP_{MELT} required to reach the melting temperature T_M is given by the ratio of melting temperature and temperature increase $\Delta T(1/\nu)$ per pulse when the next pulse is incident (time $t = \nu^{-1}$; ν being the repetition rate of the laser):

$$NP_{MELT} = \frac{T_M - T_{room}}{\Delta T(1/\nu)}, \tag{9.2}$$

where T_{room} indicates the initial temperature of the sample (typically room temperature).

NP_{MELT} scales strongly nonlinearly with the repetition rate. In Fig. 9.9 the according value for NP_{MELT} is plotted as a function of the repetition rate. For this calculation, we used values averaged within the corresponding temperature range (between T_{room} and T_M) of $\kappa = 0.217\ \mathrm{Wcm^{-1}K^{-1}}$ and $D = 0.0475\ \mathrm{cm^2s^{-1}}$ [39] for stainless steel and an absorption of $A = 0.8$ [22]. The theoretical curve reproduces the tendency of the number of pulses required for ablation properly, supporting the assumption of increased ablation efficiency due to heat accumulation and melting.

While the same trend can be observed also for higher pulse energies (Fig. 9.9b and c), slight deviations in the absolute values can be observed. This can be attributed to the fact that (9.1) is valid for low pulse energies only. Particularly, the residual thermal energy inside the sample increases with higher pulse energies [40, 41], which is not included in the model. In addition, the thermal properties of the material might change under such extreme conditions, too.

According to this discussion, the heat accumulation counterbalances particle shielding and finally even leads to a significant reduction of the number of pulses required to drill through the sample. This results in a significant increase in overall drilling speed, even stronger than expected by just taking the repetition rate increase into account. Drilling through the 0.5 mm thick stainless steel plate has been realized within few ms at 70 μJ pulse energy and 1 MHz repetition rate and even 1 mm material can be drilled within less than 10 ms [36]. However, despite the advantage of faster drilling the hole quality suffers in this regime. In fact, the main advantage of ultrashort laser pulses of quasi melt-free ablation gets lost due to the heat accumulation. As this process is incorporating melting now the holes will exhibit large burrs and resolidified material as expected from long pulse ablation (Fig. 9.10).

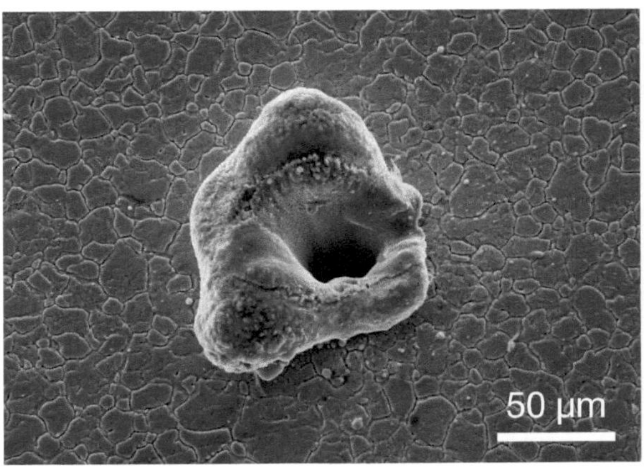

Fig. 9.10 SEM image of a hole drilled in 0.5-mm-thick stainless steel with 800 fs pulses with an energy of 30 μJ and 400 kHz repetition rate

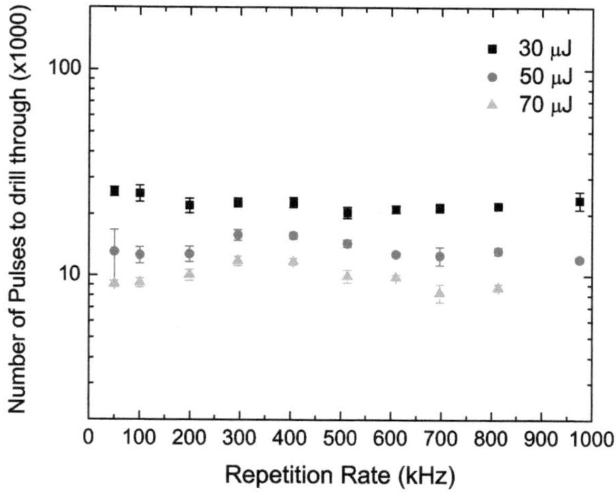

Fig. 9.11 Number of pulses to drill through 0.5-mm-thick copper sheets for various repetition rates at different pulse energies. Pulse duration was 800 fs

It has to be noted that the effect of heat accumulation is of course also strongly dependent on the material properties; particularly, the thermal conductivity and heat diffusivity play a major role (see 9.1). For example, these values are 20 times higher in copper compared to stainless steel, which means that heat is transported significantly faster. Consequently, laser percussion drilling experiments (Fig. 9.11) show completely different results compared to stainless steel (Fig. 9.9). Here, the average ablation rate is independent of the repetition rate. Thus, neither particle shielding nor heat accumulation are occurring under these experimental conditions. The SEM pictures of the drilled holes (Fig. 9.12) show for low and high repetition rates the presence of surface ripple structures, clearly indicating that melting does not play a major role. Simulations indicate that melting due to heat accumulation is expected here to occur only for repetition rates of 4 MHz and above. The explanation of negligible particle shielding is, however, not that obvious. One can probably denote the high thermal conductivity to this effect as well, since it reduces the formation of a superheated layer that originates the emission of particles due to phase explosion caused by homogeneous nucleation for longer time scales [22].

9.4.2 *Influence of the Pulse Duration*

The pulse duration has a strong influence on the ablation process as discussed in Sect. 9.1. Here, we discuss percussion drilling experiments similar to Sect. 9.4.1 for pulse durations of 800 fs, 6 ps and 19 ps in order to evaluate the influence of the pulse duration on particle shielding and heat accumulation. Pulses with an energy of up to 70 μJ are focused using a lens of 25 mm focal length onto the sample surface. The pulses originate from the same laser source and the pulse duration is adjusted by tuning the length of the pulse compressor solely, which ensures that all other laser parameters, particularly the beam profile and spot size, are unchanged and the

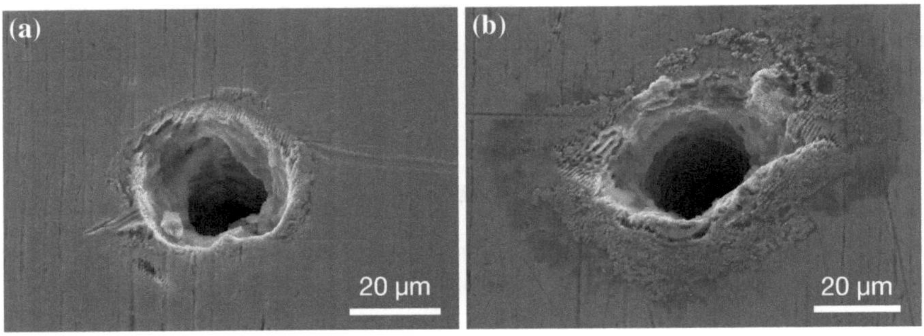

Fig. 9.12 SEM images of holes drilled in 0.5-mm-thick copper foils with 800 fs pulses with an energy of 50 μJ at **a** 50 kHz and **b** 975 kHz repetition rate

results are comparable. Figure 9.13 depicts the results as the number of pulses to drill through 0.5 mm thick stainless steel versus laser repetition rate.

It is evident that the drilling efficiency is higher for shorter pulses, especially in the range below 500 kHz. This may be explained by the fact that the shorter the pulses are the stronger is the stress generated into the target originating the phase explosion [42–44]. For longer pulses there might be an additional interaction and shielding by the generated near-surface plasma [22, 45] and potentially higher thermal losses to the surrounding. All these effects result in reduced ablation efficiency.

In the range between 200 and 400 kHz the number of pulses to drill through the sample is increased and the ablation efficiency is reduced by particle shielding. This particle shielding effect is shifted towards lower repetition rates not only with increasing pulse energy as discussed in Sect. 9.4.1 but also with increasing pulse duration. This could be explained by the ejection of a larger amount of cluster-like particles as well as a change in the ablation process itself for longer pulse durations [43]. As a consequence, the use of shorter pulse durations allows for higher repetition rates before particle shielding reduces the efficiency.

For repetition rates higher than 400 kHz one observes the decrease of the number of pulses to drill through with increasing repetition rate due to heat accumulation. This effect occurs earlier for longer pulses since the fractional power responsible for heating increases [46] and therefore heat can be accumulated more easily. As a consequence, the maximum number of pulses to drill through is shifted towards lower repetition rates for longer pulses. However, one can clearly see the detrimental effect of heat accumulation in the quality of the drilled holes. Figure 9.15 shows SEM pictures of holes fabricated at high and low repetition rates using fs and ps pulses [47]. Drilling with 19 ps pulse duration yields holes with a prominent amount of molten material inside the hole and burr at the hole edges already for a repetition rate of 300 kHz (Fig. 9.15b). Holes drilled with 800 fs at similar parameters are practically free from melting.

The use of copper instead of steel as an example for a metal with significantly higher thermal conductivity and heat diffusivity shows the same general trends irrespective of the pulse durations investigated here. For all pulse durations from

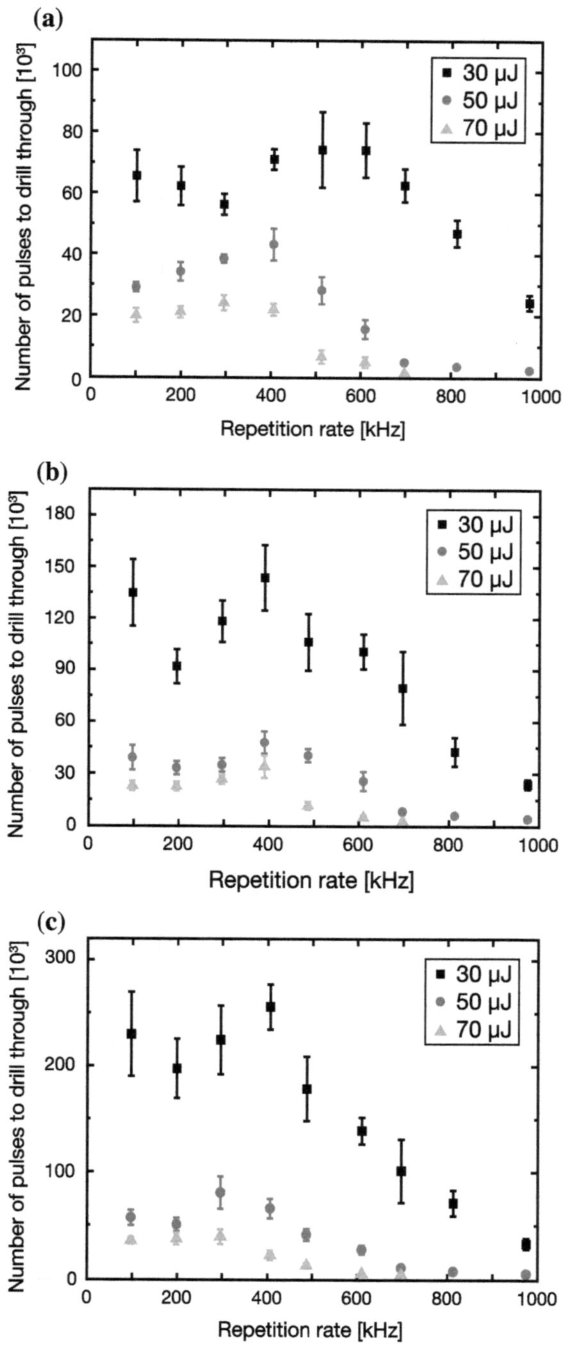

Fig. 9.13 Number of pulses to drill through 0.5-mm-thick stainless steel samples as a function of the repetition rate for different pulse durations and pulse energies: 30 μJ (*squares*), 50 μJ (*circles*), and 70 μJ (*triangles*), corresponding to fluences of 13.3, 22.1, and 30.9 J/cm^{-2}. Pulse duration was **a** 800 fs, **b** 6 ps and **c** 19 ps

Fig. 9.14 Number of pulses to drill through 0.5-mm-thick copper (99.9 %) samples as a function of the repetition rate for different pulse durations and pulse energies: 30 μJ (*squares*), 50 μJ (*circles*), and 70 μJ (*triangles*). Pulse duration was **a** 800 fs, **b** 6 ps and **c** 19 ps

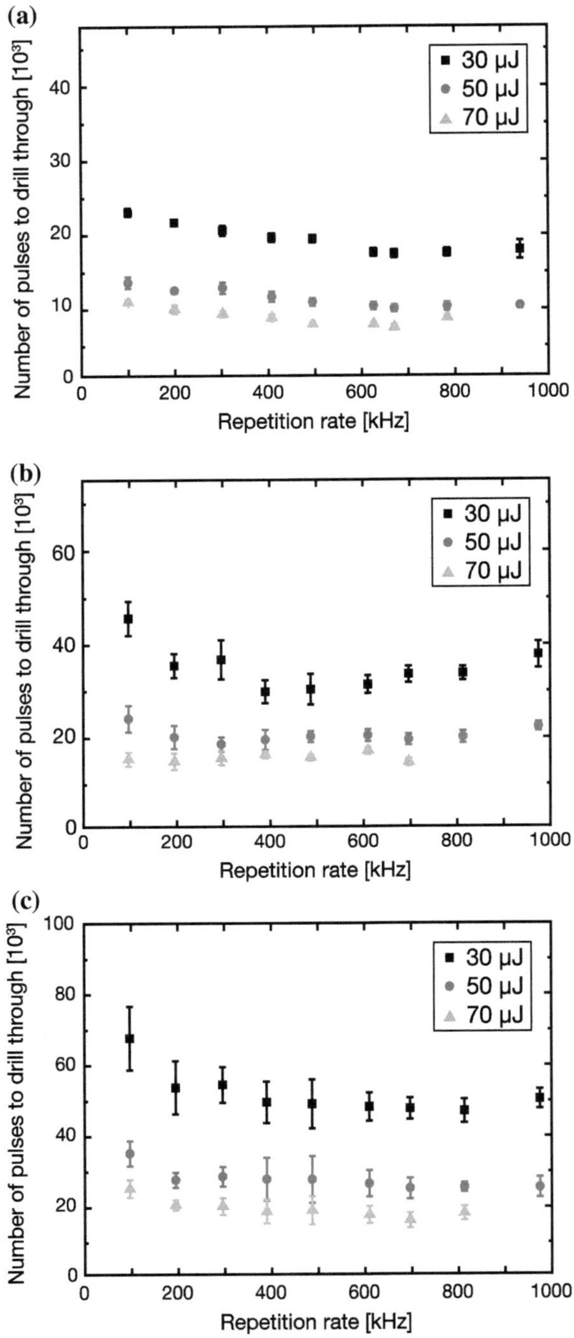

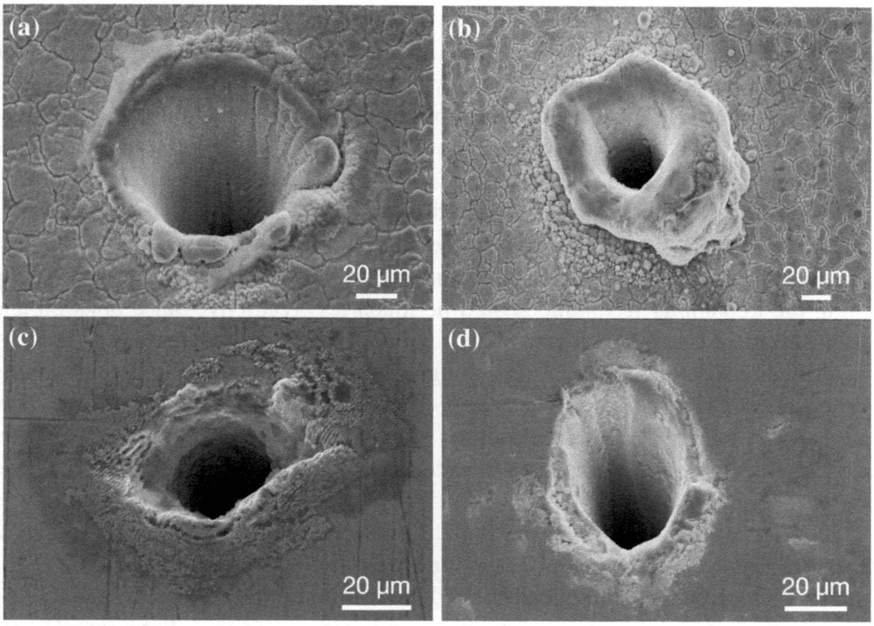

Fig. 9.15 SEM images of holes drilled with 50 µJ in stainless steel at **a** 800 fs and 200 kHz, **b** 19 ps and 300 kHz, and in copper at **c** 800 fs and 975 kHz, **d** 19 ps and 975 kHz

800 fs to 19 ps the ablation rate is almost independent of the repetition rate in the whole range from 200 kHz to 1 MHz (Fig. 9.14). However, the total number of pulses to drill through the 0.5 mm thick sample, while reduced compared to stainless steel, is significantly lower for shorter pulse durations. Moreover, a thin melt layer is already observable inside the holes drilled with 19 ps (Fig. 9.15d), while for 800 fs a pure ablation regime without signs of thermal effects is maintained (Fig. 9.15c).

Therefore, irrespective of the material, shorter pulse durations result in higher ablation efficiencies. If particle shielding and heat accumulation play a role, these detrimental effects occur later for shorter pulses allowing a scaling to higher repetition rates.

9.4.3 Trepanning

Obviously, heat accumulation and particle shielding are not only influenced by the material properties and laser parameters but depend on the energy distribution over the sample, too. In contrast to percussion drilling as discussed in Sects. 9.4.1 and 9.4.2 laser trepanning is a method, where the hole is machined using a laser beam traveling on a circular path around the center of the hole during drilling. The laser beam movement can be obtained, e.g., by applying a galvanometric scanning system or a special trepanning optics with rotating prisms. The trepanning method has the advantage that the hole geometry and the shape is mainly determined by the path on which the laser spot travels. Typically, the precision and quality can be strongly enhanced, the reliability and reproducibility become higher, and the holes have sharper and cleaner edges. Even though the processing time is longer and hole size is larger compared to percussion drilling, this method is preferred in applications, where highest accuracy is required (see Chap. 12).

Concerning the effects of particle shielding and heat accumulation limiting the scaling of the processing speed in percussion drilling as described above, drilling using the trepanning technique should be much less restrictive since ablation and laser power are distributed over a larger region. In order to prove these assumptions, we drilled holes using the laser trepanning technique with a galvanometric scanning system. The focusing optic used is an F-Theta lens with a focal length of 80 mm producing a 50 μm laser spot. In addition to the pulse energy and the repetition rate, we varied several other parameters like the trepanning radius, the rotation speed and the number of rounds.

In order to obtain melting- and burr-free structures in stainless steel we had to keep the average power below 13 W for a pulse duration of 800 fs. Figure 9.16a shows a laser trepanned hole realized with a repetition rate of 500 kHz, 25 μJ pulse energy, 75 μm trepanning radius and 265 rounds/s rotating speed. For a sample with a thickness of 0.5 mm breakthrough was achieved already after 800 ms. These results demonstrate the potential to overcome the limiting effects of particle shielding and heat accumulation for high precision drilling even in materials with low thermal conductivity.

In contrast, for copper we were able to exploit the full repetition rate of the laser system since already the percussion drilling experiments showed no influence of particle shielding or heat accumulation in this case. Figure 9.16b shows a hole trepanned with 800 fs pulse duration, 50 μJ pulse energy, 975 kHz repetition rate,

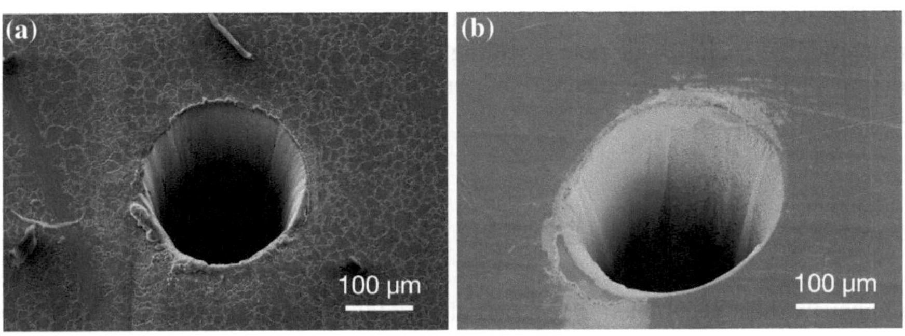

Fig. 9.16 SEM images of holes drilled with the trepanning technique in 0.5-mm thick **a** steel samples with an energy of 25 μJ at 500 kHz, using a trepanning radius of 75 μm and a rotation speed of 265 rounds/s and **b** copper sheets with an energy of 50 μJ, a repetition rate of 975 kHz, a trepanning radius of 75 μm and a rotation speed of 106 rounds/s. The time to breakthrough is 800 ms for steel and 75 ms for copper, respectively

75 μm radius and a rotation speed of 106 rounds/s. Breakthrough was obtained in this case already after 75 ms. The microstructure inside the hole as well as the absence of burr clearly demonstrate that thermal effects are negligible.

9.5 Conclusion

In this chapter we investigated fundamental limits for the scaling of the processing speed for drilling with ultrashort laser pulses by increasing the repetition rate. Therefore, we first analyzed the ablation dynamics of a single pulse. Particle emission takes place in two consecutive phases. After an initial plasma emission a second ablation phase occurs several 10 ns after the laser pulse. This second ablation phase is due to a phase explosion after heterogeneous or homogeneous nucleation. As a consequence, despite of the ultrashort laser pulse duration in the femtosecond to low picosecond domain, ablation lasts up to a microsecond time-scale. During that time subsequent pulses can be scattered or shielded by the particles and clusters present above the ablation spot.

Such a particle shielding has been proven in percussion drilling experiments in stainless steel for repetition rates of 500 kHz and below, depending on the pulse energy and pulse duration used. It leads to an increase in the number of pulses required to drill through the sample, since the ejected particles hinder consecutive pulses to hit the target directly. At even higher repetition rates the energy remaining in the sample leads to heat accumulation and finally local melting. Although this counterbalances particle shielding and increases the ablation efficiency and rate, the hole quality is significantly reduced and the advantages of precise processing by ultrashort pulses are lost.

Drilling of copper, however, did not show effects of particle shielding or heat accumulation even for repetition rates of up to 1 MHz for the pulse parameters investigated (800 fs to 19 ps, up to 68 W average power). Probably the higher thermal conductivity both reduces the superheated layer from which particle ejection is originated and prevents heat accumulation. Here, the full potential of the laser could be exploited. Consequently, it was possible to drill a hole in 1 mm thick copper in less than 1 ms without significant melting.

Higher quality could be obtained by using the laser trepanning technique. This even relaxed the limiting effects for drilling stainless steel, since the ablation and laser power is distributed over a larger area.

In all investigations, sub-picosecond pulses showed higher ablation efficiency compared to picosecond pulses. In addition, the onsets of particle shielding and heat accumulation are shifted to higher repetition rates. This means that the use of femtosecond laser pulses increases not only the ablation effectiveness but also higher repetition rates can be exploited to scale the processing speed even further. Nevertheless, still the application has to determine which pulse duration yields optimum results, not only with respect to quality and processing speed but also in terms of costs.

References

1. A.H. Zewail, Laser femtochemistry. Science **242**, 1645–1653 (1988)
2. W. Denk, J.H. Strickler, W.W. Webb, 2-photon laser scanning fluorescence microscopy. Science **248**, 73–76 (1990)
3. P. Antoine, A. L'Huillier, M. Lewenstein, Attosecond pulse trains using high-order harmonics. Phys. Rev. Lett. **77**, 1234–1237 (1996)
4. T. Brabec, F. Krausz, Intense few-cycle laser fields: Frontiers of nonlinear optics. Rev. Mod. Phys. **72**, 545–591 (2000)
5. A. Brodeur, F.A. Ilkov, S.L. Chin, Beam filamentation and the white light continuum divergence. Opt. Commun. **129**, 193–198 (1996)
6. T.A. Birks, W.J. Wadsworth, P.S. Russell, Supercontinuum generation in tapered fibers. Opt. Lett. **25**, 1415–1417 (2000)
7. T. Udem, R. Holzwarth, T.W. Hänsch, Optical frequency metrology. Nature **416**, 233–237 (2002)
8. T. Udem, J. Reichert, R. Holzwarth, T.W. Hänsch, Accurate measurement of large optical frequency differences with a mode-locked laser. Opt. Lett. **24**, 881–883 (1999)
9. B.N. Chichkov, C. Momma, S. Nolte, F. von Alvensleben, A. Tünnermann, Femtosecond, picosecond and nanosecond laser ablation of solids. Appl. Phys. A Mater. Sci. Process. **63**, 109–115 (1996)
10. D. Du, X. Liu, G. Korn, J. Squier, G. Mourou, Laser-Induced breakdown by impact ionization in SiO_2 with pulse widths from 7 ns to 150 fs. Appl. Phys. Lett. **64**, 3071–3073 (1994)
11. B.C. Stuart, M.D. Feit, A.M. Rubenchik, B.W. Shore, M.D. Perry, Laser-induced damage in dielectrics with nanosecond to subpicosecond pulses. Phys. Rev. Lett. **74**, 2248–2251 (1995)
12. T. Eidam, S. Hanf, E. Seise, T.V. Andersen, T. Gabler, C. Wirth, T. Schreiber, J. Limpert, A. Tünnermann, Femtosecond fiber CPA system emitting 830 W average output power. Opt. Lett. **35**, 94–96 (2010)
13. P. Russbueldt, T. Mans, J. Weitenberg, H.D. Hoffmann, R. Poprawe, Compact diode-pumped 1.1 kW Yb:YAG innoslab femtosecond amplifier. Opt. Lett. **35**, 4169–4171 (2010)
14. J.P. Negel, A. Voss, M.A. Ahmed, D. Bauer, D. Sutter, A. Killi, T. Graf, 1.1 kW average output power from a thin-disk multipass amplifier for ultrashort laser pulses. Opt. Lett. **38**, 5442–5445 (2013)
15. J.D. Verhoeven, Steel metallurgy for the non-metallurgist. ASM International (2007)
16. M. von Allmen, A. Blatter, *Laser-Beam Interactions with Materials* (Springer, Berlin, 1995)
17. S. Nolte, C. Momma, H. Jacobs, A. Tünnermann, B.N. Chichkov, B. Wellegehausen, H. Welling, Ablation of metals by ultrashort laser pulses. J. Opt. Soc. Am. B Opt. Phys. **14**, 2716–2722 (1997)
18. B. Neuenschwander, G. Bucher, G. Hennig, Ch. Nussbaum, B. Joss, M. Muralt, S. Zehnder, U. Hunziker, P. Schuetz, Processing of metals and dielectric materials with ps-laserpulses: results, strategies, limitations and needs. Proc. SPIE **7584**, 75840R (2010)
19. B. Jaeggi, B. Neuenschwander, M. Schmidt, M. Muralt, J. Zuercher, U. Hunziker, Influence of the pulse duration in the ps-Regime on the ablation efficiency of metals. Phys. Procedia **12**, 164–171 (2011)
20. C. Momma, B.N. Chichkov, S. Nolte, F. von Alvensleben, A. Tünnermann, H. Welling, B. Wellegehausen, Short-pulse laser ablation of solid targets. Opt. Commun. **129**, 134–142 (1996)
21. W.S. Fann, R. Storz, H.W.K. Tom, J. Bokor, Electron thermalization in gold. Phys. Rev. B **46**, 13592–13595 (1992)
22. J. König, S. Nolte, A. Tünnermann, Plasma evolution during metal ablation with ultrashort laser pulses. Opt. Express **13**, 10597–10607 (2005)
23. D. Bäuerle, *Laser Processing and Chemistry* (Springer, Berlin, 2000)
24. M.I. Kaganov, I.M. Lifshitz, L.V. Tanatarov, Relaxation between electrons and the crystalline lattice. Sov. Phys. Jetp-Ussr **4**, 173–178 (1957)

25. S.I. Anisimov, B.I. Kapeliov, T.L. Perelman, Electron-emission from surface of metals induced by ultrashort laser pulses. Zhurnal Eksperimentalnoi I Teoreticheskoi Fiziki **66**, 776–781 (1974)
26. J. König, *Dissertation "Abtragsprozesse bei Bestrahlung von Metallen mit ultrakurzen Laserpulsen"* (Friedrich Schiller University, Jena, 2006)
27. B. Le Drogoff, J. Margot, F. Vidal, S. Laville, M. Chaker, M. Sabsabi, T.W. Johnston, O. Barthélemy, Influence of the laser pulse duration on laser-produced plasma properties. Plasma Sources Sci. Technol. **13**, 223–230 (2004)
28. R.E. Russo, X.L. Mao, H.C. Liu, J.H. Yoo, S.S. Mao, Time-resolved plasma diagnostics and mass removal during single-pulse laser ablation. Appl. Phys. A **69** [Suppl.], S887–S894 (1999)
29. T.Y. Choi, C.P. Grigoropoulos, Plasma and ablation dynamics in ultrafast laser processing of crystalline silicon. J. Appl. Phys. **92**, 4918–4925 (2002)
30. L. Cerami, E. Mazur, S. Nolte, C.B. Schaffer, Femtosecond Laser Micromachining, in *Ultrafast Nonlinear Optics*, ed. by R. Thomson, C. Leburn, D. Reid (Springer, Heidelberg, 2013)
31. R. Kelly, A. Miotello, Contribution of vaporization and boiling to thermal-spike sputtering by ions or laser pulses. Phys. Rev. E **60**, 2616–2625 (1999)
32. K. Sokolowski-Tinten, J. Bialkowski, A. Cavalleri, M. Boing, H. Schüler, D. von der Linde, in *High-Power Laser Ablation*, ed. by C.R. Phipps. Dynamics of femtosecond laser induced abaltion from solid surfaces. Proc. SPIE **3343**, pp. 46–57 (1998)
33. K. Sokolowski-Tinten, J. Bialkowski, M. Boing, A. Cavalleri, D. von der Linde, in *Quantum Electronics and Laser Science Conference*. Bulk phase explosion and surface boiling during short pulse laser ablation of semiconductors, OSA Technical Digest, Optical Society of America, Washington, pp. 231–232 (1999)
34. G. Mie, Beiträge zur Optik trüber Medien, speziell kolloidaler Metallösungen. Ann. der Physik **330**, 377–445 (1908)
35. F. Röser, T. Eidam, J. Rothhardt, O. Schmidt, D.N. Schimpf, J. Limpert, A. Tünnermann, Millijoule pulse energy high repetition rate femtosecond fiber chirped-pulse amplification system. Opt. Lett. **32**, 3495–3497 (2007)
36. A. Ancona, F. Röser, K. Rademaker, J. Limpert, S. Nolte, A. Tünnermann, High speed laser drilling of metals using a high repetition rate, high average power ultrafast fiber CPA system. Opt. Express **16**, 8958–8968 (2008)
37. A. Weck, T.H.R. Crawford, D.S. Wilkinson, H.K. Haugen, J.S. Preston, Laser drilling of high aspect ratio holes in copper with femtosecond, picosecond and nanosecond pulses. Appl. Phys. A Mater. Sci. Process. **90**, 537–543 (2008)
38. S.M. Eaton, H.B. Zhang, P.R. Herman, Heat accumulation effects in femtosecond laser-written waveguides with variable repetition rate. Opt. Express **13**, 4708–4716 (2005)
39. R.S. Graves, T.G. Kollie, D.L. Mcelroy, K.E. Gilchrist, The thermal-conductivity of AiSi 304 l stainless-steel. Int. J. Thermophys. **12**, 409–415 (1991)
40. A.Y. Vorobyev, V.M. Kuzmichev, N.G. Kokody, P. Kohns, J. Dai, C.L. Guo, Residual thermal effects in Al following single ns- and fs-laser pulse ablation. Appl. Phys. A Mater. Sci. Process. **82**, 357–362 (2006)
41. A.Y. Vorobyev, C.L. Guo, Enhanced energy coupling in femtosecond laser-metal interactions at high intensities. Opt. Express **14**, 13113–13119 (2006)
42. J.P. Colombier, P. Combis, A. Rosenfeld, I.V. Hertel, E. Audouard, R. Stoian, Optimized energy coupling at ultrafast laser-irradiated metal surfaces by tailoring intensity envelopes: consequences for material removal from Al samples. Phys. Rev. B **74**(22), 224106 (2006)
43. P. Lorazo, L.J. Lewis, M. Meunier, Thermodynamic pathways to melting, ablation, and solidification in absorbing solids under pulsed laser irradiation. Phys. Rev. B **73**, 134108 (2006)
44. J. Yang, Y. Zhao, X. Zhu, Transition between nonthermal and thermal ablation of metallic targets under the strike of high-fluence ultrashort laser pulses. Appl. Phys. Lett. **88**, 094101 (2006)

45. B. Salle, O. Gobert, P. Meynadier, M. Perdrix, G. Petite, A. Semerok, Femtosecond and picosecond laser microablation: ablation efficiency and laser microplasma expansion. Appl. Phys. A **69** [Suppl.], S381–S383 (1999)
46. D. Breitling, A. Ruf, F. Dausinger, Fundamental aspects in machining of metals with short and ultrashort laser pulses. Proc. SPIE **5339**, 49–63 (2004)
47. A. Ancona, S. Döring, C. Jauregui, F. Röser, J. Limpert, S. Nolte, A. Tünnermann, Femtosecond and picosecond laser drilling of metals at high repetition rates and average powers. Opt. Lett. **34**, 3304–3306 (2009)

Chapter 10
Processing Techniques and System Technology for Precise and Productive Microdrilling in Metals

Martin Kraus, Dmitrij Walter, Andreas Michalowski and Jens König

Abstract The fabrication of microstructures in metals employing ultrashort laser pulses has been under study since the middle of the 1990s. Over many years, extraordinary precision and machining quality have been demonstrated in surface patterning, cutting and drilling applications. However, apart from some exceptions, micromachining applications with ultrashort laser pulses could not be used in industrial production because of insufficient throughput until the late 2000s. This was the case particularly for microdrilling applications, where processing times easily reached several minutes per hole. This situation has changed fundamentally in recent years. In 2008, industry standard picosecond laser sources with average powers of 50 W combined with repetition rates up to 1000 kHz have been launched onto the market. In the present chapter, it is shown that these newly available laser specifications can be utilized for efficient and high-quality drilling techniques, the basic principles of which will be explained in combination with the necessary engineering expertise. A beam steering system for helical drilling with rotating beam profile will be presented. Finally, various diagnostic methods will be discussed with regard to visualization and fundamental understanding of ablation processes, but also as a basis for process monitoring and control.

10.1 Introduction

The fabrication of high-precision microholes in metals is required in various industrial applications. Increasingly small orifices with high aspect ratios and extraordinarily restrictive tolerances are used particularly in the fields of injection, dosing, optical, medical and sensing technologies [1–6]. In these applications, the

M. Kraus (✉) · A. Michalowski · J. König
Robert Bosch GmbH, Postbox 300240, 70049 Stuttgart, Germany
e-mail: Martin.Kraus5@de.bosch.com

D. Walter
Manz AG, Steigäckerstraße 5, 72768 Reutlingen, Germany

S. Nolte et al. (eds.), *Ultrashort Pulse Laser Technology*,
Springer Series in Optical Sciences 195, DOI 10.1007/978-3-319-17659-8_10

borehole diameter typically ranges from 50 to 100 μm. The material thickness usually lies between 0.5 and 1 mm, yielding aspect ratios between 5:1 and 20:1. Depending on the ratio between the inlet and the outlet diameter, the geometrical shape of the capillary can be either cylindrical or conical. In many applications, the laser drilled holes are used as spray holes which are designed to create specific flow patterns of fluid media. Consequently, the inlet and outlet orifices are generally expected to have a high contour accuracy and sharp edges. In order to avoid post processing, the holes have to be free of burr and recast. Besides borehole geometry and machining quality, processing time is an important factor. To ensure that the laser drilling process meets the cycle time requirements of mass production, the processing time for the fabrication of a high-precision hole in a 1 mm steel sheet should not exceed 15 s. Finally, it should be noted that the drilling process must have a high reproducibility, as well as a high reliability.

It has been shown in a variety of publications that microholes complying with the criteria set out above can be manufactured making use of ultrashort laser pulses. This is possible due to a combination of short interaction times with high power densities, leading to material ablation via immediate evaporation and reducing melt generation to a negligible extent [1, 5]. It must be emphasized, however, that until the middle of the 2000s, the majority of the published results were obtained on the basis of laser sources operating at average powers less than 5 W and repetition rates between 1 and 10 kHz. These parameters lead only to a relatively low level of laser-plasma interaction and a minor thermal load of the workpiece, finally resulting in extraordinary machining quality, but also in insufficient process efficiency and excessively long processing times [5, 6].

As already mentioned in the abstract, powerful picosecond laser sources have been made available by several laser manufacturers in 2008 [7, 8]. This new generation of ultrashort pulsed lasers has found broad use in surface processing and cutting applications because in many cases, the available laser power can be converted into productivity by simply increasing the feed rate. In high-aspect-ratio microdrilling however, the use of high pulse energies and high repetition rates causes a strong ionization of the ambient gas, the formation of melt due to heat accumulation and, consequently, a deterioration of the borehole quality.

This chapter describes how to make use of the available laser power in order to raise process efficiency without quality loss. It focuses on processing techniques, system technology and monitoring concepts for the efficient production of high-precision boreholes in metals or metal alloys with picosecond laser pulses.

10.2 System Engineering for Helical Drilling

A principal condition for the production of high-precision microholes with well-defined geometry is a specialized optical system creating a beam rotation on the workpiece [5, 6]. In order to achieve high aspect-ratios (>5:1), cylindrical or even negatively conical borehole shapes or small diameters (<50 μm), helical laser drilling should be performed using purpose-built drilling optics which permit an independent and precise adjustment of path diameter and beam inclination angle, the latter being an essential requirement for the fabrication of microholes with defined tapering.

Optical systems for helical drilling usually generate a rotation of a laser beam which is incident on an aberration corrected focusing objective. In order to control the beam inclination angle and the path diameter on the workpiece, the laser beam is rotated on the focusing objective with a defined lateral offset and a defined angle of incidence. Consider a collimated laser beam entering a focusing objective at a distance r from and an angle α to the optical axis and being focused at a focal distance f'. It can easily be seen from ray optics that in the focal plane, the distance of the beam from the optical axis is given by $f' \cdot \alpha$, whereas its angle to the optical axis becomes $\alpha - r/f'$. In most cases one can assume that $\alpha \ll r/f'$, so that the beam inclination angle can be approximated by r/f'. Hence, the diameter of the circular beam path on the workpiece surface and the beam inclination angle relative to the surface normal can be adjusted independently by controlling r and α on the focusing objective.

In order to meet these requirements several optical concepts have been developed in the past. Most of these concepts are based on rotating wedges [5]. Other systems are based on a rotating Dove prism [9, 10], tilted or shifted lenses [11] or galvanometer scanners. The concept described in this section is based on a rotating telescope comprising two cylindrical lenses [12]. As shown in Fig. 10.1, the collimated laser beam eccentrically hits the first cylindrical lens (ZL) at a given distance from and angle to the optical axis, and is then focused at the centre of the telescope. The second cylindrical lens collimates the beam. After free space propagation along a distance L, the beam is focused on the workpiece by an aberration corrected objective (SL). The cylindrical lenses rotate around their optical axes, causing also the transmitted laser beam to rotate.

A major advantage of the cylindrical lens concept compared to the wedge-based concept is the simultaneous rotation of both spot position and beam intensity profile. This enables the fabrication of holes with excellent roundness even if the beam profile is not perfectly symmetric. In addition, the optical setup can be designed in such a way that the rotating components, i.e. the cylindrical lens telescope, are mechanically separated from the optics used for setting the desired

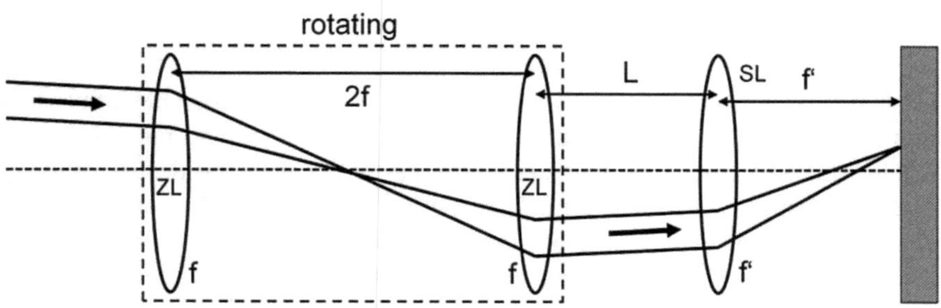

Fig. 10.1 Scheme of a rotating telescope drilling optics

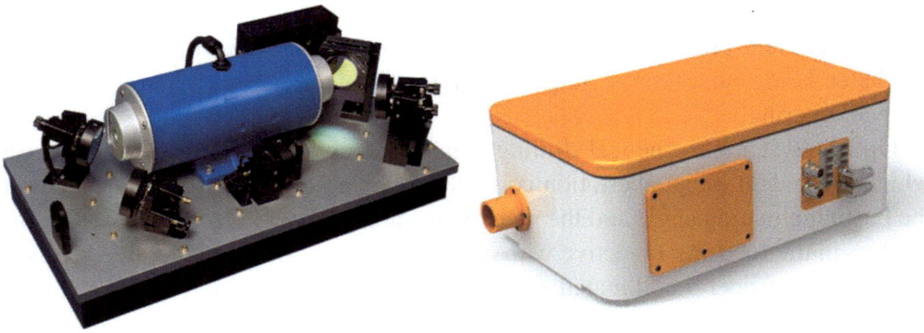

Fig. 10.2 *Left* Prototype of the helical drilling optics based on rotating cylindrical lenses. The two mirrors which provide the adjustment of the path diameter and the inclination angle are located at the front *right-hand corner* of the base plate and just beside the *blue* housing of the hollow shaft motor which rotates the telescope. The focusing objective is not shown. *Right* Industrial implementation of the prototype shown on the *left* [13]

path diameter and the beam inclination angle. This allows for smooth, precise and high-speed beam rotation. A prototype realization of this concept is shown in Fig. 10.2 (left image).

Based on the depicted prototype, a technically mature helical drilling system including a rotating cylindrical lens telescope has been developed. The GL.trepan (see right image in Fig. 10.2) allows to rotate the beam with frequencies as high as 30.000 rpm. Unlike most other helical drilling systems, the GL.trepan can be inserted into the beam path of a micromachining station at any position between the laser source and the focusing optics. Due to the adjustable beam inclination angle, the hole conicity can be varied continuously from positive to negative tapers. In combination with a superimposed lateral movement of the laser spot, the presented system enables precise cutting with right-angle cutting edges. Figure 10.3 shows the inlet and the outlet of a high-quality microhole drilled in a 300 μm steel sheet using the GL.trepan.

10.3 Generation and Effects of Plasma at Picosecond Microdrilling in Metals

The high peak intensity of ultrashort laser pulses causes a considerable influence of laser-induced plasma on the ablation process. In high-aspect-ratio microdrilling of metals, ablated particles accumulate in the gas atmosphere inside the capillary. Under irradiation with subsequent laser pulses, they emit high-energy electrons which are capable of initiating an impact ionization avalanche, thus causing an instantaneous low threshold breakdown of the surrounding gas species [14–20]. Since the propagation velocity of the ionization front is 10^9 cm/s equivalent to

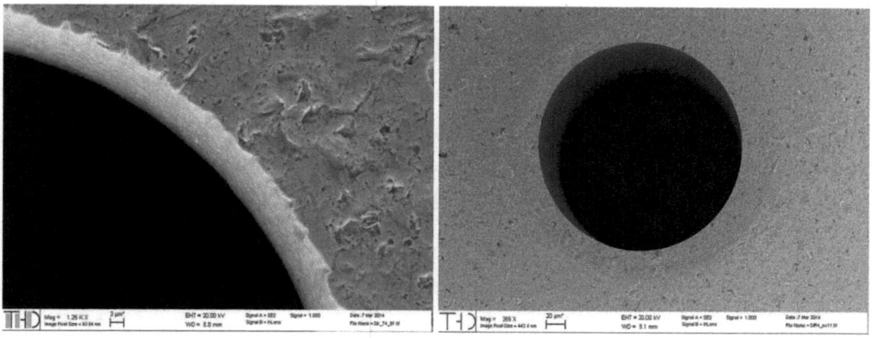

Fig. 10.3 Inlet (*left*) and outlet (*right*) of a microhole in a 1.4125 Cr steel sheet [13]. The depicted borehole was manufactured using the GL.trepan helical drilling optics presented in Fig. 10.2 (*right image*). The SEM images illustrate the excellent roundness and machining quality which can be achieved with beam rotation optics. Inlet diameter 200 μm, outlet diameter 217 μm, sheet thickness 300 μm, wavelength 1030 nm, pulse duration 0.9 ps, assist gas compressed air, processing time 1.7 s

10 μm/ps [19], the so-called particle-ignited plasma expands over the entire capillary cross-section in a few ps.

At picosecond microdrilling in metals, the threshold for particle-induced breakdown amounts to a few 10 J/cm^2 or approximately 10^{11} W/cm^2. These values are commonly exceeded in industrial microdrilling applications, whereas the threshold for optical breakdown without the presence of a workpiece is beyond the relevant energy density regime [1, 17]. The energy of the laser field is transferred to the plasma mainly via inverse bremsstrahlung (IB), which is the dominating mechanism for the absorption of photon energy by free electrons in laser processing at power densities up to 10^{16} W/cm^2 [1, 18–20]. As stated in [21], the laser-plasma interaction leads to plasma heating for pulse durations starting from 5 ps.

Due to the ionization of the gas atmosphere inside and above the capillary, a considerable part of the pulse energy is transferred to the comparatively long-living laser-induced plasma [22] and subsequently redistributed to the workpiece on a nanosecond timescale. This leads to a radial expansion of the capillary, the formation of melt due to prolonged surface heating [23–25] and, consequently, a deterioration of the borehole quality. A smoothing effect of the plasma cloud as it has been observed at ns drilling [1, 26] could not be confirmed in the ps regime. On the contrary, burr and eroded inlet edges, grooves and melt ridges on the borehole walls as well as insufficient widening of the outlet orifice are observed particularly if strong ionization has occurred during the drilling process. Hence, a successful control of ionization is to be regarded as the key to both borehole quality and productivity.

10.4 Core Techniques for High-Aspect-Ratio Microdrilling with Ultrashort Laser Pulses

As described in Sect. 10.1, the technical utilization of high-precision microholes is associated with a variety of requirements concerning borehole geometry, machining quality and the productivity of the drilling process. These requirements can be met by employing specific processing techniques, the most effective of which are explained in the following sections.

10.4.1 Pulse Energy Versus Repetition Rate

In high-aspect-ratio microdrilling in thick workpieces, high productivity strictly requires high energy densities. Previous investigations have shown that at percussion drilling in steel using 125 fs pulses, the volume ablation rate, which is the decisive factor for processing speed, grows extraordinarily strong with energy density in the range between 10 and 100 J/cm^2 [5, 17]. In fact, a tenfold increase in energy density was found to cause a 100 fold improvement in drilling velocity, yielding a tenfold increase in process efficiency per unit pulse energy. The implementation of a corresponding efficiency gain via an enhancement of the repetition rate would require considerably more average power. At the same time, however, it must be taken into account that high energy densities lead to a vigorous ionization of the surrounding gas species and to a degradation of the borehole due to radial etching and melt production.

Figures 10.4 and 10.5 illustrate the investigation of the optimum division of a fixed average power of 10.3 W at helical drilling in 1 mm CrNi steel sheets. During

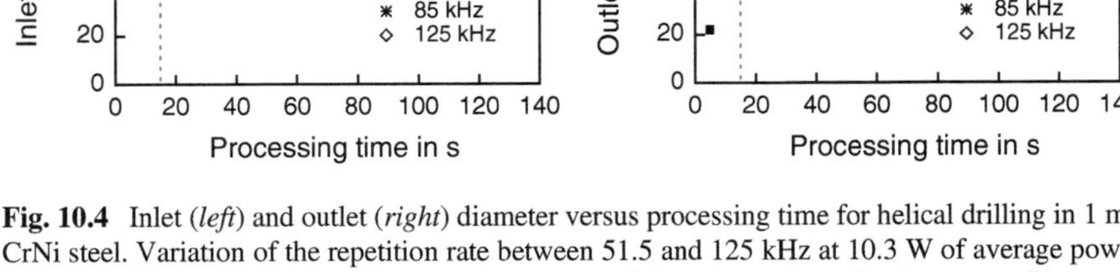

Fig. 10.4 Inlet (*left*) and outlet (*right*) diameter versus processing time for helical drilling in 1 mm CrNi steel. Variation of the repetition rate between 51.5 and 125 kHz at 10.3 W of average power. The depicted diameters are each mean values averaged from the major and minor axes of the cross-section. Wavelength 1030 nm, pulse duration 6 ps, focal diameter 24 µm, assist gas He 4.6, feed gas pressure 5 bar, linear pulse energy ramp over 5 s, circular polarization

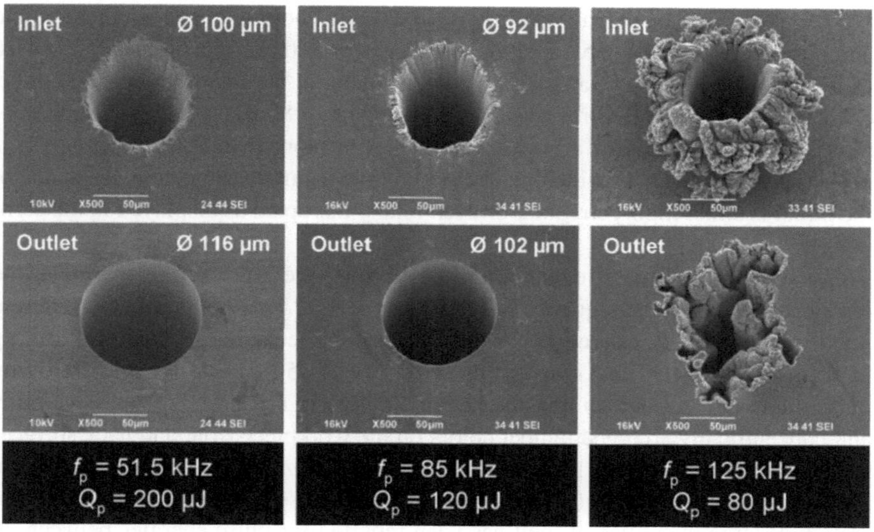

Fig. 10.5 Scanning electron microscope (SEM) images of microholes in 1 mm CrNi steel, drilled at constant average power of 10.3 W and with repetition rates of 51.5 kHz (*left*), 85 kHz (*middle*) and 125 kHz (*right*) [28, 29]. The depicted diameters are each mean values averaged from the major and minor axes of the cross-section. Processing time 15 s, other parameters as indicated in Fig. 10.4

the experiments the repetition rate was set to 51.5, 85 and 125 kHz, yielding pulse energies of 200, 120 and 80 µJ and focal energy densities of 44, 26 and 17 J/cm^2, respectively. The experiments were carried out under Helium atmosphere, and the pulse energy was gradually increased in the form of a linear ramp starting from 5 µJ. Figure 10.4 shows that the borehole geometry is completed most efficiently at a repetition rate of 51.5 kHz, while an inefficient and irregular widening behavior is observed at 125 kHz. At 51.5 kHz, the inlet (outlet) diameter is widened to the target value of 100 µm (115 µm) already after a processing time of 15 s. In this context, it is important to note that, although a gradual growth in diameter is recorded in the further course of the process, the slope gradient at 15 s is small enough to guarantee high reproducibility (compare [27]). At 85 kHz, the desired inlet (outlet) diameter is reached only after 30 s (45 s). At 125 kHz, eventually, the expansion of the capillary proceeds in such an inefficient way that the development of the borehole dimensions stagnates at roundly 90 % of the target values. Moreover, it can be seen that an enhancement of the repetition rate results in an extension of the time period which is needed to fully penetrate the workpiece. At 51.5 kHz, the first outlet orifice can be detected after 5 s compared to 7.5 s at 85 kHz and 12.5 s at 125 kHz.

Figure 10.5 reveals that preference should be given to high pulse energies instead of high repetition rates not only in terms of process efficiency, but also in terms of machining quality. From the depicted scanning electron microscope (SEM) images, it can be seen that at 51.5 kHz ($Q_p = 200$ μJ), both inlet and outlet are round and absolutely free of visible melt deposits or debris. If the repetition rate is raised to 85 kHz ($Q_p = 120$ μJ), a small burr is generated at the inlet, whereas the outlet cross-section tends to become oval. At 125 kHz ($Q_p = 80$ μJ), a large production of melt and the formation of a burr at the inlet with more than 100 μm in height is observed. As an efficient widening of the borehole is not possible here, the shape of the outlet is either highly irregular or even totally disrupted as displayed in the SEM image.

If the repetition rate is enhanced while keeping the pulse energy constant, the deterioration of borehole quality is even more severe. This holds true for different pulse energy regimes as well as for different drilling techniques. At helical drilling in 1 mm CrNi steel with a constant pulse energy of 150 μJ (wavelength 1030 nm, pulse duration 6 ps, energy density 32 J/cm^2), heavily eroded inlets, massive burrs and concentrically arranged temper colors on the workpiece front and back surfaces are formed for repetition rates greater than 100 kHz. At percussion drilling in 500 μm CrNi steel sheets (wavelength 1030 nm, pulse duration 800 fs), the generation of large quantities of melt has been reported for pulse energies between 30 and 70 μJ, the critical repetition rates here lying between 400 and 200 kHz [30].

Overall, it appears that the use of pulse energies up to 200 μJ is absolutely appropriate in high-precision picosecond microdrilling in steel. At high repetition rates, however, a considerable part of the incident laser energy does not contribute to the desired material removal in vapor form, but leads to an unwanted thermal load of the bulk material, resulting in extensive melt production and inacceptable drawbacks in machining quality and borehole shape. The critical repetition rate depends on laser parameters, material properties, borehole dimensions and processing technology and is thus specific for each application.

A possible reason for the limitation of the repetition rate at high-aspect-ratio microdrilling in metals is heat accumulation in the interaction zone [30]. Despite the ultrashort pulse duration and the fast laser-material interaction, a small amount of residual energy remains in the workpiece after each pulse [1, 17]. If the repetition rate is so high that the time between the pulses is too short to allow substantial energy transfer into the bulk, the residual energy can accumulate in the irradiated area and eventually heat up the workpiece to the melting point. The observed behavior could also be caused by the above-mentioned post heating effect of the particle-ignited plasma. The magnitude of plasma absorption and heating depends on the particle density inside the capillary at the moment of laser irradiation [14]. The higher the repetition rate, the shorter is the period of time which is available to lower the particle density by precipitation on the borehole walls or removal via convection. A higher density of sources for starting electron emission, in turn, yields a more intensive gas breakdown and causes a larger amount of laser energy to be redistributed from the ultrashort to the nanosecond domain.

10.4.2 Time-Controlled Drilling Strategies

In percussion and helical drilling of metals with ultrashort laser pulses, the highest ablation rate and the highest gas breakdown intensity are generally observed at the very beginning of the process [1, 15, 31]. Consequently, also deteriorating phenomena such as plasma etching or melt production are not uniformly distributed over the machining time, but mainly affect the process during the starting phase. This idea is confirmed by the finding that at high-aspect-ratio microdrilling, the strongest growth of burr at the borehole inlet occurs during the first 1000 pulses [32]. A promising option to defuse the initial phase of the process is to gradually ramp up the governing parameters to their final values. Accordingly, time-controlled drilling strategies were investigated for a variety of process parameters provided by the laser source, the helical drilling optics, the motion system and the feed gas supply, utilizing various ramp patterns, starting values and ramp-up times in each case [29]. As pulse energy and path diameter ramps were found to be most effective, they will be discussed in the following section.

Figure 10.6 illustrates the effect of different pulse energy ramps on machining quality and ionization behavior at helical drilling in 1 mm CrNi steel sheets. Whereas SEM images of the obtained borehole inlets and outlets are shown in the top and middle rows, the bottom row shows the intensity evolution of the optical emissions of the corresponding processes. The optical process emissions, comprising continuum emission from laser-induced plasma and line emission from hot vapor, were recorded using a fast Si photodiode, compare Sect. 10.5.2 and [31]. The borehole depicted on the left was fabricated employing a standard process without ramp, the pulse energy here being 120 μJ from the beginning of laser irradiation. The holes displayed in the middle and on the right of the figure were manufactured utilizing an exponential and a linear pulse energy ramp, respectively. Both ramps started from 3 μJ and were applied over the first 10 s of the process. Since the laser control did not allow the pulse energy to be shifted continuously, the desired ramps were approximated by step functions, thus providing the photodiode signal with a time base. In accordance with the standard process, the final value of the pulse energy ramps was 120 μJ.

It can be seen from the SEM images that drilling with a standard process results in poor inlet quality characterized by an oval cross-section, burr and spatter on the workpiece surface. In contrast, a round inlet orifice with nearly no widening as well as clean edges with only minor nicks are produced with linear pulse energy ramps. Employing exponential ramp patterns, the formation of burr and spatter can be avoided, but the inlet zone is widened in a funnel-shaped manner. In addition, the borehole edge is damaged because of the sharp rise of the pulse energy at the end of the exponential ramp. Also logarithmic ramps are not suitable to adequately enhance machining quality, since their pulse energy development differs only slightly from the standard process. Finally, it is worth noting that the positive effect of pulse energy ramps on machining quality is not restricted to the borehole inlets.

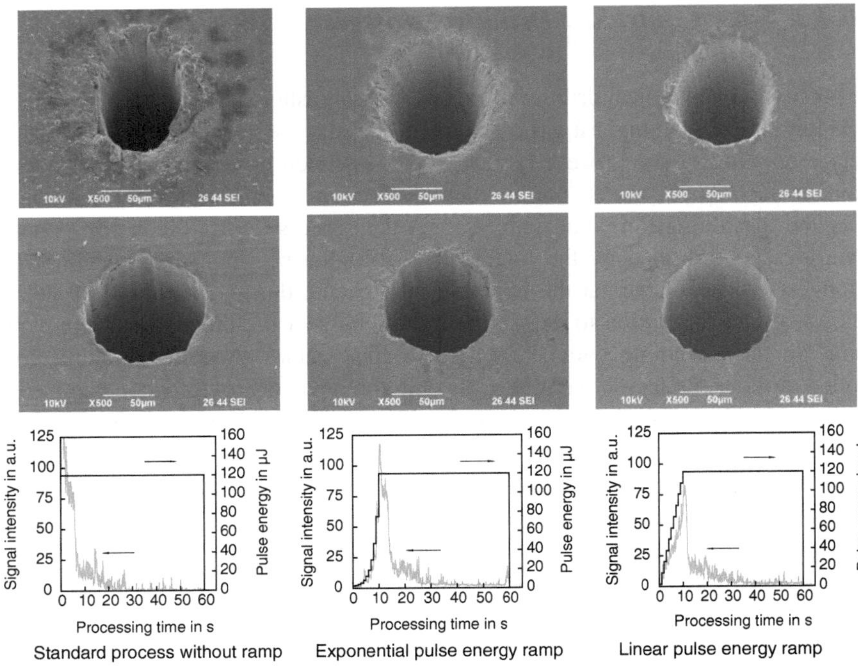

Fig. 10.6 Helical drilling in 1 mm CrNi steel employing different pulse energy ramps with a starting value of 3 μJ and a ramp-up time of 10 s [29]. *Top row* Scanning electron microscope (SEM) images of the borehole inlets. *Middle row* SEM images of the borehole outlets. *Bottom row* Pulse energy (*black*) and intensity evolution of the optical process emissions (*grey*) versus processing time. Wavelength 1030 nm, pulse duration 6 ps, pulse energy 120 μJ, repetition rate 85 kHz, average power 10.3 W, focal length 100 mm, focal diameter 24 μm, energy density 27 J/cm^2, circular polarization, processing time 60 s

In fact, particularly linear energy ramps lead to a considerable improvement of the outlet contour and the wall smoothness.

From the plots of the optical process emissions, it is apparent that plasma ignition during the initial phase of the process is effectively minimized by gradually ramping up the pulse energy. Regardless of the ramp pattern, the intensity of the optical process emissions is proportional to the pulse energy. The standard process, however, starts with heavy ionization and highly intensive optical emissions [31]. In all depicted plots, the moment of breakthrough is indicated by a sharp drop of the photodiode signal, followed by a smooth decrease of the signal intensity during the widening phase. Generally, starting values for pulse energy ramps should be chosen in the order of a few μJ. This corresponds to the energy range which is typically employed in surface processing and provides for negligible melt generation in the inlet region. Adequate ramp-up times usually lie between 15 and 50 % of the total machining time.

Provided that helical drilling is carried out employing a fully automated helical drilling system, also the path diameter can be shifted during the manufacturing process. A successful implementation of a spiral-shaped path of the laser beam at the beginning of irradiation has already been demonstrated in nanosecond drilling [33]. At microdrilling with ultrashort pulses, the effectiveness of path diameter ramps strongly depends on the desired borehole geometry, the processing parameters and the material properties. At drilling in 0.5 mm steel sheets with frequency-doubled radiation, for instance, outstanding quality can be achieved using linear path diameter ramps. Also at drilling in 1 mm sheets with infrared radiation, linear path diameter ramps generally reduce burr and spatter. Yet, intense ionization of the surrounding gas atmosphere during the starting phase of the process cannot be avoided. This causes a widened borehole inlet with eroded edges, accompanied by insufficient widening of the outlet region due to reduced machining time at the final value of the drilling diameter.

A variation of the beam inclination angle during the starting phase has only minor influence on machining quality. At helical drilling with inclined beam, the beam axis performs a precessional motion about the surface normal of the workpiece. Since the pivot of the precession is usually located near the workpiece front surface, the motion of the beam axis undergoes only negligible change in the inlet region if an inclination angle ramp is applied to the drilling process. Thus, inlet quality is not improved in comparison to the standard process, whereas the outlet orifice shows insufficient widening and poor roundness.

Altogether it is shown that by gradually ramping up the pulse energy or the path diameter during the initial phase of a helical drilling process, the borehole quality can be significantly improved. A direct enhancement of process efficiency is usually not obtained by means of time-controlled drilling strategies. However, they allow high pulse energies to be used without quality loss, thus enabling a better exploitation of the available laser power. The correlation between the intensity of the optical process emissions and the magnitude of the deteriorating effects underlines the relevance of laser-induced atmospheric plasma regarding the machining quality and, in addition, supports the assumption of plasma-induced post heating. A further reduction of processing time and a further enhancement of borehole quality can be achieved by making use of assist gases or by drilling with frequency-doubled radiation, as shown in the following sections.

10.4.3 Gas-Assisted Microdrilling

The basic influence of the surrounding gas species on the interaction of ultrashort pulses with matter has been under study since the 1970s. Concerning the role of the gas atmosphere in metal processing with femtosecond pulses, it has been shown in [34, 35] that employing appropriate assist gases, plasma-induced effects such as beam distortion, defocusing or plasma shielding can be significantly reduced, whereas precision and ablation rate can be enhanced. The following section discusses the potential of assist gases with regard to ablation velocity, borehole morphology and machining quality at picosecond microdrilling in 1 mm CrNi steel sheets.

The corresponding experiments were carried out in ambient air at normal pressure and under various assist gases which are typically used in laser applications, including compressed air, nitrogen, argon and helium. The assist gases were delivered by a conical single-hole nozzle which was installed coaxially with the laser beam and between the focusing optics and the workpiece. The nozzle diameter was fixed to 1 mm, whereas the distance of the nozzle outlet to the workpiece surface and the feed gas pressure were optimized experimentally. Depending on the gas type and the desired borehole dimensions, the optimum nozzle distance lies between 0.5 and 1.5 mm, whereas the optimum pressure range is from 3 to 6 bar. Insufficient nozzle distances and high pressure levels usually lead to unwanted widening of the inlet area. Also large nozzle distances can lead to the formation of funnel shaped inlet geometries and, additionally, produce spatter on the workpiece front surface. Insufficient gas pressures result in poor roundness of the borehole orifices and in inefficient widening of the outlet region. It is interesting to note that the described dependencies are exclusively observed at drilling under compressed air, nitrogen and argon, while in case of helium, the influences of the nozzle distance and the feed gas pressure were found to be negligible in the investigated parameter range.

Figure 10.7 depicts five holes manufactured under different atmospheric conditions in a 1 mm steel sheet by means of helical drilling. Using a pulse energy of 200 μJ, the desired borehole geometry could be completed in a processing time of 15 s. It is apparent that the use of helium enables the fabrication of microholes with outstanding machining quality. As can be seen from the transverse section on the far right, the capillary has a regular shape, smooth walls and round inlet and outlet orifices. With an inlet diameter of 100 μm and an outlet diameter of 116 μm, the borehole has a negatively conical geometry, as it is frequently required in fuel injection technology. In contrast, extensive plasma etching and melt generation are

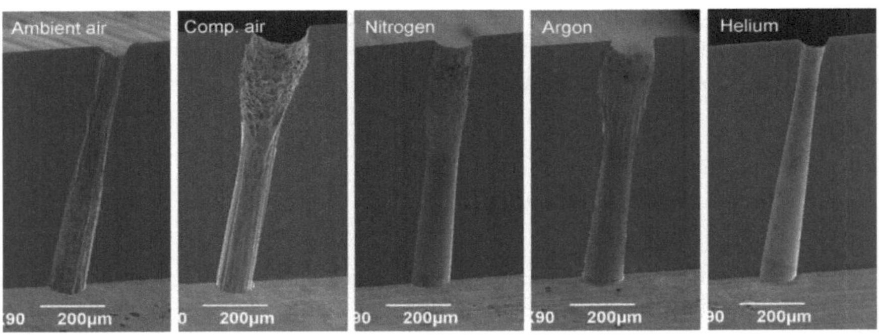

Fig. 10.7 Transverse sections of microholes in 1 mm CrNi steel, drilled in ambient air at normal pressure and under compressed air, nitrogen, argon and helium atmosphere [28]. Wavelength 1030 nm, pulse duration 6 ps, repetition rate 51.5 kHz, pulse energy 200 μJ, focal length 100 mm, focal diameter 24 μm, energy density 44 J/cm^2, power density 7.4•10^{12} W/cm^2, focal position −100 μm, circular polarization, processing time 15 s, feed gas pressure 5 bar, linear pulse energy ramp over 5 s

recorded in the inlet zone of the boreholes manufactured under compressed air, nitrogen and argon. As the plasma cloud builds up in the zone of highest energy density, radial widening is most evident at the beam waist position at 100 μm below the workpiece surface. Drilling in ambient air at normal pressure actually induces only minor plasma etching, but yields eroded inlet edges, an irregular outlet contour as well as melt ridges and grooves stretching out over the entire length of capillary.

Figure 10.8 illustrates the intensity evolution of the optical plasma emissions recorded during the fabrication of the boreholes displayed in Fig. 10.7. It can be seen that drilling under argon atmosphere causes by far the highest signal intensity, i.e. the strongest ionization. Employing nitrogen and helium results in a massive reduction of the signal intensity and a faster breakthrough. Although the helium curve is congruent with the nitrogen curve during the first five seconds of the process, it is remarkable that in case of helium, the breakthrough already occurs before the pulse energy has reached its highest level. Drilling in ambient air leads to continuous plasma ignition over the entire irradiation period. The compressed air curve, which is not shown in the figure, is congruent with the ambient air curve during the energy ramp, while afterwards a persistent signal with about twice the intensity of the ambient air curve is produced until the breakthrough is reached at roundly 14 s.

Particular attention should be given to the fact that at drilling under argon, nitrogen and helium atmosphere, the optical process emissions stop entirely shortly after breakthrough. This can be seen as a confirmation of the assumption that the low threshold gas breakdown which is commonly observed in high-aspect-ratio

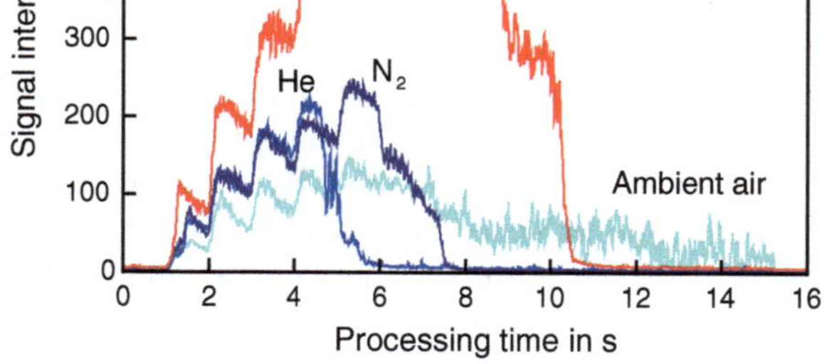

Fig. 10.8 Intensity evolution of the optical process emissions for helical drilling in CrNi steel in ambient air at normal pressure and under nitrogen, argon and helium atmosphere [28]. The irradiation of the workpiece starts at time zero and ends after 15 s. In order to protect the photodiode, the signal intensity of the argon curve has been attenuated to half the original value by inserting a neutral density filter into the monitoring beam path. As already described in Sect. 10.4.2, the pulse energy ramp was approximated by a step function, thus enabling a time-dependent study of the process emissions. Processing parameters as indicated in Fig. 10.7

picosecond microdrilling is facilitated by residual particles inside the capillary. As soon as the outlet orifice is wide enough to allow an effective gas flow, the ablated material is expelled from the borehole. With the gas atmosphere inside and above the capillary now being mostly free of particles, the breakdown threshold increases by one order of magnitude [14, 15], thus exceeding the available energy density. A further ignition of plasma is inhibited, the breakdown spark which was visible at the borehole inlet disappears, and the intensity of the photodiode signal drops to zero. If plasma is continuously ignited as it is the case for drilling in ambient air and under compressed air, the borehole walls are deteriorated along the entire length of the capillary. If the outlet region is widened to its final dimensions after breakthrough has occurred and ionization has stopped, as observed for argon and nitrogen, clean and regular borehole walls can be produced in the lower half of the sheet. Overall, it is evident again that intense process emissions correlate with strong ionization, plasma etching and melt formation.

The major difference between the depicted borehole morphologies and the corresponding optical process emissions confirms the essential role of the atmospheric conditions at high-aspect-ratio microdrilling with ultrashort pulses. Assist gases strongly influence the characteristics of the dissociation, ionization and recombination processes which are involved in the formation and the decay of laser-induced plasma. In addition, their thermal conductivity and their flow pattern affect the heat dissipation from the borehole and from the plasma cloud. Finally, assist gases enable or inhibit chemical reactions such as oxidation processes. Due to the major impact of laser-induced plasma on the drilling process, the influence of the assist gas on plasma generation is of particular interest. However, discussing the ionization behavior of different gas types, it is certainly not sufficient to exclusively consider the first ionization energies. The electron number density in gas plasmas ignited during picosecond laser ablation of metals with power densities in the range of 10^{12}–10^{13} W/cm^2 was measured to be as high as 10^{20} cm^{-3} [19, 36]. This means that investigating the atmospheric breakdown at picosecond microdrilling, one has to take into account multiple ionization of gas atoms and, consequently, higher ionization energies. This approach might contribute to explain the poor performance of argon compared to nitrogen.

In summary, it can be seen that the application of adequate assist gases not only enables a substantial improvement of machining quality and borehole shape, but also a significant reduction of processing time. By allowing the use of high pulse energies and controlling the influence of laser-induce plasma, assist gases particularly accelerate the widening process of the borehole outlet, which is the most time-consuming part of the entire drilling process. In this context, it has been shown that with a total processing time of 15 s for the completion of a negatively conical borehole in a 1 mm CrNi steel sheet, the picosecond drilling process has become compatible to cycle times in mass production. Besides helium, also nitrogen is well suited as an assist gas. High-precision microdrilling should not be performed in low-pressure atmosphere since although it supports an efficient completion of the basic borehole geometry, large burrs with up to 100 μm in height are generally formed under vacuum conditions.

10.4.4 Microdrilling with Frequency-Doubled Radiation

Modern picosecond laser sources provide average powers up to 50 W combined with repetition rates of several 100 kHz and excellent beam quality. In addition, they feature harmonic generation units which convert the fundamental infrared (IR) radiation into the second, third or even fourth harmonic by frequency doubling and sum frequency generation in nonlinear crystals. Employing frequency-doubled radiation for microdrilling applications can be expected to be beneficial from both productivity and quality aspects.

First of all, frequency-doubled radiation offers better focusability than IR radiation [27, 37]. This provides for a larger energy density in the interaction zone and enables a potential increase in productivity [5, 15, 17]. Furthermore, the optical absorption length is proportional to the wavelength. Since in machining of metals with ultrashort laser pulses, the thickness of the heated material layer is determined by the absorption length [23], it can be assumed that the pulse energy is deposited in a smaller volume for frequency-doubled radiation. This, in turn, causes a decline of the threshold fluence [38] and leads to a more efficient widening of the borehole by the wings of the Gaussian profile, which is decisive for the formation of the borehole shape in the final phase of the drilling process. As regards the wavelength dependence of absorptivity, the calculated values for pure iron at room temperature and perpendicular incidence are roundly 44 % at 532 nm compared to 36 % at 1064 nm [39]. However, it is not clear whether the absorptivity is significantly higher for the second harmonic than for the IR radiation in the application discussed here. This is because the absorptivity of laser radiation in metal surfaces does not only depend on wavelength, but on a range of influencing variables, such as workpiece temperature, angle of incidence, alloying elements, surface roughness and oxidation (see [39] for a general discussion). These variables, if at all, are only known within wide limits in the investigated regime.

Since the transfer of energy from the laser pulses to the workpiece is largely affected by laser-generated plasma, it is of particular interest that the plasma intensity can be minimized by reducing the processing wavelength. As stated above, inverse bremsstrahlung (IB) is the principal mechanism for the absorption of photon energy by free electrons in picosecond laser drilling, and the IB absorption coefficient varies with the third power of the wavelength ($\alpha_{IB} \sim \lambda^3$) in the investigated regime [39]. This means that collisional heating and impact ionization can be significantly reduced by employing the second harmonic. Moreover, the use of frequency-doubled radiation reduces scattering and beam distortion due to nonlinear effects occurring in the focal region [40].

Figure 10.9 illustrates the productivity improvement which can be achieved with the second harmonic. It can be seen that using frequency-doubled instead of IR radiation for the fabrication of a 100 µm hole in 1 mm CrNi steel, the completion time can be reduced by a factor of four. At 532 nm, the outlet is widened to the desired diameter as early as 60 s after process initiation compared to 240 s at

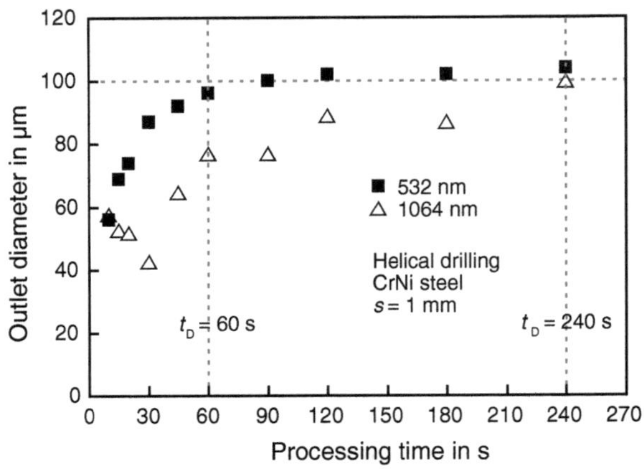

Fig. 10.9 Outlet diameter versus drilling time for helical drilling in 1 mm CrNi steel at 1064 and 532 nm with equal pulse energies and equal beam diameters on the focusing optics [27, 37]. 1064 and 532 nm: repetition rate 40 kHz, pulse energy 80 µJ, focal length 100 mm, no assist gas, no ramps, circular polarization. 1064 nm: pulse duration 12 ps, focal diameter 19.8 ± 0.6 µm, energy density 26 J/cm^2. 532 nm: pulse duration 10 ps, focal diameter 15.4 ± 1 µm, energy density 43 J/cm^2

1064 nm. Moreover, it is shown that utilizing the second harmonic, the drilling process has a considerably higher reproducibility due to steadier progression.

In addition to the reduction of processing time, drilling with frequency-doubled radiation instead of infrared radiation enables a substantial enhancement of machining quality. Figure 10.10 compares two holes manufactured at 1064 and 532 nm, respectively, in a 1 mm CrNi steel sheet by means of helical drilling in ambient air at atmospheric pressure. The microhole fabricated with the second harmonic shows a regular shape, smooth borehole walls and an absence of recast. Contrary to this, at fundamental wavelength, melt ridges are formed on the borehole walls, and a burr is generated at the inlet. The presence of melt at 1064 nm explains the bumpy widening behavior as it is displayed in Fig. 10.9. At 532 nm however, with the drilling process not being affected by the randomly distributed melt structures inside the capillary, a smooth development of the outlet diameter takes place. The depicted holes can also be distinguished by geometrical criteria. Although the machining time was 30 s longer at 1064 nm than at 532 nm, the outlet diameter of the hole fabricated with IR radiation remains at 85 µm, while the second harmonic widens the outlet to 95 µm, thus producing a negative conical shape. At 1064 nm, the capillary is widened in the upper zone, whereas at 532 nm, the borehole shape seems to reproduce the beam waist.

As the threshold intensity for a particle-ignited breakdown of the air inside the capillary is exceeded for 1064 and 532 nm, laser-induced plasma is observed at both wavelengths. At fundamental wavelength, a considerable part of the pulse energy is absorbed by the plasma, leading to radial etching, the formation of melt due to prolonged surface heating as well as insufficient widening of the outlet,

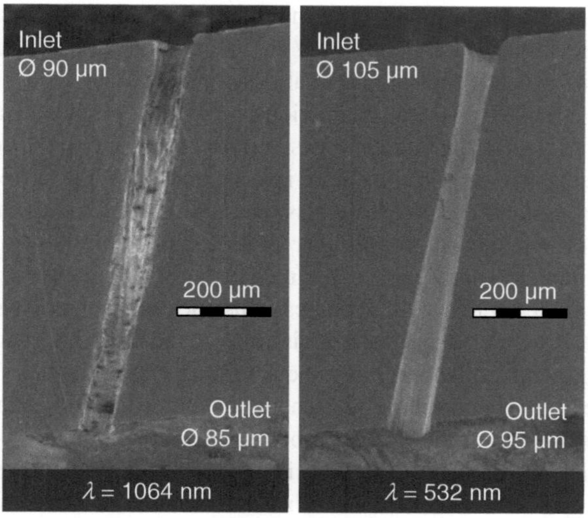

Fig. 10.10 Transverse sections of microholes in 1 mm CrNi steel, drilled at 1064 nm (*left*) and 532 nm (*right*) with equal beam diameters on the focusing optics [27, 37]. The depicted diameters are each mean values averaged from the major and minor axes of the cross-section. 1064 and 532 nm: repetition rate 30 kHz, pulse energy 80 µJ, focal length 100 mm, focal position 200 µm below workpiece surface, helical drilling, no assist gas, no ramps, circular polarization. 1064 nm: processing time 120 s. 532 nm: processing time 90 s. Pulse durations, focal diameters and energy densities see Fig. 10.9

accompanied by poor roundness. As described in [15], the plasma cloud builds up in the zone of highest energy density. Figure 10.10 confirms that the etching zone corresponds to the position of the beam waist, here being located 200 µm below the workpiece surface. At 532 nm, a smaller fraction of the pulse energy is transferred to the plasma, causing less radial expansion in the upper region of the capillary and leaving more energy available in the outlet zone. Moreover, the reduced plasma absorption at 532 nm leads to less post heating of the bulk material, which finally results in reduced melt generation.

Altogether, employing frequency-doubled radiation ensures a high quality level even without sophisticated drilling methods or extra processing technology. A further improvement of process efficiency and precision can be expected from combining the second harmonic with the abovementioned techniques and from using higher harmonics in the UV range.

10.5 Online Quality Assurance at Microdrilling of Metals

As stated above, microdrilling with ultrashort laser pulses (USP) enables the fabrication of high-precision boreholes with outstanding machining quality and high contour accuracy. Independent of this, current quality standards frequently require the implementation of real-time online monitoring systems. This section discusses different concepts for process monitoring at USP microdrilling and their feasibility regarding online process control and quality assurance.

So far, the complex physical interactions between an incident laser beam, the irradiated material and the ambient gas atmosphere have not been understood entirely. Consequently, only rudimentary information about the correlation between the drilling process and the process emissions is available. As a simple transfer of existing monitoring concepts from longer pulse durations to ultrashort pulses is not possible, the development and experimental verification of new monitoring approaches is needed. First of all, technically usable and analyzable signals have to be identified and extracted from the emissions of the USP microdrilling process. Furthermore, particular attention has to be paid to the interpretation of the recorded signals. As the borehole morphology is changing and the ablation characteristics are rapidly fluctuating during the drilling process, also the emitted process signals are highly dynamic in space and time and, in addition, subject to enormous fluctuations. Finally, specific demands concerning the location of the sensor units within the machining setup have to be met. Since the non-irradiated side of most workpieces is not accessible for installing sensors, the monitoring systems have to be installed above the workpiece.

10.5.1 Process Monitoring via Laser-Induced Shock Waves

Ultrafast laser-induced heating of solids and ceramics in an ambient gas produces a hemispherical shock wave leaving the target surface and propagating into the ambient atmosphere at ultrasonic speed. This shock wave is caused by the explosive evaporation of the ablated material. Using a high-speed schlieren photography technique, the propagation characteristics of shock waves from deep capillaries can be analyzed during the drilling process. The acquired images show the morphology and the temporal evolution of the shock waves, thus allowing the observation of different stages of the microdrilling process [41, 42]. Figure 10.11 depicts schlieren

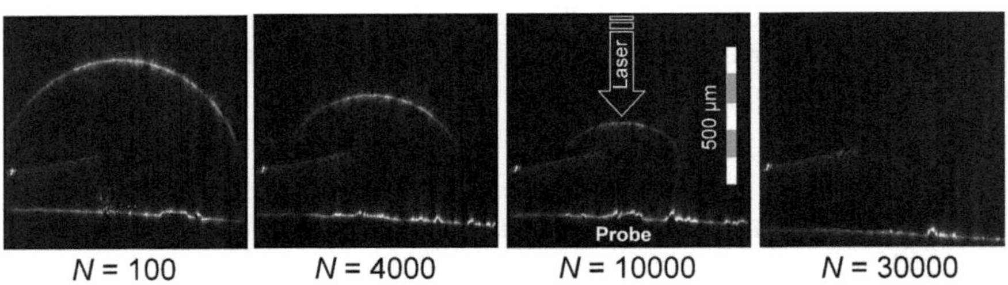

Fig. 10.11 Schlieren photographs showing the propagation of laser-induced shock waves in ambient atmosphere from different depths of the borehole. N denotes the number of pulses. The delay time with respect to the processing pulse was 500 ns in all cases

photographs of shock waves originating from different borehole depths. In order to achieve information about the borehole depth, all images are taken at a fixed time delay after the interaction between the laser pulse and the target material.

It can be stated that the shock wave radius significantly decreases with increasing capillary depth. One of the reasons for this behavior is a decrease of the ablation rate with growing borehole depth, leading to a decrease of the energy content of the shock wave. This, in turn, causes a reduction of the velocity of the shock wave front, according to the theory of gas-dynamic blast wave expansion. A further effect which has to be taken into account is the extended propagation time of the shock front from the tip of the drilling capillary to the target surface with increasing borehole depth. The combination of both effects leads to the observed propagation behavior of laser induced shock waves.

In order to judge whether the observed correlation is suitable for process monitoring, the shock wave radius and the depth of the generated boreholes can be plotted against the number of processing pulses (see Fig. 10.12). Over roughly 2500 pulses, a strong decrease of the shock wave radius from 550 to 350 μm is obtained. At the same time, the borehole depth increases to 500 μm, and breakthrough occurs. Figure 10.13 displays the same data as Fig. 10.12, but with the shock wave radius directly plotted against the borehole depth. The linear correlation between the two parameters provides a precise and non-contact measurement method for the real-time detection of the drilling depth.

Analogue experiments were carried out using various pulse durations in the range between 100 fs and 5 ps. Figure 10.14 displays the results acquired for drilling in 500 μm steel sheets at three different pulse durations and at a constant energy density of 80 J/cm^2. Irrespective of the pulse duration, the linear dependence between the shock wave radius and the capillary depth is clearly visible. These results confirm that for microdrilling in metals using ultrashort laser pulses, the borehole depth can be monitored online from the front side of the target.

The results presented above exclusively refer to the percussion drilling technique. As shown in previous chapters, however, the borehole quality can be significantly enhanced by using the helical drilling technique. Figure 10.15 illustrates the evolution of the shock wave radius versus the borehole depth during helical

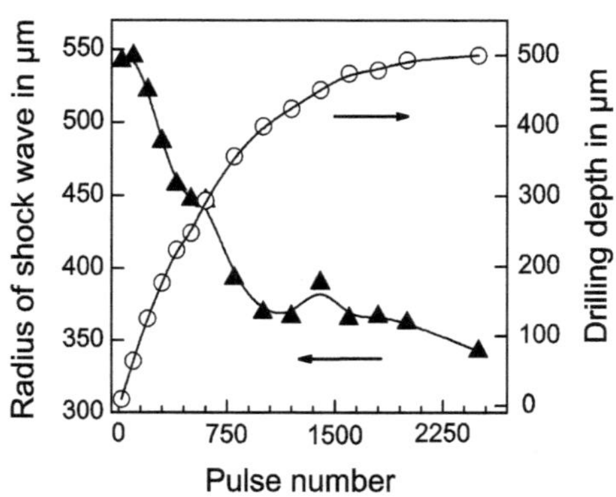

Fig. 10.12 Evolution of shock wave radius and borehole depth as a function of the pulse number for drilling in a 0.5 mm steel sheet

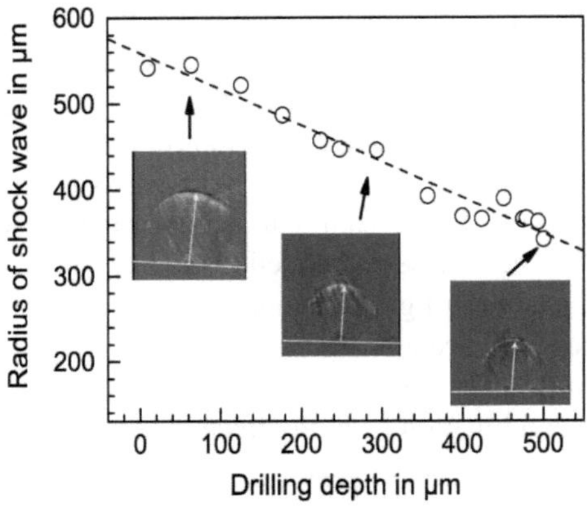

Fig. 10.13 Linear relationship between shock wave radius and borehole depth. The shock wave radius was measured from schlieren photographs which were taken at a fixed delay of 500 ns after each laser pulse

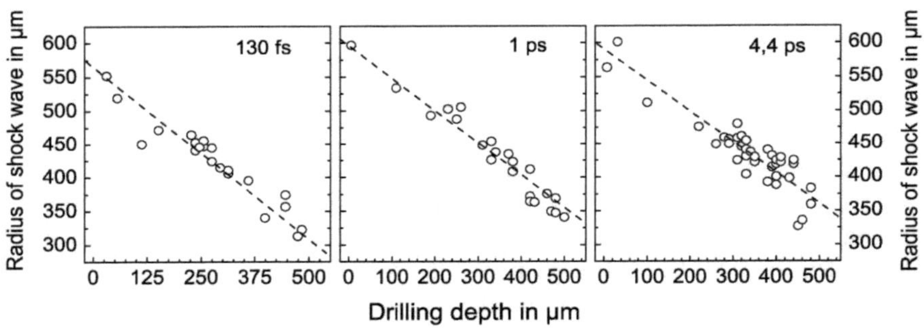

Fig. 10.14 Linear relationship between shock wave radius and borehole depth for different pulse durations. Wavelength 800 nm, repetition rate 1 kHz, focal diameter 18 µm, sheet thickness 500 µm, delay time 500 ns

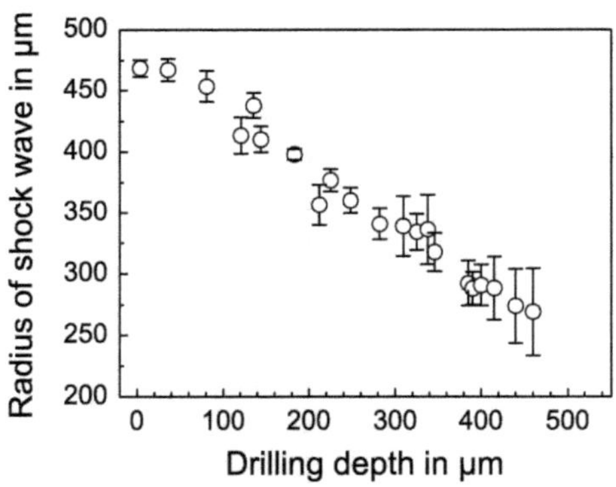

Fig. 10.15 Linear relationship between shock wave radius and borehole depth for drilling with the helical drilling technique. Wavelength 800 nm, repetition rate 1 kHz, focal diameter 24 µm, sheet thickness 500 µm, delay time 270 ns

drilling in steel. In analogy to percussion drilling, the radius of the shock wave gradually decreases with increasing capillary depth. Also, a linear dependence can be observed over the entire depth of the workpiece. This is evidence that the presented monitoring method can be applied not only at different pulse durations, but also at different drilling techniques.

In addition to the borehole depth, this detection method predicts the termination of the drilling process, since the shockwave radius decreases significantly. However, the detection of breakthrough at drilling in metals only works securely in case of the helical drilling technique. For drilling in ceramics, both helical drilling and percussion drilling allow the detection of breakthrough. CrNi steel does not allow any breakthrough detection because of an irregular opening mechanism at the workpiece backside.

Apart from detecting the borehole depth and the process termination, the presented monitoring system is able to monitor the outlet diameter. As the outlet opening widens, the radius of the shock wave decreases proportionally. With this linear correlation, it is possible to measure the outlet diameter, which is a decisive geometric feature for many applications.

10.5.2 Real-Time Monitoring and Control Using the Optical Process Emissions

The emissions of the plasma which is generated during USP microdrilling have been studied by an optical method. It has been found that the analysis of the optical plasma emissions during the process permits not only the estimation of the plasma temperature and the free electron density, but can also be of significant interest as a possible approach for process control [43].

The instantaneous ablation rate was found to correlate with the temporal integral of the optical process emissions. Tracking the temporal change of these emissions combined with subsequent algorithmic analysis allows to pinpoint the breakthrough as well as the completion of the outlet widening phase. A coaxial photodiode was used in order to monitor the optical process emissions. The coaxial alignment is a convenient option for monitoring process signals emitted during the fabrication of microholes, even at high aspect ratios. Due to the high reproducibility and validity of the inspected process emissions, a real-time monitoring system based on coaxial detection could be developed and implemented.

Figure 10.16 shows the graphical interface of the evaluation unit which was part of the monitoring system. The signals of the coaxial sensor and a reference photodiode are displayed in separate windows. The reference photodiode was positioned on the rear side of the workpiece in order to acquire the laser radiation which is transmitted through the open capillary after the penetration of the workpiece. The corresponding signal is shown in the top window. It was used as a reference to identify the moment of breakthrough and to measure the widening dynamics of the

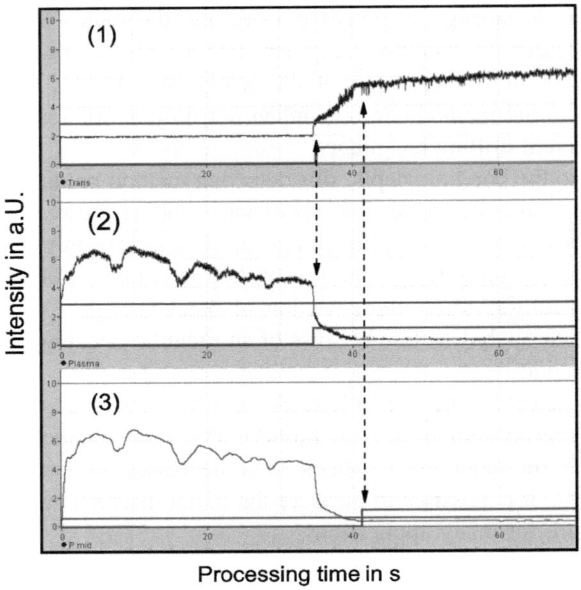

Fig. 10.16 Characteristic evolution of time-integrated photodiode signals acquired with an evaluating unit during microdrilling in CrNi steel with ultrashort laser pulses. *Channel 1* Reference signal of a photodiode positioned at the back side of the sample. *Channel 2* Signal of the sensor photodiode positioned in coaxial alignment to the laser beam, used for the detection of breakthrough. *Channel 3* Same signal as displayed in channel 2, but used to define the completion of the process. Wavelength 800 nm, pulse duration 5 ps, repetition rate 1 kHz, focal diameter 18 μm, energy density 170 J/cm^2, sheet thickness 500 μm

borehole outlet. The moment of breakthrough can be identified via the rise of the signal intensity of the reference photodiode at about 34 s.

The signal of the sensor photodiode is shown in the middle and in the bottom windows. Depending on the monitored process phase, different analyzing algorithms can be applied to the sensor signal. Channel 2 is used to determine the moment of breakthrough. The breakthrough is clearly visible from the sharp drop of the signal. Having detected this drop, the unit generates a control signal which is shown as a step function in the middle window. The rise of the control signal value from 0 to 1 denotes the detection of breakthrough. The comparison of the real moment of breakthrough (channel 1) and the time which was detected by the evaluating unit (channel 2) shows a good correlation. Broad experimental results confirm the high accuracy of this real-time monitoring approach for breakthrough detection.

As can be seen from the bottom window, the sensor photodiode can also be employed for the detection of the end of the drilling process. Figure 10.16 shows that the control signal generated by channel 3 is in good temporal agreement with the attainment of the peak intensity by the signal of the reference diode (channel 1) and the beginning of saturation of the transmitted laser intensity. The saturation of

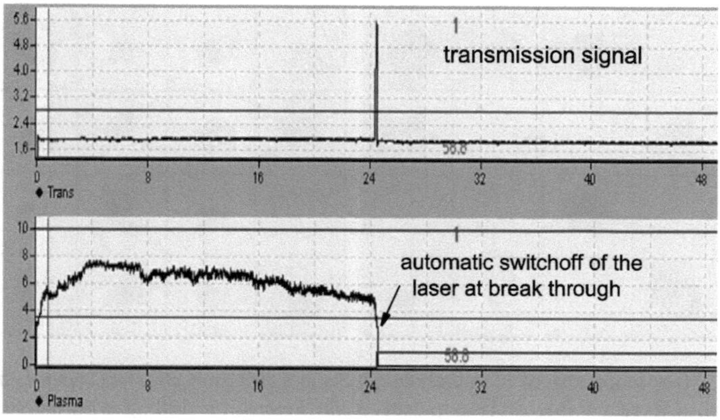

Fig. 10.17 Evolution of time-integrated photodiode signals acquired during microdrilling in CrNi steel with automatic process termination. *Top window* Reference photodiode. *Bottom window* Coaxial photodiode used for breakthrough detection. Wavelength 800 nm, pulse duration 5 ps, repetition rate 1 kHz, focal diameter 18 μm, energy density 170 J/cm^2

the laser transmission indicates the termination of the outlet widening phase, i.e. the end of the drilling process.

In order to control the drilling process, the evaluating unit can be programmed to generate a 24 V control signal immediately after the detection of breakthrough via the coaxial photodiode. The 24 V signal can be used to stop the laser emission and, coupled with this, the drilling process. Figure 10.17 shows that as soon as the defined breakthrough criterion is met, the drilling process is stopped, and the optical emissions recorded by the coaxial photodiode drop to zero (see bottom window). At the same moment, the signal of the reference photodiode shows a narrow intensity peak (see top window). This means that the laser emission was stopped immediately after the penetration of the workpiece, and that an automatic process termination was established.

Figure 10.18 demonstrates the effect of the automatic process termination on percussion drilling in 500 μm CrNi steel. The microscope images on the left side show the outlets of boreholes fabricated with a fixed number of laser pulses and without using the monitoring system. It is apparent that drilling with a fixed number of pulses leads to a highly irregular machining result. While in some cases, the workpiece is not penetrated at all, the borehole outlets are widened to large diameters in other cases. Using the monitoring system to stop the laser emission after breakthrough, however, the drilling process can be terminated reproducibly (see right image in Fig. 10.18). Apart from one borehole which has not been completed because of a temporary drop of signal intensity before the actual breakthrough, the depicted borehole outlets have similar characteristics. Their small diameter and the substantial melt deposits near the outlet orifices can be interpreted as an indicator for process termination just after the moment of breakthrough.

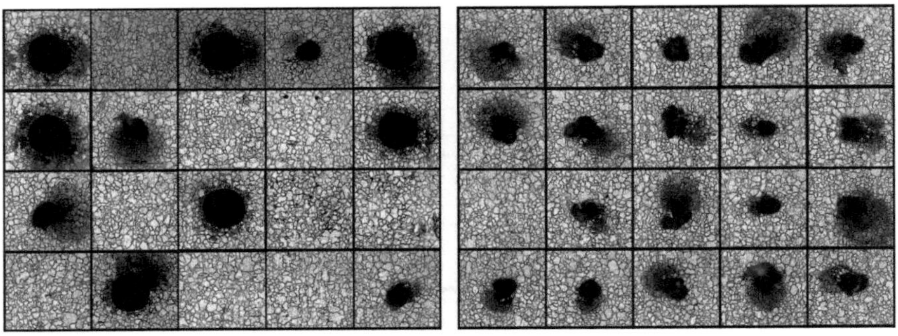

Fig. 10.18 Microscope images of microhole outlets in a 500 μm CrNi steel sheet. *Left* Percussion drilling with a fixed number of laser pulses. *Right* Percussion drilling with automatic process termination after breakthrough. Wavelength 800 nm, pulse duration 5 ps, repetition rate 1 kHz, focal diameter 18 μm, energy density 170 J/cm^2, focal position −200 μm

10.5.3 Other Methods for Online Monitoring

In addition to camera-based imaging of laser-induced shock waves and coaxial monitoring of optical process emissions, other techniques were developed and tested [44]. Measuring the ratio of ionized particles within the plasma plume using a conducting electrode as well as capturing the airborne and structure-borne acoustic process emissions are quite promising means for detecting the breakthrough or for monitoring the widening of the borehole outlet. The first experimental results demonstrate the potential for the development of further monitoring approaches for USP microdrilling and ablation processes. Yet, this field of research is still at its onset, and further experimental work is necessary.

10.6 Industrial Microdrilling of Diesel Injector Nozzles

A prominent application of drilling with ultrashort laser pulses is the fabrication of precise and specifically shaped boreholes in diesel fuel injector nozzles (see illustration in Fig. 10.19). This application is currently pushed by the perspective to fulfill future emission standards by optimizing the combustion process in diesel engines of cars and trucks. Such an inner-engine optimization is a possible alternative to the treatment of the combustion products in the exhaust line using e.g. particle filters or catalyzers that reduce nitrogen oxide (NO_x) emissions.

From these boundary conditions, the functional requirements of future diesel systems and diesel nozzles can be deduced. In the next step, the functional requirements have to be transferred into geometrical requirements so that finally the borehole dimensions and the necessary precision can be defined. However, a potential use of laser drilling with ultrashort laser pulses in mass production does

Fig. 10.19 Section of a diesel engine with glow plug and diesel injector

not only depend on the fulfillment of the functional requirements, but also on the manufacturing costs which must not exceed a certain target price. In the case of the diesel nozzle, the maximum additional costs for an improved combustion are determined by the costs of a corresponding exhaust gas treatment system. These conditions yield a processing time of not more than 12 s per borehole, provided that post processing such as deburring is not necessary.

The material of a diesel nozzle is hardened chromium-nickel (CrNi) steel. Depending on the nozzle type, the spray hole depth is about 1 mm. The hole position at the surface of the geometrically complex nozzle tip has to be adjusted with a tolerance of only a few microns. The requirements concerning the spray hole shape are equally high, which means that the standard deviations of hole diameter and straightness have to be below 1 μm. Another challenging issue is the protection of the inner walls of the nozzle main bore, which can be damaged by laser radiation transmitted through the spray holes after breakthrough, i.e. during the widening phase of the drilling process. This problem is illustrated by the cross-section of a typical diesel nozzle displayed in Fig. 10.20.

With the comprehensive understanding of process fundamentals and processing techniques presented in this chapter, it was possible to transfer the experimental results from metal sheets to a demonstrator diesel nozzle. The highest challenges included the protection of the inner walls of the diesel nozzle and the reduction of the processing time to less than 12 s while maintaining the required precision. As can be seen from Fig. 10.21, the required specifications have successfully been implemented.

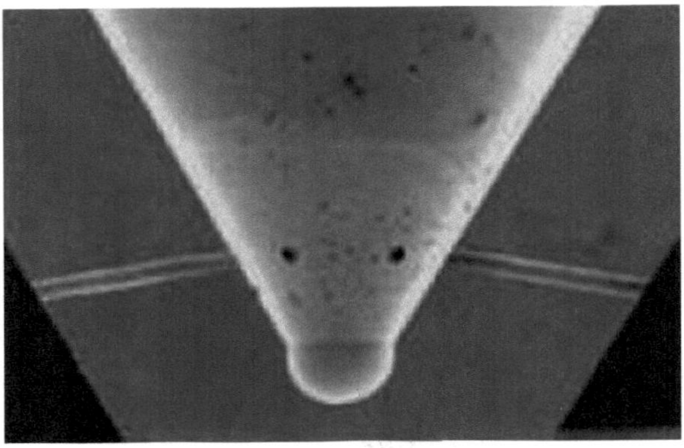

Fig. 10.20 Cross-section of the head of a diesel nozzle. The inner wall of the depicted nozzle does not show any defects on the opposite side of the borehole outlets

Fig. 10.21 *Left* Outer side of the spray hole (entrance of laser beam). *Middle* Inner side of spray hole (exit of laser beam). *Right* View inside the nozzle

10.7 Conclusions

This chapter shows that high precision microholes in metals can be fabricated reproducibly using picosecond laser pulses together with the helical drilling technique. Cylindrical and negatively conical boreholes with diameters in the range of 100 μm and excellent machining quality have been manufactured in 1 mm CrNi steel sheets (see Fig. 10.22) and in demonstrator diesel fuel injector nozzles. The productivity of the helical drilling process could be enhanced to an extent that the processing time of the fabricated spray holes meets the cycle time requirements of serial production. Several characteristics of the manufactured microholes and the drilling process such as the capillary depth, the moment of breakthrough or the completion of the widening phase could be monitored or even controlled via the observation of laser-induced shock waves or the optical process emissions.

The results of the reported experiments suggest that the use of high pulse energies instead of high repetition rates is beneficial from both quality and efficiency aspects. The quality and the roundness of the borehole inlet and outlet as well as the wall smoothness can be significantly improved by gradually increasing

Fig. 10.22 Transverse sections of microholes in a 1 mm CrNi steel sheet. Inlet diameter 90 μm, outlet diameter 110 μm, wavelength 1030 nm, pulse duration 6 ps, helical drilling, processing time 15 s. The combination of the helical drilling technique with focal diameters <10 μm allows to reduce the borehole diameter far below the depicted range. Microholes with a diameter of 25 μm and an aspect ratio of 7:1 have recently been manufactured in spinning nozzles consisting of 240 μm AuPt sheets [45]

the pulse energy in the form of a linear ramp over the first seconds of the drilling process. Path diameter ramps can be employed to avoid the formation of burr and spatter. A benefit of repetition rate or inclination angle ramps in burr height, machining quality or widening behaviour could not be identified in this analysis. Also a focal shift during the drilling process has not been found to cause any positive effects. Instead, it could be examined that a focal shift into the workpiece leads to an unintended enlargement of the inlet zone. Assist gases with high ionization energies have a significant impact on laser-material-atmosphere interaction. They can be utilized to raise productivity and to enhance machining quality. Drilling with frequency doubled radiation ensures a high quality and productivity level, even without the use of complementary processing techniques. This effect can be explained by a substantial reduction of laser-plasma interaction compared to drilling with infrared radiation.

Finally, it can be stated that borehole quality and process efficiency are decisively determined by the magnitude of ionization of the gas atmosphere inside and above the capillary. The observed correlation between the production of melt and the intensity of the optical process emissions supports the theory of post heating caused by laser-induced plasma. Plasma control is thus the key objective for further process development in the field of high aspect-ratio picosecond microdrilling in metals.

References

1. F. Dausinger, F. Lichtner, H. Lubatschowski, Femtosecond technology for technical and medical applications. Top. Appl. Phys. **96** (2004)
2. H. Tönshoff, C. Momma, A. Ostendorf, S. Nolte, G. Kamlage, Microdrilling of metals with ultrashort laser pulses. J. Laser Appl. **12**(1), 23–27 (2000)
3. E. Weynant, D. Antonov, J. Tremblay, P. Zivojinovic, E. Mottay, High aspect ratio deep micromachining in a mono crystalline copper-based shape memory alloy using a high-energy femtosecond laser, in *Proceedings of the Fourth International WLT-Conference on Lasers in Manufacturing*, ed. by F. Vollertsen, C. Emmelmann, M. Schmidt, A. Otto (AT-Fachverlag, Stuttgart, 2007), pp. 635–638
4. T. Bauer, J. König, Applications and perspectives of ultrashort pulsed lasers, *presented at the Stuttgart Laser Technology Forum*, Stuttgart, Germany, 8–10 June 2010
5. F. Dausinger, Femtosecond technology for precision manufacturing: Fundamental and technical aspects. Proc. SPIE **4830**, 471–478 (2003)
6. C. Föhl, F. Dausinger, High precision deep drilling with ultrashort pulses. Proc. SPIE **5063**, 346–351 (2003)
7. R. Knappe, A. Nebel, High-speed micromachining with high-power picosecond ultraviolet lasers. Proc. SPIE, Solid State Lasers XVII: Technol. Devices **6871,** 687121 (2008)
8. J. Kleinbauer, D. Eckert, S. Weiler, D. Sutter, 80 W ultrafast CPA-free disk laser. Proc. SPIE, Solid State Lasers XVII: Technol. Devices **6871**, 68711B (2008)
9. W. Wawers, Präzisions-Wendelbohren mit Laserstrahlung, *Aachen University, Dissertation* (Shaker, Aachen, 2008)
10. G. Kühnle, N. Streibl, Optische Vorrichtung zum Bohren mittels Laserstrahls, Offenlegungsschrift DE19741029A1 (1997)
11. Laser- und Medizin-Technologie GmbH, Vorrichtung zum Führen eines Lichtstrahls, Gebrauchsmusterschrift DE202008017745U1 (2008)
12. A. Michalowski, Untersuchungen zur Mikrobearbeitung von Stahl mit ultrakurzen Laserpulsen, *University of Stuttgart, Dissertation* (Herbert Utz, München, 2014)
13. Courtesy of GFH GmbH, Deggendorf, Germany
14. S. Klimentov, T. Kononenko, P. Pivovarov, S. Garnov, V. Konov, A. Prokhorov, D. Breitling, F. Dausinger, The role of plasma in ablation of materials by ultrashort laser pulses, in *Proceedings First Intl. WLT-Conference on Lasers in Manufacturing* (AT-Fachverlag, Stuttgart, 2001), pp. 273–283
15. T. Kononenko, S. Klimentov, V. Konov, P. Pivovarov, S. Garnov, F. Dausinger, D. Breitling, Propagation of short-pulsed laser radiation and stages of ablative deep-channel formation. Proc. SPIE **4274**, 248–257 (2001)
16. S. Klimentov, S. Garnov, V. Konov, T. Kononenko, P. Pivovarov, O. Tsarkova, D. Breitling, F. Dausinger, Effect of low-threshold air breakdown on material ablation by short laser pulses. Phys. Wave Phenomena **15**(1), 1–11 (2007)
17. D. Breitling, A. Ruf, F. Dausinger, Fundamental aspects in machining of metals with short and ultrashort laser pulses. Proc. SPIE **5339**, 49–63 (2004)
18. X. Mao, W. Chan, M. Shannon, R. Russo, Plasma shielding during picosecond laser sampling of solid materials by ablation in He versus Ar atmosphere. J. Appl. Phys. **74**(8), 4915–4922 (1993)
19. S. Mao, X. Mao, R. Greif, R. Russo, Initiation of an early-stage plasma during picosecond laser ablation of solids. Appl. Phys. Lett. **77**(16), 2464–2466 (2000)
20. R. Russo, Laser ablation. Appl. Spectrosc. **49**(9), 14A–28A (1995)
21. B. Le Drogoff, J. Margot, F. Vidal, S. Laville, M. Chaker, M. Sabsabi, T. Johnston, O. Barthélemy, Influence of the laser pulse duration on laser-produced plasma properties. Plasma Sour. Sci. Technol. **13**, 223–230 (2004)
22. T. Kononenko, D. Walter, V. Konov, F. Dausinger, Optical spectroscopy of laser plasma in a deep crater. Quantum Electron. **39**(4), 328–332 (2009)

23. T. Kononenko, V. Konov, S. Garnov, R. Danielius, A. Piskarskas, G. Tamoshauskas, F. Dausinger, Comparative study of the ablation of materials by femtosecond and pico- or nanosecond laser pulses. Quantum Electron. **29**(8), 724–728 (1999)
24. J. Chang, B. Warner, E. Dragon, M. Martinez, Precision micromachining with pulsed green lasers. J. Laser Appl. **10**(6), 285–291 (1998)
25. M. Jackson, W. O'Neill, Laser micro-drilling of tool steel using Nd:YAG lasers. J. Mater. Process. Technol. **142**, 517–525 (2003)
26. F. Dausinger, T. Abeln, D. Breitling, J. Radtke, V. Konov, S. Garnov, S. Klimentov, T. Kononenko, O. Tsarkova, Bohren keramischer Werkstoffe mit Kurzpuls-Festkörper-lasern. LaserOpto **31**(3), 78–85 (1999)
27. M. Kraus, S. Collmer, S. Sommer, A. Michalowski, F. Dausinger, Microdrilling in steel with ultrashort laser pulses at 1064 nm and 532 nm, in *Proceedings Fourth International WLT-Conference on Lasers in Manufacturing*, ed. by F. Vollertsen, C. Emmelmann, M. Schmidt, A. Otto (AT-Fachverlag, Stuttgart, 2007), pp. 639–644
28. M. Kraus, C. Markmann, A. Michalowski, R. Weber, T. Graf, Gas-assisted microdrilling in steel with ultrashort pulsed laser radiation, *presented at the 11th International Symposium on Laser Precision Microfabrication*, Stuttgart, Germany, 7–10 June 2010
29. M. Kraus, B. Neubig, A. Michalowski, R. Weber, T. Graf, Time-controlled strategies for microdrilling in steel with ultrashort pulsed laser radiation, *presented at the Stuttgart Laser Technology Forum*, Stuttgart, Germany, 8–10 June 2010
30. A. Ancona, F. Röser, K. Rademaker, J. Limpert, S. Nolte, A. Tünnermann, High speed laser drilling of metals using a high repetition rate, high average power ultrafast fiber CPA system. Opt. Express **16**, 8958–8968 (2008)
31. D. Walter, A. Michalowski, F. Dausinger, Real-time monitoring and control of micro-drilling process using optical emission of material-vapor plasma, in *Proceedings Fourth International WLT-Conference on Lasers in Manufacturing*, ed. by F. Vollertsen, C. Emmelmann, M. Schmidt, A. Otto (AT-Fachverlag, Stuttgart, 2007), pp. 569–573
32. A. Michalowski, D. Walter, F. Dausinger, T. Graf, Melt dynamics and hole formation during drilling with ultrashort pulses. J. Laser Micro/Nanoeng. **3**, 211–215 (2008)
33. G. Rutherford, D. Karnakis, A. Webb, M. Knowles, Optimization of the laser drilling process for fuel injection components, in *Proceedings Advanced Laser Applications Conference ALAC 2005*, Ann Arbor, MI, 19 September 2005
34. J. Sun, J.P. Longtin, Inert gas beam delivery for ultrafast laser micromachining at ambient pressure. J. Appl. Phys. **89**, 8219–8224 (2001)
35. L. Walker, R. Maynard, W. Clark, Atmospheric affects on ultrashort-pulsed material processing, in *Proceedings 21st International Congress on Applications of Lasers and Electro-Optics*, ed. by E. Beyer, R. Patel (Laser Institute of America, Orlando, FL, 2002)
36. R. Russo, X. Mao, H. Liu, J. Yoo, S. Mao, Time-resolved plasma diagnostics and mass removal during single-pulse laser ablation. Appl. Phys. A **69**, 887–894 (1999)
37. M. Kraus, S. Collmer, S. Sommer, F. Dausinger, Microdrilling in steel with frequency-doubled ultrashort pulsed laser radiation. J. Laser Micro/Nanoeng. **3**, 129–134 (2008)
38. S. Sommer, M. Kraus, F. Dausinger, Surface structuring of technical metal surfaces with ultrashort laser pulses—enhancement of quality and efficiency, in *Proceedings Fourth Int. WLT-Conference on Lasers in Manufacturing*, ed. by F. Vollertsen, C. Emmelmann, M. Schmidt, A. Otto (AT-Fachverlag, Stuttgart, 2007), pp. 625–628
39. H. Hügel, F. Dausinger, Fundamentals of laser-induced processes, in *Laser Physics and Applications*, ed. by W. Martienssen (Springer, Berlin, 2004), pp. 3–71 (Landolt-Börnstein: Numerical Data and Functional Relationships in Science and Technology—New Series VIII/1C)
40. S. Klimentov, P. Pivovarov, V. Konov, D. Walter, M. Kraus, F. Dausinger, Spectral dependences of conical emission in gases: minimization of scattering for ultra-short pulsed laser ablation. Laser Phys. **19**(6), 1282–1287 (2009)

41. D. Walter, A. Michalowski, R. Gauch, F. Dausinger, Monitoring of the micro-drilling process by means of laser-induced shock waves, in *Proc. Fourth Int. WLT-Conference on Lasers in Manufacturing*, ed. by F. Vollertsen, C. Emmelmann, M. Schmidt, A. Otto (AT-Fachverlag, Stuttgart, 2007), pp. 557–562
42. D. Walter, A. Michalowski, R. Gauch, F. Dausinger, Monitoring of the micro-drilling process by detection of laser-induced shock waves in air, in *Proc. 26th Int. Congress on Applications of Lasers & Electro-Optics* (Orlando, Florida, 2007), paper M805
43. D. Walter, A. Michalowski, F. Dausinger, Real-time monitoring and control of micro-drilling process using optical emission of material-vapor plasma, in *Proc. Fourth Int. WLT-Conference on Lasers in Manufacturing*, ed. by F. Vollertsen, C. Emmelmann, M. Schmidt, A. Otto (AT-Fachverlag, Stuttgart, 2007), pp. 569–573
44. D. Walter, Online-Qualitätssicherung beim Bohren mittels ultrakurz gepulster Laserstrahlung, *University of Stuttgart, Dissertation* (Herbert Utz, München, 2010)
45. A. Feuer, C. Kunz, M. Kraus, V. Onuseit, R. Weber, T. Graf, D. Ingildeev, F. Hermanutz, Influence of laser parameters on quality of microholes and process efficiency. Proc. SPIE **8967**, 1–10 (2014)

Chapter 11
Surface Structuring of Metallic Materials

Steffen Sommer and Sven Döring

Abstract By the use of ultrashort laser pulses in surface structuring, the achievable quality and precision can be improved compared to pulse durations in the nanosecond regime. Further post processing to eliminate burr can be saved. Nevertheless, the process efficiency for picosecond pulses is much lower compared to the nanosecond regime. Therefore, spatial and temporal strategies need to be applied to achieve economic ablation rates. This chapter demonstrates the enhancement of productivity by factor of 3–10 by spatial beam shaping and about several orders of magnitude by temporal division of the pulses. Furthermore, the precision and the efficiency can be increased by the use of a shorter wavelength. Finally selected tools for processing assurance are introduced to enable a higher reliability within industrial production.

11.1 Introduction

Surface structuring with pulsed lasers enables the fast and precise generation of arbitrary geometries. Compared to mechanical technologies like turning and milling, with laser ablation very small dimensions of several microns can be achieved very easily and the machining of hard materials is possible without increased wear of the tool or losses in precision. Compared with technologies like etching, the laser ablation is quite flexible. Due to these advantages many processes like marking, engraving of injection molds [1] and structuring of tribological surfaces [2] have been established in industrial production since many years with pulse durations in

S. Sommer (✉)
Dausinger + Giesen GmbH, Rotebühlstraße 87, 70178 Stuttgart, Germany
e-mail: sommer@dausinger-giesen.de

S. Döring
Institute of Applied Physics, Abbe Center of Photonics, Friedrich-Schiller-Universität Jena, Max-Wien-Platz 1, 07743 Jena, Germany

S. Nolte et al. (eds.), *Ultrashort Pulse Laser Technology*,
Springer Series in Optical Sciences 195, DOI 10.1007/978-3-319-17659-8_11

the nanosecond regime. Especially for the ablation of metals, these pulses form burr due to thermal effects and need additional cleaning steps afterwards. By reduction of the pulse duration in the regime of a few picoseconds it is possible to ablate with nearly no heat affected zone. This can save further cleaning steps and leads to a higher precision of the ablated structures. But the process efficiency is very poor due to limitations in the applicable energy density [3].

There are several ways to enhance the productivity of ablation with ultrashort laser pulses and to establish this technology for industrial production. The applied average power can be increased by spatial or temporal division at constant energy density without losses in quality by raising thermal effects. The use of a reduced wavelength compared to the state of the art near infrared enables better absorption and benefits in focusing conditions. Further it is necessary to implement tools for processing assurance to provide stable and reliable production processes.

11.2 Productive Processing Strategies

Surface ablation of metals with laser irradiation needs to heat up, melt and vaporize the work piece. Thereby volume expansion appears and a shock wave is induced. Due to the impact of this shockwave liquid material is released out of the interaction zone. After the ablation process recast of molten, but not fully detached material remains in the processed area, thereby the moved and solidified material bulges and forms burr beside the interaction zone, see Fig. 11.1a.

To reduce heat conduction within the process, the pulse duration τ_H has to be in the regime of the timescales for reaching the thermal equilibrium state τ_{eq} between electron system and atomic lattice, which is typically between hundreds of femtoseconds and a few picoseconds for metallic materials. Nevertheless, the formation of a liquid phase cannot be avoided completely. The amount of generated melt primarily depends on the applied energy density H, also referred to as fluence [4]. For an energy density, which is much higher than the ablation threshold H_{th},

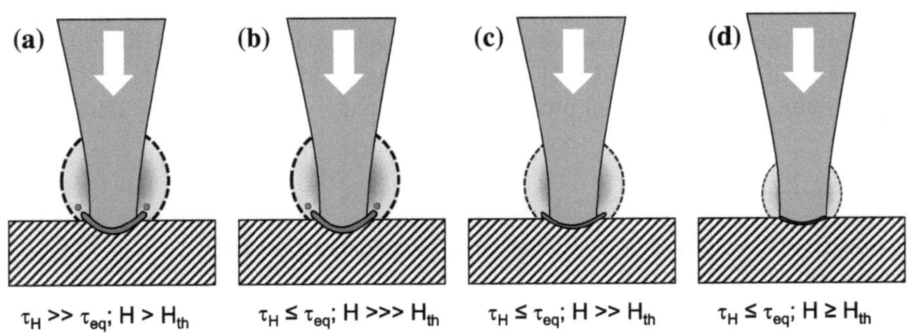

Fig. 11.1 Behavior at ablation of metallic materials for short ($\tau_H >> \tau_{eq}$) and ultrashort ($\tau_H < \tau_{eq}$) laser pulses with different levels of energy density H

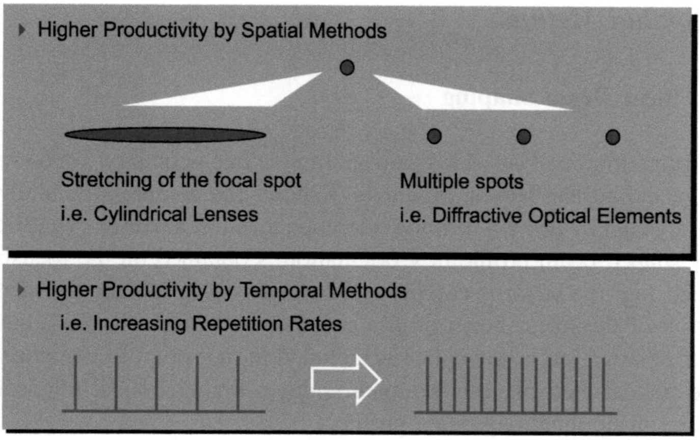

Fig. 11.2 Strategies to improve the productivity of ablation with ultrashort laser pulses

a significant melt volume is generated and thermal effects like bulging can be observed, see Fig. 11.1b. Furthermore, the pressure of the shockwave becomes high enough to move molten material out of the laser interaction zone, similar to the ablation behavior for long pulse durations, Typical values for this regime are 50–100 J/cm^2. To obtain a negligible amount of melt and avoid agitation in the molten material, the energy density for irradiation needs to be close to the ablation threshold, typically 0.05–0.2 J/cm^2 [5], depending on the material, see Fig. 11.1d. However, only a low ablation rate of several nm/pulse can be achieved in this case. An energy density in the range of 1–10 J/cm^2, see Fig. 11.1c, offers in most cases the best compromise between the achievable quality and efficiency for high precision ablation without post processing.

In conclusion, it is possible to ablate metals with ultrashort laser pulses without burr and at very high precision, but this quality requires a limited energy density of the irradiating pulses and reduces the volume ablation rate for the process, because of relatively low applicable average power. Productivity directly scales with average power. Concepts with high average power at limited energy density are spatial and temporal division (see Fig. 11.2). Using spatial beam shaping the ablation zone can be enlarged, either by transforming the spot into a line geometry or due to parallelization by division of the laser beam [6]. As a temporal method, the division of the laser power within a higher number of pulses by an increase of the repetition rate should enable a faster processing velocity. A successful use of both concepts strongly depends on the application.

11.2.1 Spatial Methods

11.2.1.1 Linear Beam Shaping

The transformation of a Gaussian beam spot into a line geometry can be solved by a telescope of cylindrical lenses. Figure 11.3 shows the compression along one axis of the initial beam by using a forming telescope made of a pair of cylindrical lenses. A linearly shaped beam profile in y-orientation is obtained on the focusing optics. During focusing, the beam is compressed stronger along its larger dimension, and the direction of the linear beam profile switches to x-direction. To adjust the orientation of polarization, a waveplate is included in the setup. In comparison with a diffractive optical element, this setup provides a higher optical efficiency and is independent of the incoming beam profile.

In order to show the influence of beam stretching on the achievable ablation efficiency, in Fig. 11.4 ablation results of beam profiles with three different stretching ratios of the shaped beam are compared to the unshaped focal spot. The energy density and the relative overlapping of the pulses on the target are kept constant, therefore pulse energy and the machining speed have been adapted in proportion to the aspect ratio.

The graph demonstrates a linear growth of the ablation rate with increasing aspect ratio of the stretched focal spot. At constant energy density, higher pulse energy and resulting higher average power can directly be transferred into an increase of processing speed. SEM-pictures of the resulting grooves in Fig. 11.5 show some periodic profile within the structure for higher aspect ratios, but even for the stretching factor of 15, there is no accumulation of burr beside the ablated area.

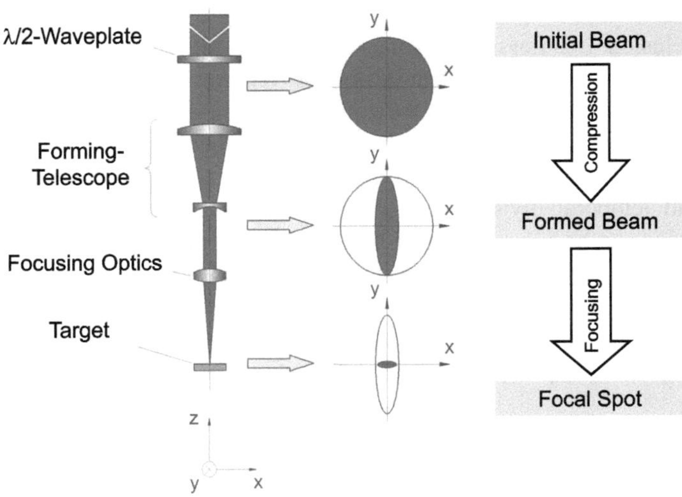

Fig. 11.3 Experimental setup for generation of a lined focal spot by the use of a telescope of cylindrical lenses

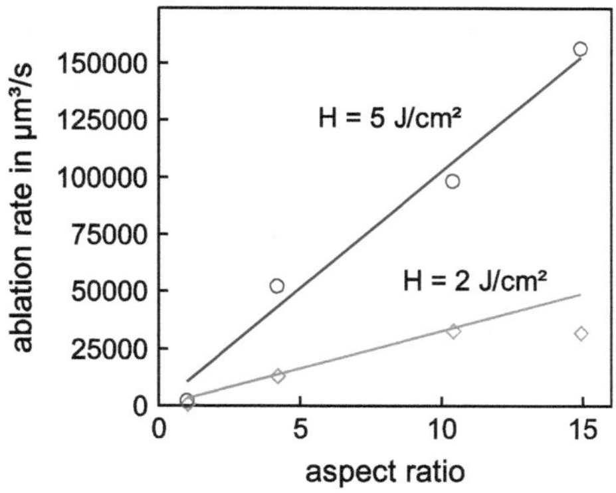

Fig. 11.4 Volume ablation rate by structuring grooves in grey cast iron dependent on the stretching ratio of the formed beam. $\tau_H = 5$ ps; $f_p = 1$ kHz

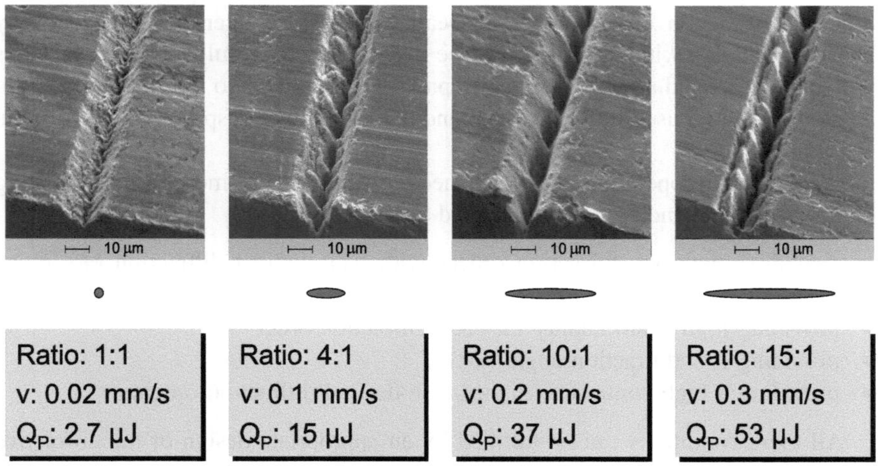

Fig. 11.5 Grooves structured in grey cast iron with different stretching ratio of the formed beam. $\tau_H = 5$ ps; $f_p = 1$ kHz; $H = 2$ J/cm^2

11.2.1.2 Multi Spot Beam Shaping

An alternative approach to increase the overall ablation zone per pulse is parallel processing. The fundamental concept is based on the splitting of the laser beam into several sub-beams with preferably equivalent power and quality, while the beam shape itself remains unchanged. In the simplest configuration this can be achieved by means of a partially reflecting and partially transmitting beam splitter providing beams for different workstations.

For some applications, especially in the micromachining regime, parallel processing at the same workstation is preferable to realize structures with a separation

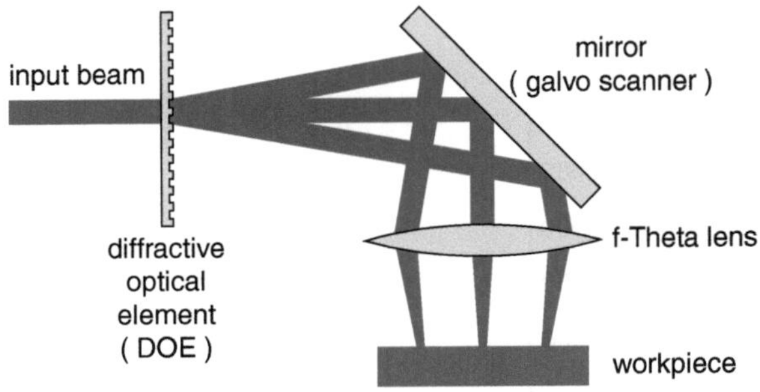

Fig. 11.6 The diffractive optical element (DOE) splits the beam into multiple diffraction orders, which can then be focused to the workpiece by using a single optic (e.g. galvo scanner including an f-Theta lens)

of a few tens of micrometers in a single step. This can be achieved by inserting a diffractive optical element (DOE) into the beam path, which splits the beam into different diffraction orders along the beam path due to interference induced by phase modulation within the element (see Fig. 11.6). This results in multiple beams with clearly defined arrangement and separation angle. Due to the small diffraction angle they can be used in front of one and the same optic despite the limited input aperture.

For successful operation, the DOEs need to fulfill one or more of the following requirements, depending on the intended application:

- providing multiple diffraction orders with well-defined diffraction efficiency, e.g. equivalent splitting of the incident pulse energy
- providing high beam quality for each diffraction order
- providing low diffraction angle (<<1°)
- providing a high contrast to higher, non-desired diffraction orders

All these conditions can be fulfilled by an appropriate design of the diffractive optical element.

In the following a commercially available, triple spot DOE is used to demonstrate the capability of such a system. This element provides three beams (zero and ± 1st diffraction order) with a diffraction angle of 0.135°. In combination with a scanner system and an f-Theta lens with 100 mm focal length, these beams are projected into the same focal plane with a spot separation of about 225 μm. As can be seen from geometrical optics considerations, the spot separation only depends on the diffraction angle and the focal length used, irrespective of the distance between DOE and lens or workpiece. Therefore, the system is resistant to misalignment. Measurements of the intensity profile and M^2-parameter showed that the beam quality of the laser system is preserved for each diffraction order. The element has a high efficiency of 90 % and a uniform power splitting ratio of 1:1:1.

As a practical example, grooves with a width of 25 and 20 μm depth were structured in stainless steel. A galvo scanner in front of the focusing optic was used

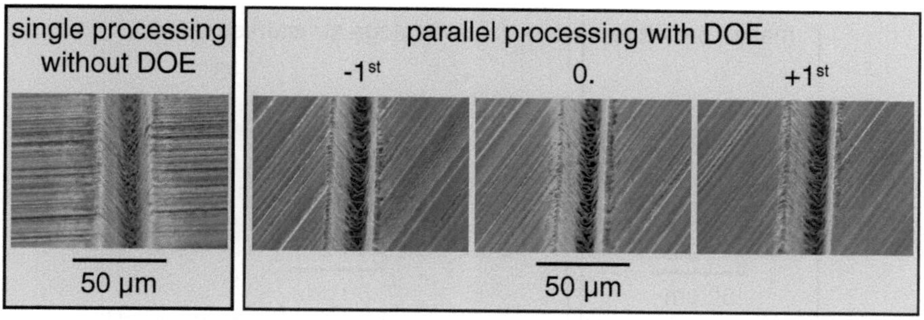

Fig. 11.7 Ultrashort pulse laser structuring with a triple spot DOE produces three grooves, one for each diffraction order. They show uniform quality and dimensions. The quality is also comparable to a line generated without using the DOE. The feed rate is the same in both cases, while for application of the DOE the pulse energy is adapted appropriately. 7.5 μJ pulse energy per line, corresponding to $H = 4\ J/cm^2$, effective feed rate of 5 mm/s

to move the beams on the sample surface, see Fig. 11.6. Figure 11.7 shows scanning electron microscope images of the line structures obtained with an ultrashort pulse laser system (8 ps pulse duration, 200 kHz repetition rate). A single line processed without using the DOE is given for comparison. When using the DOE, the total pulse energy has been scaled up to match the conditions for ablation without the DOE while the feed rate remains constant. For both cases, the grooves show the same nearly burr- and melt-free quality as well as the same dimensions. The triple grooves generated using the DOE are uniform in their quality, size and shape. The roughened ground level is a typical feature of the structuring of stainless steel with ps laser pulses. Consequently, parallel processing by means of a DOE significantly increases the process efficiency without a reduction in quality of the produced structures.

It has to be mentioned, that non-desired, higher diffraction orders can significantly damage the workpiece depending on the actual process parameters and the threshold fluence of the material to be machined. Figure 11.8 shows the damage generated by the +2nd and +3rd diffraction order in comparison to the master structure (zero or 1st diffraction order). Typically, the total power in the higher diffraction orders needs to be suppressed below 1 % of the entire transmitted power to reduce the fluence at these spots below the threshold. The microstructure of a DOE can be specially designed to meet these prerequisites.

In conclusion, diffractive optical elements are able to provide optimal beam splitting for parallel processing at micrometer distance. The splitting pattern, respectively number and distance of generated spots, and the splitting ratio are determined by the design of the element, which has to be realized properly to meet the application requirements. Hence, the elements are inadaptable to changing processing demands, but on the other hand are essentially insensitive to misalignment. Thus, DOEs show a high potential for parallel processing in combination with standard optics for micromachining.

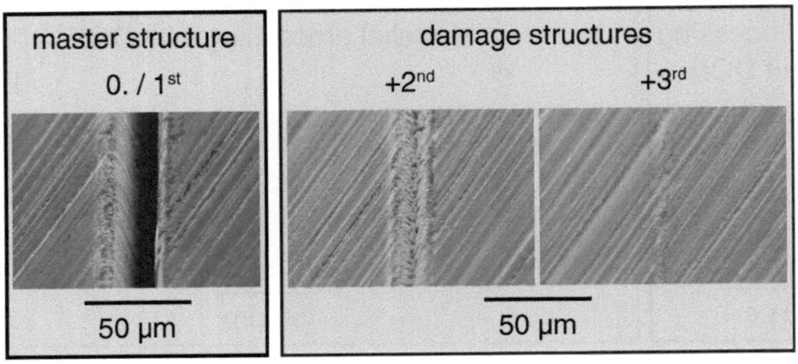

Fig. 11.8 Non-desired higher diffraction orders can significantly damage the workpiece (+2nd) or induce a ripple structure on the surface (+3rd). The effective process speed in this example was 2 mm/s

Fig. 11.9 Influence of the repetition rate on the cross section and the volume ablation rate for structuring grooves at constant pulse overlap into grey cast iron

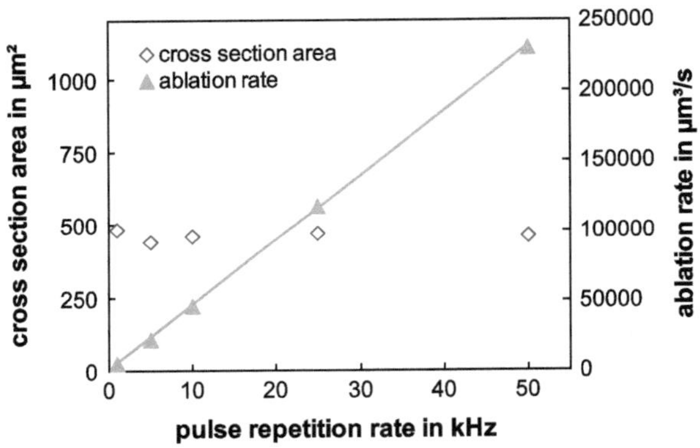

11.2.2 Temporal Method

Apart from energy density the pulse overlap, or progressive feed per pulse, is a main parameter for structuring grooves. To increase the processing speed, while keeping theses two parameters constant, a higher repetition rate can be applied. For the following investigations an Yb:YAG-laser with a pulse duration of about 5 ps was used. The regenerative amplifier featured a thin disc laser. This laser continuously worked at 50 kHz while an external modulator divided the repetition rate. This enables working with constant beam parameters even when changing the pulse frequency. For constant energy density and a fixed pulse overlap, grooves structured in grey cast iron with different pulse repetition rates can be compared.

The obtained cross sectional area shown in Fig. 11.9 is constant for frequencies between 1 and 50 kHz. Due to the increasing machining speed, the ablation rate grows linearly with increasing repetition rates.

With regard to the quality of the structured grooves in Fig. 11.10, there are no changes such as increased burr or melt layers detectable towards higher repetition

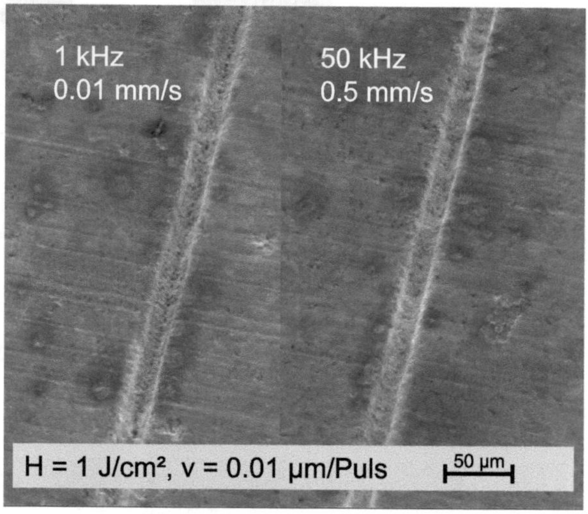

Fig. 11.10 Grooves ablated into grey cast iron at constant pulse overlap and energy density with 1 and 50 kHz

rates. Hence, increasing the pulse repetition rate is an adequate means for providing higher process efficiency while maintaining good ablation quality. Limitations strongly depend on the material, the focusing conditions and the feed rate per pulse. For typical applications like surface structuring metals with a depth of up to several microns no interaction effects can be observed till 1 MHz. For higher repetition rates interaction between the laser beam and plasma and material vapor generated during the pulse can influence the process significantly. Nevertheless, for ablation of (metallic) thin films of several nanometer thickness with very low energy density and a low overlap per pulse, these effects can be kept small and it can be possible to apply repetition rates even higher than 1 MHz.

11.3 Influence of the Wavelength

In laser material processing, the processing wavelength generally affects the absorption as well as the focusing properties. The shorter penetration depth for shorter wavelength is leading to a smaller interaction volume and a better absorption of the irradiated laser beam. Figure 11.11 depicts the achievable ablation rates for structuring grooves in grey cast iron with wavelengths in the near infrared, the visible regime and with ultraviolet light. At the same fluence and spot size a higher volume ablation rate with reduced wavelength can be demonstrated.

Considering the rate of the increase of the process efficiency, it is nearly linearly with the rate of frequency reduction. However, taking into account the conversion losses at frequency-conversion, the productivity for ablation of ferrous metals cannot be improved significantly. Depending on the application, a fabrication of

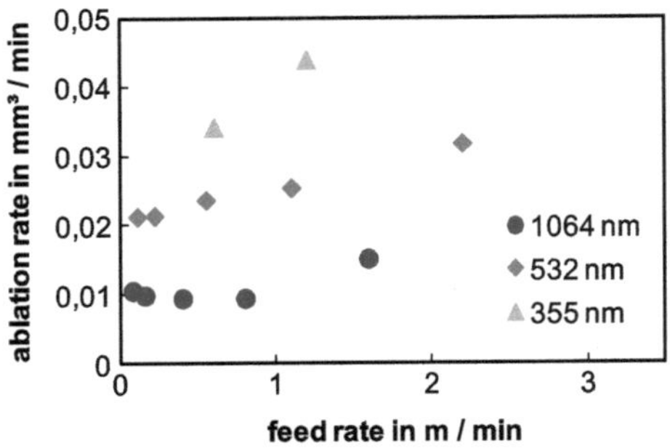

Fig. 11.11 Volume ablation rate at structuring grooves in grey cast iron for different processing wavelength. $\tau_H = 12$ ps; $H = 1$ J/cm^2; $f_P = 100$ kHz

Fig. 11.12 Grooves in grey cast iron ablated with 1064 nm (*upper*) and 532 nm (*lower*) for machining speeds from 1.6 m/min (*left*) to 0.15 m/min (*right*). $\tau_H = 12$ ps; $H = 2$ J/cm^2; $f_P = 100$ kHz

deeper geometries at constant machining speed or an enhancement of the feed rate at constant structure depth is possible.

The grooves displayed in Fig. 11.12 illustrate a higher quality at reduced wavelength. At 1064 nm (upper row), a burr is generated at an energy density of 2 J/cm^2 even at a high feed rate of 1.6 m/min (left). Contrary to this, the grooves manufactured with 532 nm (lower row) are burr free. The internal structure of the grooves is much smoother at 532 nm than at 1064 nm. In addition, the wavelength of 532 nm produces sharper edges and ablates the graphite lamellae within the grey cast iron more efficiently (2nd column from right).

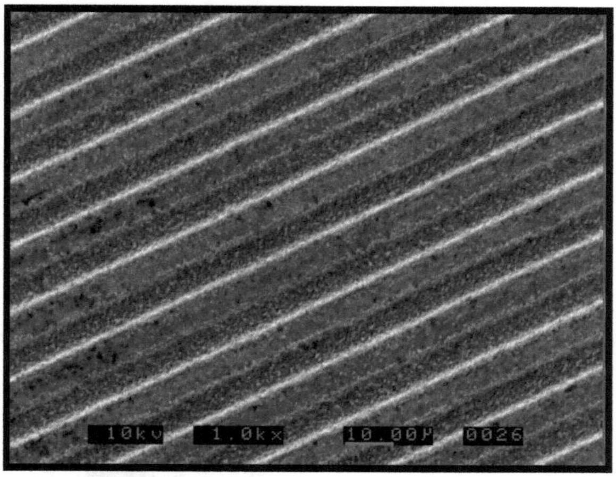

Fig. 11.13 Laser manufactured grating in copper. Pitch: 10 μm; Width of grooves: 6 μm; λ = 532 nm; fp = 100 kHz; v = 30 mm/s

It can be seen that the change to shorter wavelength does not increase productivity directly, but rather the quality. Besides this main advantage, the second harmonic has higher focusability, which enables either smaller focal spots for precise machining or an enlargement of the processing window by using optics with higher focal length or an increase of the field size at a constant spot size by the use of a galvo scanner system. Thus, new applications in metals can be enabled.

For non-ferrous metals the increase of productivity due to the reduction of the wavelength can be higher than the losses at frequency conversion [7]. In copper the use of 532 nm enables a quality useable for optical gratings as shown in Fig. 11.13.

11.4 Processing Assurance

Within industrial production, especially of automotive parts, beside productivity the surveillance of new process technology is very important for a successful implementation. Therefore, the quality and risk management needs tools to control the processes. Beside post measurement an inline control would be preferable. In Fig. 11.14a method for detecting the processing emission is displayed. A photo detector captures the coaxial light of the ablation zone. Filters block the wavelength of the laser [8]. Further the impact sound in the work piece is detected.

The graph in Fig. 11.15 left shows the variation of the width and depth of the groove among the variation of the focal position (upper). There is a correlation with the optical processing emission (middle) that is minimized at the focus position as well as between the impact sound that has a maximum in the focal position (lower). The variation of the pulse energy shown in Fig. 11.15 right has an increase of width, depth, optical emission and impact sound for an increasing pulse energy. Due to this behavior an increase or decrease of optical and acoustic emission

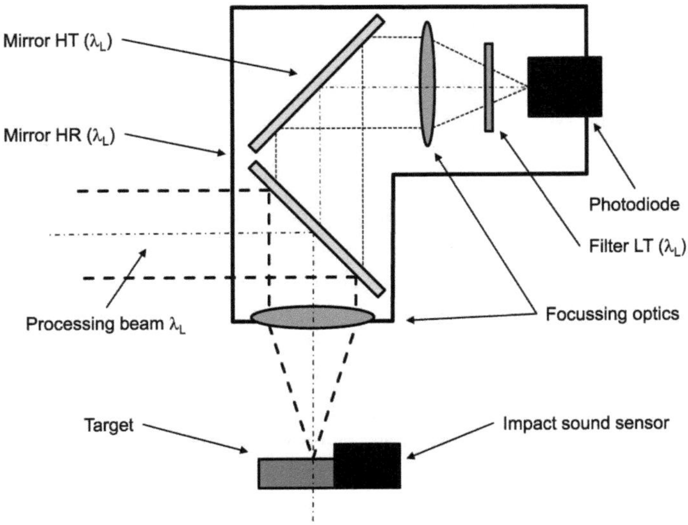

Fig. 11.14 Setup for coaxial detection of the optical processing emission and the recording of the impact sound

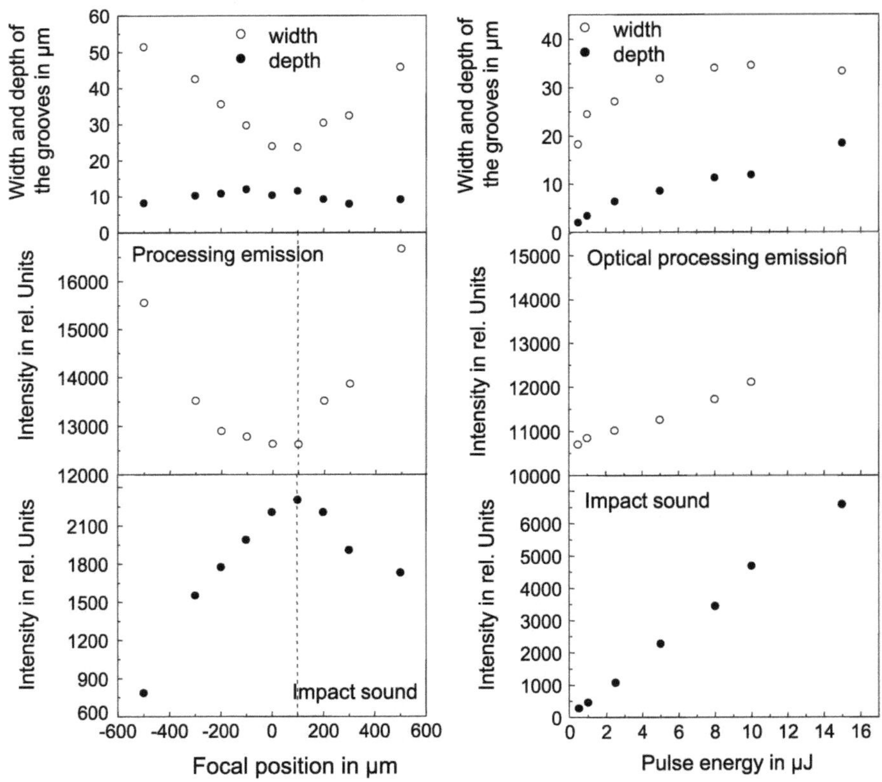

Fig. 11.15 Detection of the processing emission and the impact sound in correlation with the width and depth of the resulting grooves for a variation of the focal position (*left*) and the pulse energy (*right*) [9]

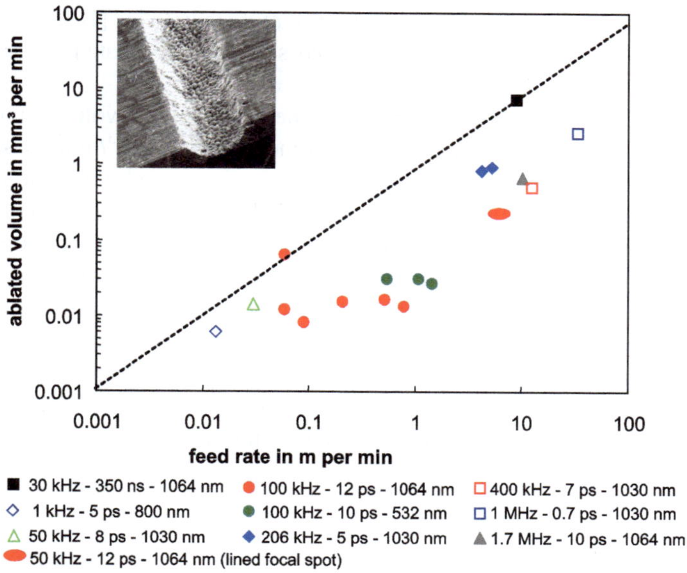

Fig. 11.16 Enhancement of the productivity during the recent years by the use of spatial and temporal beam shaping together with the reduction of the wavelength displayed for the volume ablation rate at structuring grooves in grey cast iron

implicates a change in pulse energy, while an increase of optical emission at decreasing impact sound a change in the focal position implies. This shows the high potential for processing surveillance methods for ablation with ultra short pulses, even it seems to be necessary to detect several signals to get a conclusion which parameter changes within the process [9].

11.5 Conclusion

The enhancement of the productivity is the core topic to bring the technology of ultra short pulsed lasers into industrial applications. Figure 11.16 summarizes experimental achieved ablation rates for laser structuring of metals. Grooves in grey cast iron in the dimension of 20 μm width and 5 μm depth are chosen as benchmark application. The value for 1 kHz was state of the art at the begin of this millennium. The item 30 kHz–350 ns is the productivity of the laser ablation with nanosecond pulses at which the burr has to be removed in a final cleaning step. The measurement values in between summarize the spatial and temporal strategies together with the reduction of the wavelength. The recent availability of stable high power lasers with pulse durations in the sub-picosecond regime enables further potential for the enhancement of the process efficiency. So the change from 7 ps to 0.7 ps even in metals raises the ablation rate per pulse more than twice [10]. For significant shorter pulses effects like air breakdown and scattering at ablated material appear and reduce the quality and efficiency.

Based on these different concepts within the last years the volume ablation rate could be raised from 0.006 mm^3/min up to more than 1 mm^3/min. Based on the availability of high power laser systems emitting picosecond pulses at a high stability it is possible to combine methods of parallelizing or modification of the beam profile at high repetition rates with reduced wavelengths. There are different industrial applications realized through the main benefit for the efficiency resulting of higher pulse repetition rates. Today the speed of many processes is limited due to the dynamic of the motion system so it becomes more and more important to use the spatial methods to become more productive. Not every application allows the implementation of parallelization so still every process has to be designed in consideration of all the mentioned tools for improvement of productivity and quality.

References

1. F.J. Gruber, in *Technologische Punktlandung—Laser-Formbohren. EuroLaser 1* (Schlütersche Verlagsgesellschaft, Hannover, 2006) pp. 30–33
2. H.W. Mayer, W. Pester, Laser lässt dieselmotoren sparsamer fahren. VDI Nachrichten **49**, 12 (2005)
3. F. Dausinger, F. Lichtner, H. Lubatschowski, Femtosecond Technology for Technical and Medical Applications, in *Topics in Applied Physics 96* (Springer, Berlin, 2004)
4. D. Breitling, *Gasphaseneinflüsse beim Abtragen und Bohren mit ultrakurz gepulster Laserstrahlung* (Herbert Utz Verlag, München, 2010)
5. S. Sommer, *Effizienzsteigerung beim präzisen Oberflächenabtragen mit ultrakurzen Laserpulsen* (Shaker Verlag, Aachen, 2010)
6. Z. Kuang, W. Perrie, D. Liu, S. Edwardson, J. Cheng, G. Dearden, K. Watkins, Diffractive multi-beam surface micro-processing using 10 ps laser pulses. Appl. Surf. Sci. (2009). doi:10.1016/j.apsusc.2009.06.089
7. S. Sommer, M. Kraus, F. Dausinger, Surface struc-turing of technical metal surfaces with ultrashort laser pulses—enhancement of quality and efficiency, in *Proc. Fourth Intl. WLT-Conference on Lasers in Manufacturing 2007* (München) ed. by Wissenschaftliche Gesellschaft Lasertechnik (WLT) e.V. (AT-Fachverlag, Stuttgart, 2007), pp. 625–628
8. D. Walter, *Online-Qualitätssicherung beim Bohren mittels ultrakurz gepulster Laserstrahlung* (Herbert Utz Verlag, München, 2010)
9. D. Walter, A. Michalowski, F. Dausinger, Real-time monitoring and control of micro-drilling process using optical emission of material-vapor plasma, in *Proceedings of the Fourth International. WLT-Conference on Lasers in Manufacturing 2007* (München) ed. by Wissenschaftliche Gesellschaft Lasertechnik (WLT) e.V (AT-Fachverlag, Stuttgart, 2007), pp. 569–573
10. S. Sommer, F. Dausinger, Flexible and precise material processing with femtosecond disk lasers, in Proc. of LPM2011 (Takamatsu) by Japan Laser Processing Society (JLPS)

Chapter 12
Beam Guidance, Focal Position Shifting and Beam Profile Shaping in Ultrashort Pulsed Laser Materials Processing

Peter Bechtold, Maik Zimmermann, Stephan Roth, Ilya Alexeev and Michael Schmidt

Abstract In ultrashort pulsed (USP) laser micro-processing several optical and optomechanical components are needed to enable the desired process. Besides the USP laser itself, systems to guide and focus the laser beam are mandatory in almost all applications. Furthermore beam profile shaping, e.g. the creation of top-hat or super-Gaussian beam profiles, results in beneficial effects for many applications, such as increased efficiency and superior quality. In this chapter an overview of both established and up-to-date technologies regarding beam guiding, focal position shifting and beam profile shaping will be presented. For beam guidance and focal position shifting comparable key parameters will be stated to enable easy comparability. As an outlook on probable future applications, both temporal and spatial polarization profile shaping will be presented.

12.1 Introduction

As stated in other chapters of this book, ultrashort pulsed lasers are increasingly favored compared to other laser sources and non-laser technologies in specific fields of micro manufacturing, particularly structuring, drilling and cutting of μm-sized features. Optimization of the laser-material interaction of an ultrashort pulsed laser beam regarding selected key features, such as volume of burr and recast, roughness

P. Bechtold (✉) · I. Alexeev · M. Schmidt
Institute of Photonic Technologies, Friedrich-Alexander-Universität Erlangen-Nürnberg, Paul-Gordan-Str. 3, 91052 Erlangen, Germany
e-mail: Peter.Bechtold@lpt.uni-erlangen.de

M. Zimmermann · S. Roth · M. Schmidt
Bayerisches Laserzentrum GmbH, Konrad-Zuse-Str. 2-6, 91052 Erlangen, Germany

P. Bechtold · M. Zimmermann · S. Roth · I. Alexeev · M. Schmidt
Erlangen Graduate School in Advanced Optical Technologies (SAOT), Friedrich-Alexander-Universität Erlangen-Nürnberg, Paul-Gordan-Str. 6, 91052 Erlangen, Germany

S. Nolte et al. (eds.), *Ultrashort Pulse Laser Technology*,
Springer Series in Optical Sciences 195, DOI 10.1007/978-3-319-17659-8_12

of the generated structure and overall process efficiency, is heavily dependent on and partly limited by the main parameters of the applied beam guiding, focal position shifting and beam profile shaping technology. It is therefore of upmost importance to know and understand up-to-date technologies and developments in these fields of technology to ensure best use of an ultrashort pulsed laser.

In a vast majority of ultrashort pulsed laser applications well-known and established galvanometer scanners with quasi-fixed focal position and Gaussian beam profile are used, though usage of further advanced technologies would result in beneficial changes of the laser-material interaction. To state more specific examples, beam guiding with faster angular speed will result in lower pulse overlap, which is beneficial for structuring and cutting quality [1]. Usage of focal position shifting technologies allows three-dimensional processing and beam profile shaping —reshaping to top-hat distribution for instance—increases processing efficiency and decreases heat impact in surface structuring [2].

Within this chapter the authors will state an up-to-date overview of systems for beam guidance by beam deflection, focal position shifting and beam profile shaping. After covering the basics of each topic, both established and novel technologies are explained in detail. To ease the selection of a specific technology for each field, comparable key parameters for the most important technologies will be stated. Each section will also include an outlook on upcoming technologies, which may be of interest in future.

12.2 Systems for Beam Guidance by Beam Deflection

Beam guidance is to be understood as angular deflection of the beam in one (single axis beam guiding) or two axes (dual axis beam guiding) perpendicular to the beam propagation and onto each other. In micro-processing and other applications with USP lasers, the laser beam typically passes the focusing optics after angular deflection.[1] Thus angular deflection is transformed into lateral focus spot displacement by the focusing optics, see Fig. 12.1a. With idealized paraxial approximation[2] the equivalent lateral displacement Δx can be calculated by (12.1), where f is the effective focal length of the focusing optics and α the beam deflection angle.

$$\Delta x = f \tan \alpha \approx f \alpha \tag{12.1}$$

In case of reflective beam guidance (e.g. galvanometer, piezo and MEMS scanners) the achievable deflection angle may be multiplied by the principle of

[1]In some macro-processing applications the collimated laser beam passes the focusing optics before angular deflection. This enables larger working areas in remote processes, such as remote welding. As focal distance and thus spot size is increased, this technique is uncommon in micro-processing.

[2]Paraxial approximation assumes that deflection angle is low, so that $\sin \alpha = \tan \alpha = \alpha$, and beam height is negligible, i.e. all beams are near to optical axis.

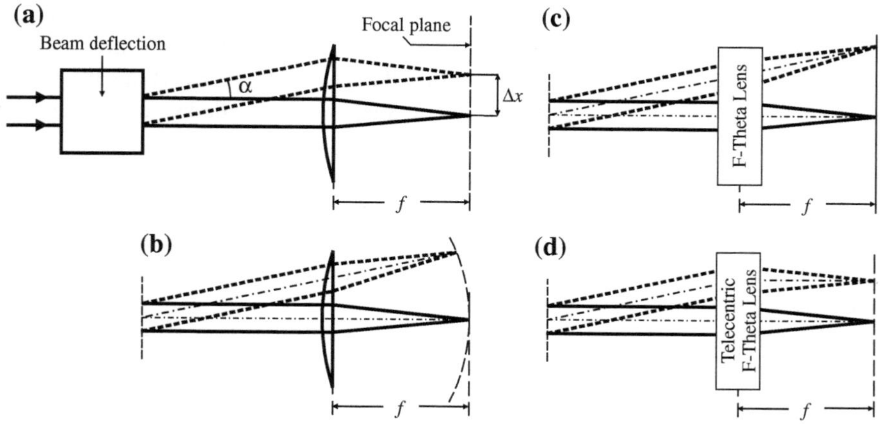

Fig. 12.1 Transformation of beam deflection into lateral focus shift; **a** idealized, paraxial approximation; **b** curvature of focus plane for single lens in non-paraxial (real) case; **c** planarization of focal plane by F-Theta lens; **d** equaling angle of incidence to 0° by telecentric F-Theta lens

optical beam folding, i.e. the laser beam hits the same tilted mirror surface more than once by use of an appropriate optical setup (folding mirrors). Optical beam folding is a well-known technique, which is often used in relation to gratings. As the deflection angle is multiplied by the folding factor, the same accounts to the values of maximum angular speed, repeatability and resolution.

Before explaining most established and widely used optical systems for beam deflection in detail, a succinct overview on focussing options of the ultrashort pulsed laser beam after beam deflection will be given.

12.2.1 Focussing Methods for Ultrashort Pulsed Lasers

Different lenses and lens combinations are used as focussing optics in ultrashort pulsed laser processing. If no beam deflection is needed—e.g. the workpiece is translated via linear axes—a single lens (plano-convex, best-form or aspherical lens) or lens combination (achromat or so-called 'focussing optic') is used most commonly. However, if beam deflection takes place, such lenses will focus the beam not onto a planar area, as shown in Fig. 12.1a, but on a curved area, see Fig. 12.1b. Planarization of the focal plane is done by F-Theta lenses[3] (Fig. 12.1c), which are lens combinations of mostly three or more single lenses. If aspect ratio of

[3]The term 'F-Theta' historically relates to the optimization of such lens combination towards linear dependence of lateral beam height after focussing ('F') to deflection angle before focussing ('Theta'). In modern scanning systems, this relation is optimized by software, so that planarization of the focal plane is the main task of F-Theta lenses today.

the feature to be produced is high, telecentricity, i.e. the value of the angle of incidence of the focused beam onto the focal plane, has to be taken into account. Telecentric F-Theta lenses (Fig. 12.1d) keep the telecentricity low (typically below 2°), whereas the telecentricity is comparable to the deflection angle for non-telecentric F-Theta lenses.

In application, telecentric F-Theta lenses are to be favored in most ultrashort pulsed laser processes. When the feature to be produced is of low aspect ratio (approximately 0.1 and below), non-telecentric F-Theta lenses are a viable, more cost-efficient option. Only if there is no beam deflection (e.g. the workpiece is moved), single lenses are viable.

When using any lens system in conjunction with ultrashort pulsed lasers two further aspects have to be taken into account. Firstly, some lens combinations (F-Theta lenses in particular) produce additional foci due to back-reflection at concave optical surfaces, which might result in damaging the lens system or optical elements prior to the lens system (e.g. components of the beam deflection). Therefore, lens systems optimized to avoid such additional foci are to be increasingly favored when pulse duration is ultrashort (~10 ps and less) and pulse energy is moderate or high (~1 μJ and above). Secondly, some glass types used in lens combinations induce a significant amount of group velocity dispersion (GVD).[4] In case of ps-laser pulses (pulse duration approximately 10 ps) the spectral bandwidth is rather low (~0.1 nm), so that the impact of GVD is negligible for almost all established glass materials. When pulse duration is shorter (~100 fs and below), spectral bandwidth is increasing (for 100 fs: ~10 nm) and thus GVD will significantly lengthen the pulse[5] [4]. Additionally, lateral chromatic aberration takes place, which can lead to significant deformation of the beam profile shape at high deflection angles. Low-dispersion glasses, such as fused silica, and color-corrected (achromatic) lens systems are to be favored for almost all ultrashort pulsed lasers, especially if pulse durations are below 1 ps.

12.2.2 Galvanometer-Based Optical Scanners

Galvanometers, first reported by Johann Schweiger in 1820 [5], are instruments for detecting electric currents. A rotary deflection of a coil (or magnet) surrounded by a magnet (or coil) is produced in response to an electric current flowing through the coil, as depicted schematically in Fig. 12.2a). The pointer, which is used to visualize the rotary deflection, was exchanged by a mirror later on allowing higher

[4]For more details on GVD and impact on pulse duration please refer to [3].

[5]As a numeric example a bandwidth-limited pulse of 50 fs duration and 800 nm wavelength will be lengthened to 111 fs (54 fs) after passing through 10 mm SF14 (10 mm fused silica) [4].

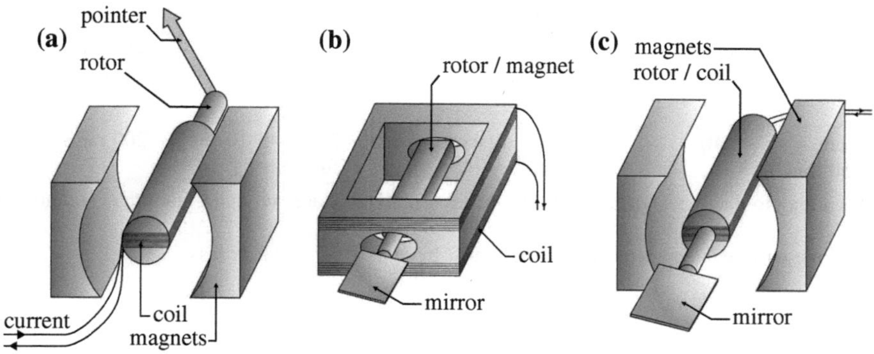

Fig. 12.2 Schematic layout of **a** galvanometer with pointer, **b** galvanometer scanner with moving magnet actuator and **c** galvanometer scanner with moving coil actuator

resolution and representing an early form of a Galvanometer-based optical scanner, commonly referred to as galvo scanner. Today the laser beam is typically reflected by a pair of galvanometer-driven mirrors to enable dual axis beam guiding, where the rotary deflections in both axes are controlled via the applied currents. The mirrors are each attached to the so-called rotor, which is either fixed to the freely moving magnet surrounded by the fixed coil (moving magnet actuator, Fig. 12.2b) or fixed to the freely moving coil surrounded by the fixed magnet (moving coil actuator, Fig. 12.2c). Where the moving magnet actuator enables higher positioning speeds due to superior rigidity and stiffness (higher torque), the moving coil actuator offers better positioning accuracy due to high magnetic flux (smaller coil area). Today both technologies are still in use [6].

The position of the mirror is detected by different techniques (e.g. capacitive or optical position detection) enabling closed-loop control and thus maximizing accuracy and speed of galvanometer scanners. The mirror itself is one of the key elements in galvanometer scanners. It has to offer a suitable size to reflect the beam without disturbance over the whole angular range. Adequate stiffness is needed to withstand deformation caused by the rotor torque. In parallel it must not increase the scanner's inertia to an unacceptable level by too much weight. Overall the mirror design (thickness, cross section, profile) and material (typically fused silica, silicon, silicon carbide) has to meet the trade-off between speed and optical aperture size. The second mirror in a two-axis galvanometer scanner is typically larger than the first one as it has to cover both angular deflections, and thus limits the overall maximum speed [6].

In summary, galvanometer-based optical scanners offer typical parameters as stated in Table 12.1.

Table 12.1 Typical key parameters for galvanometer-based optical scanners[a]

Aperture	Beam deflection up to	Positioning repeatability better	Maximum sinusoidal frequency	Maximum angular speed
5 mm… > 50 mm	±40°	20 µrad (short term)[b]	~2 kHz @ 10 % of full step	200 rad/s

[a]According to datasheets of Scanlab AG, Arges GmbH, Raylase AG and Cambridge Technology Inc

[b]On long term, galvanometer mirrors may drift by appr. 0.1 mrad/h or less

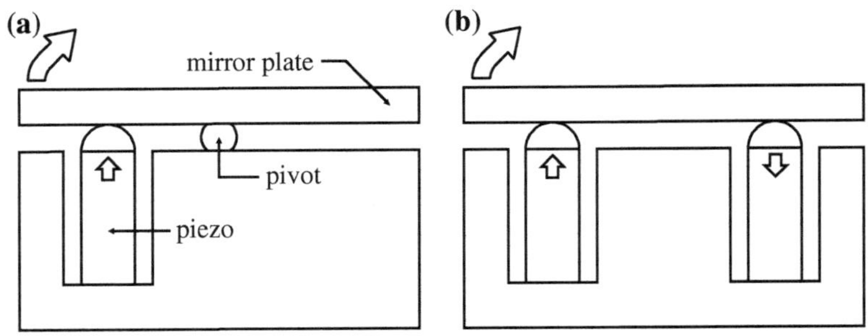

Fig. 12.3 Schematic sectional view of single-axis piezo-driven steering mirror using **a** a single piezo actuator and **b** two piezo actuators

12.2.3 Piezo-Driven Steering Mirror

Piezo-driven steering mirrors, commonly related to as piezo scanners, consist of a mirror attached to one or more linear piezo actuators and flexures to enable tilting of the mirror.

In case of single-axis beam deflection a single piezo actuator in combination with a single flexure may be used, as depicted in Fig. 12.3a. Here, the flexure serves as pivot point. Alternatively, two piezo actuators may be used, which support the mirror directly without the need of an additional flexure, see Fig. 12.3b. In this case the piezos are actuated oppositional, i.e. one piezo is pulling, while the other one pushes the platform. Whereas the single piezo solution offers highest cost efficiency and smallest system size, the mirror tilting angle is sensitive to temperature changes, as the piezo actuator and flexure possess different thermal expansion coefficients. In contrast, the two piezo solution offers stable and precise positioning over a wide temperature range, as a temperature change will only cause a linear movement of the attached mirror. Also faster and larger deflection is achieved.

In most beam guiding tasks the laser beam has to be deflected in both axes. Two single-axis tilting platforms may be used, one for the first axis deflection and the second one in distance to the first one for the second axis deflection. As the same problems arise as with galvanometer scanners (the second mirror has to be larger)

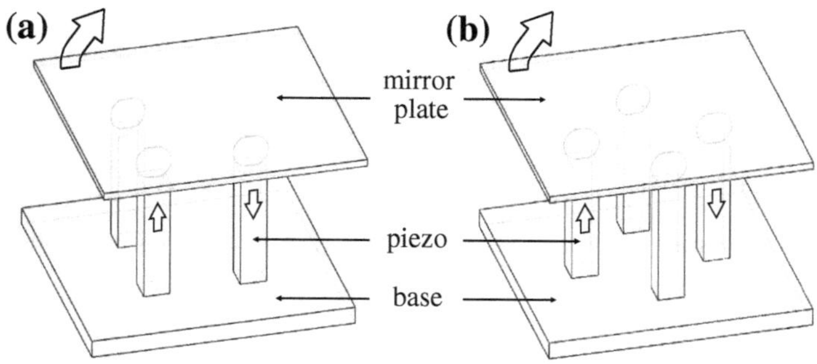

Fig. 12.4 Schematic view of dual-axis piezo-driven steering mirror using **a** a three piezo actuator arranged in L-shape and **b** four piezo actuators arranged in cross-shape

dual- and multi-axis tilting platforms are used much more often. In contradiction to galvanometer-actuated mirrors (rotary actuation is transferred to tilting), dual-axis tilting is realized much easier using piezo actuators, as a linear actuation is transferred to tilting.

Dual axis tilting is typically realized using three or four linear piezo actuators directly attached to the mirror, see Fig. 12.4a, b. Tilting platforms using three linear piezo actuators allow an additional linear movement of the mirror, e.g. for path-length adjustments in applications such as phase-shifting. In case of four piezo actuators each pair of piezo actuators is actuated in opposite directions (push/pull), similar to the single-axis tilting platform using two piezo actuators. In this case each piezo actuator is attached to the mirror in such way, that the distance to the middle pivot point is the same and an angle of 90° is arranged between the two piezo actuator pairs. As before, the setup with four piezo actuators typically exceeds the one with three piezo actuators in terms of maximum speed and deflection [7].

As a major part of beam guiding is related to dual-axis deflection of a laser beam, the summary in Table 12.2 comprises typical parameters of dual-axis piezo-driven optical scanners solely [7].

12.2.4 Acousto-Optical Deflection

Acousto-optical devices consist of a parallelepiped of transparent material (e.g. fused silica, germanium, lithium niobate) polished to optical quality, and a piezo transducer bonded to one side of that cuboid. The transducer induces an acoustic wave into the material, which in reverse induces an according periodical density distribution, see Fig. 12.5. As the index of refraction is depending linearly on the density in first approximation, the cuboid acts as a diffraction grating as soon as an acoustic wave is coupled into the material. The diffraction angle θ is linearly

Table 12.2 Typical key parameters for dual-axis piezo-driven steering mirrors[a]

Aperture	Beam deflection up to	Positioning repeatability better	Maximum sinusoidal frequency	Maximum angular speed
5 mm… > 30 mm	±3°	5 μrad	~1 kHz @ ± 1.5°	50 rad/s

[a]According to datasheets of Physik Instrumente (PI) GmbH & Co. KG and Piezosystem Jena GmbH

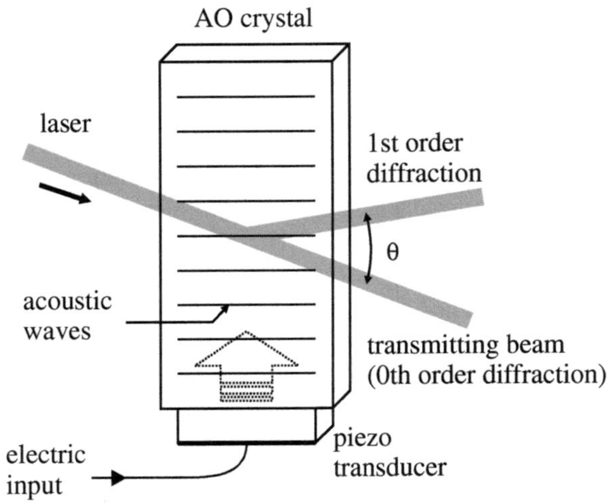

Fig. 12.5 Schematical layout and working principle of the acousto-optical deflector

proportional to the acoustic frequency, so beam guiding is achieved by varying acoustic frequency f (so called “chirping”) [8].

$$\theta = \frac{\lambda f}{V_a} \tag{12.2}$$

For small angles the diffraction angle θ can be estimated by (12.2), where λ is wavelength, f acoustic frequency and V_a the acoustic velocity. Larger diffraction angles and thus deflection ranges can be achieved for longer wavelengths. Typical values for f are 50–500 MHz and for V_a 600–6000 m/s.

As the acoustic frequency can be varied very fast (typical response times in the range of μs), angular speeds exceeding 1000 rad/s are achieved. Though, several challenges arise when acousto-optical deflection (AOD) is used. The diffraction efficiency amounts to appr. 80 %, so at least 20 % of laser power is lost into the zero order (i.e. non-diffracted) beam. As the acoustic frequency is chirped in time-domain when laser beam deflection takes place, the frequency and thus deflection is not constant over the whole aperture volume. This results in a behaviour comparable to a cylindrical lens, which is stronger as angular speed increases. At an angular speed of appr. 1000 rad/s, the focal length of the so-called cylinder lensing effect is between 0.6 m and 6 m depending on the acoustic velocity V_a. Thus, the laser beam is strongly astigmatic after passing the acousto-optical deflector [8, 9].

Of all beam guidance devices reviewed in this chapter, the acousto-optical deflector is the only one where the beam passes material (others are reflective). As a result, chromatic dispersion will cause group-velocity dispersion (GVD), lengthening and chirping the pulse. Moreover, the deflection angle is dependent on the wavelength as can be seen in (12.2). Hence, ultrashort laser pulses of high bandwidth will be laterally dispersed after deflection (lateral chromatic aberration). These effects have to be taken into account at pulse durations below appr. 200 fs [10].

As an acousto-optical deflector enables one-dimensional scanning of the laser-beam, 2-D scanning can be achieved by the use of two deflectors in close vicinity and at an angle of 90° between the scanning axes. However, in this case the lost laser power adds up to at least 50 %. Therefore, acousto-optical deflectors are not widely used for beam guidance in micro-processing to date. Instead they are applied in other technologies such as imaging and optical tweezers, where laser power is typically not an issue [11, 12].

The summary in Table 12.3 comprises typical parameters of 2-D optical scanners using two acousto-optical deflectors [9].

12.2.5 Micro-Electro-Mechanical System Scanners

In MEMS (micro-electro-mechanical systems) and MOEMS (micro-opto-electro-mechanical systems) technology, promising new scanner devices, so-called MEMS scanners, are developed. MEMS scanners mainly consist of a silicon mirror (optionally coated) attached to silicon hinges, so that the mirror may rotate in either one or two dimensions. There are different ways used to drive the mirror, in summary either electrostatic (differing potential of mirror and base) or electromagnetic (similar to galvanometer) forces are utilized. Figure 12.6 depicts an exemplary layout of a 2-D MEMS scanner system with electrostatic actuation schematically.

The development of MEMS scanners started appr. three decades ago and key parameters are reaching values comparable to piezo and galvanometer scanners today. Up to now, most commercially available MEMS scanners are used in

Table 12.3 Key parameters for acousto-optical deflectors[a]

Aperture	Beam deflection up to	Positioning repeatability better	Maximum sinusoidal frequency	Maximum angular speed
0.1 mm… > 5 mm	±25 mrad (eq. ± 1.4°)	1 µrad	~100 kHz… 1 MHz @ full step	>5000 rad/s[b]

[a]According to datasheets of Gooch & Housego Ltd. and Brimrose Corp

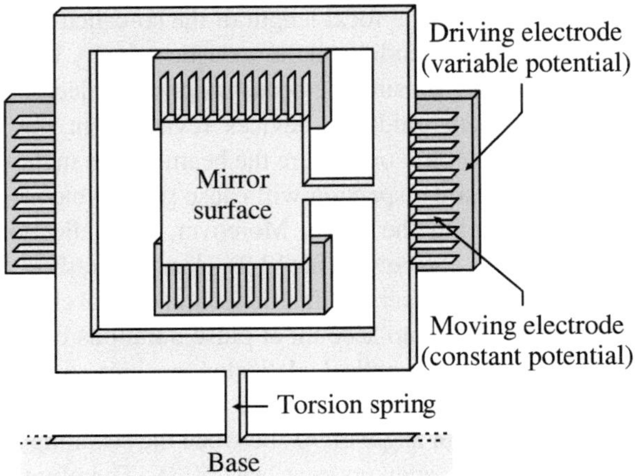

Fig. 12.6 Schematical layout of 2-D MEMS scanner with electrostatic actuation

resonant mode (resonant MEMS scanners), i.e. the mirror is constantly moving at its own eigenfrequency. As mirror weight and moment of inertia are designable to some limit, the eigenfrequency can be chosen up to appr. 50 kHz. The production of MEMS scanners is very cost-effective compared to other scanner technologies (galvanometer, piezo and acousto-optical scanners) as semiconductor technology is used mainly. I.e. many MEMS scanners are fabricated in parallel and monolithically on a wafer. In bar-code-scanners MEMS scanners are already industrially established (generation of laser line via 1-D resonant scanning). In future, the same will probably account to display and projector technology, laser scanning microscopy (both 2-D resonant scanning) and other possible applications [13–15].

Regarding micro-processing with USP lasers, MEMS scanners are not yet established. Resonant scanning is not desired to a large part, therefore non-resonant MEMS scanners may be used. Though, as MEMS scanners can basically be compared to miniaturized galvanometer scanners (in case of electromagnetic actuation), aperture size is limited. Ongoing research and especially low costs may lead to an application in micro-processing in future. Table 12.4 states the key parameters of such non-resonant MEMS scanners, which could be applied in micro-processing [16].

12.2.6 Electro-optical Scanners

Comparable to using the acousto-optical effect for acousto-optical deflection, the electro-optical effect (i.e. the index of refraction is dependent on the electric field strength) is also used for deflection. Typically, electro-optical crystals are shaped in prism-like geometries and an externally applied electrical field changes the index of refraction and thus the deflection angle. Though maximum angular speed exceeds

Table 12.4 Key parameters for non-resonant MEMS scanners

Aperture	Beam deflection up to	Positioning repeatability better	Maximum sinusoidal frequency	Maximum angular speed
1 mm… > 5 mm	±30°	0.1 mrad	~1 kHz @ 10 % of full step	200 rad/s

any other technology (typically > 100,000 rad/s), electro-optic scanners lack usable aperture size, often being smaller than 100 µm [17, 18]. Also, investigations on whether this technology is applicable to ultrashort pulsed lasers were not conducted to date. Therefore, electro-optical scanners are still of pure scientific interest mostly.

Another relatively new technology uses an extension of electro-optic scanning. Free space-charges are induced into a crystal ($KTa1-xNbxO3$, abbreviated as KTN), which deform the externally applied electrical field in such way that scanning is achieved without the crystal being in prism shape necessarily. In result, high-speed and large-angle scanning is reached (±5° @ 40,000 rad/s). With approximately 0.5 mm, the aperture size is bigger than other electro-optical scanners, but still rather small [19]. Regarding the deflection of ultrashort laser pulses, the free space charges interact with the laser pulse in an unfavorable way, which diminishes the scanning effect to a part, if pulse energy and repetition rate is too high [20]. To date, the applicability in ultrashort pulsed laser-processing is therefore limited to low-power regimes. Also, random-access scanning is very limited, as certain KTN-scanner properties are only stable for high-speed sinusoidal scanning. However, the KTN-scanner is undergoing extensive development and could become more viable mid-term.

12.2.7 Summary and Outlook

To date a large part of micro-processing applications with USP lasers utilizes established galvanometer scanners for beam deflection. Piezo and galvanometer scanners achieve comparable beam guiding parameters, though 2-D piezo scanners may consist of a single steering mirror guiding the beam in both axes. Typically piezo scanner costs exceed that of galvanometer scanners and offer lower maximum deflection angle (which could be increased by beam folding). Acousto-optical deflection offers highest angular speeds exceeding 1000 rad/s, though the maximum beam deflection is limited to a very low value. If acousto-optical deflectors are used, dispersive effects have to be considered. Above that, astigmatism is caused at high angular speeds due to the working principle (chirped acoustic wave). MEMS scanners achieve parameters comparable to galvanometer and piezo scanners but lack a big aperture. As MEMS scanners offer best cost-effectiveness and further development is ongoing, they will probably become more established in future. KTN-scanners are very promising and will certainly become more viable for

Table 12.5 Summarized key parameters of devices for beam deflection

Scanner technology	Aperture	Beam deflect-tion up to	Positioning repeatability better	Maximum sinusoidal frequency	Maximum angular speed
Galvano-meter	5 mm… > 50 mm	±40°	20 µrad (short term)	~2 kHz @ 10 % of full step	200 rad/s
Piezo- based	5 mm… > 30 mm	±3°	5 µrad	~1 kHz @ ± 1,5°	50 rad/s
Acousto-optic	0.1 mm… 5 mm	±20 mrad (eq. ± 0.05°)	1 µrad	~100 kHz… 1 MHz @ full step	> 5000 rad/s
MEMS	1 mm… > 5 mm	±30°	0.1 mrad	~1 kHz @ 10 % of full step	200 rad/s
KTN	0.5 mm	±5°	n.a.	100 kHz @ full step	>5000 rad/s

ultrashort pulsed laser applications in future, if the interaction of the laser pulse with the free space charges is lowered to an insignificant level and random-access scanning is possible without limitation. Table 12.5 comprises the beam guiding parameters of above technologies, including the KTN-scanner (though being only suitable for a rather limited selection of USP laser applications).

12.3 Systems for Focal Position Shifting

For the majority of ultrashort pulsed laser applications, the optimal position of the focal plane varies due to advancing ablation (drilling), variation of workpiece geometry (different workpieces, geometric tolerances) or the process itself (e.g. 3-D waveguide-writing with ultrashort laser pulses). For some of these factors it is suitable to reposition the focussing optic or the workpiece, which both is typically slow because of high amounts of moving masses. If faster repositioning of the focal plane is necessary to meet the application-driven demands, focal position shifting systems are viable.

Focal position shifting is achieved by varying divergence of the laser beam typically in front of the focusing optics.[6] The resulting focal position shift Δz can be calculated by (12.3), where f_{FO} is the focal length of the focusing optics.

[6]Alternatively, divergence and thus focal position along the beam propagation axis is varied without using focusing optics. In this case, the divergence variation is used for both the compensation of the deviation from a planar focal position induced by beam guiding and focal position shifting. This technique is sometimes used in remote processing, e.g. remote welding. In micro-processing additional focusing optics are standard.

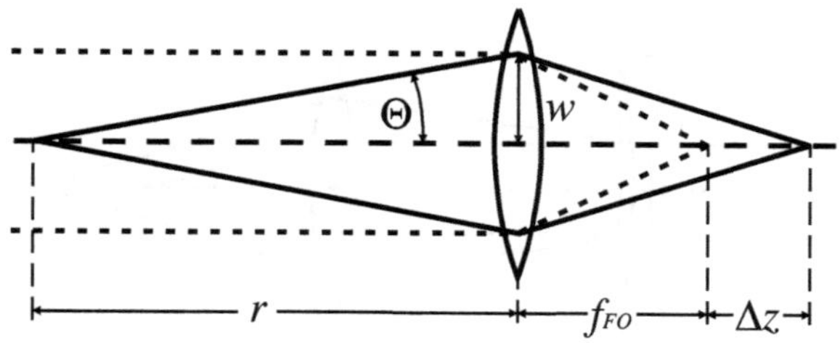

Fig. 12.7 Position of variables for focal position shift calculation due to divergence variation

$$\Delta z = \left(\frac{1}{f_{FO}} - \frac{1}{r}\right)^{-1} - f_{FO} \text{ with } r \cong \frac{w}{\Theta} \tag{12.3}$$

The variable r is the radius of the light wavefront incident on the focusing optics, i.e. the distance of the (virtual) focus to the focusing optics which is varied by the divergence angle Θ, see Fig. 12.7. For small values of divergence Θ and beam radius on the focusing optics w (paraxial approximation), the light wave radius r can be calculated by (12.3). In many cases w can be assumed to be constant. The sign and position of variables r, f_{FO}, Δz, w and Θ are depicted in Fig. 12.7.

Typically, the supplied data for most focal position shifting devices are stating either the refraction power range $D_{min} \ldots D_{max}$ or the focal length range $f_{min} \cdots f_{max}$ as a substitute to divergence variation. In conjunction with focusing optics these values can be used to calculate the resulting focal position shifts Δz_{min} and Δz_{max} according to the following (12.4) and (12.5). Figure 12.8 shows the position and sign of all variables. Typically f_{FO} is positive and much smaller then both $|f_{min}|$ and $|f_{max}|$.

$$\Delta z_{min} = \frac{f_{FO}^2}{d - f_{min} - f_{FO}} \quad \text{with} \quad f_{min} = \frac{1}{D_{min}} \tag{12.4}$$

$$\Delta z_{max} = \frac{f_{FO}^2}{d - f_{max} - f_{FO}} \quad \text{with} \quad f_{max} = \frac{1}{D_{max}} \tag{12.5}$$

Equations (12.4) and (12.5) point to the fact that positioning of the focal position shifting device is crucial. The distance d between focusing optics and focal position shifting device should be chosen in accordance with the following remarks (pre-condition: $0 < f_{FO} << |f_{min/max}|$).

- Only accounts to positive values of f_{min} and f_{max}:
  - d smaller than appr. $2f_{min}$ and $2f_{max}$, otherwise focal position shifting capability is lower than possible and
  - d not in proximity of f_{min} to $f_{min} + f_{FO}$ and f_{max} to $f_{max} + f_{FO}$, otherwise focus inside focusing optics or divergent beam after focusing optics may result.

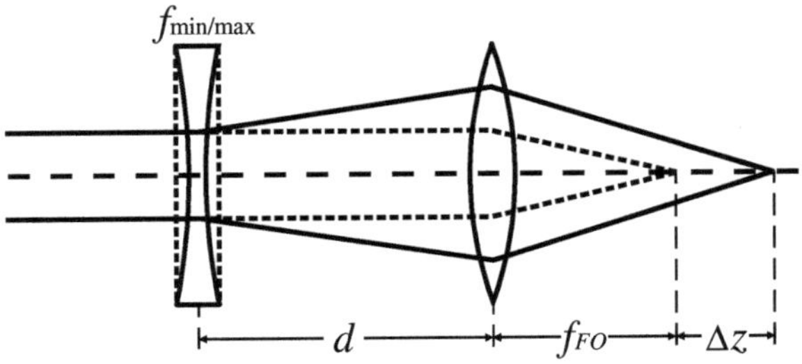

Fig. 12.8 Position of variables for the focal position shift calculation with a substitute focal length

- Only accounts to negative values of f_{min} and f_{max}:
  - d as small as possible, otherwise focal position shifting capability is lower than possible.

With regard to compact system size and high focal position shifting capability, d is typically chosen to be much smaller than $|f_{min}|$ and $|f_{max}|$ and in range of f_{FO}. In the following a selection of both industrially established and focal position shifting devices in development will be reviewed.

12.3.1 Moving Lens

The most common way to vary the focal position is the use of a telescope setup (in general Galilei type) in front of the focusing optics. One of the lenses is mounted moveable and translated by an actuator, see Fig. 12.9. The actuator has to realize high-speed linear movement of the lens at low rates of angular displacement, hysteresis of displacement, jiggling or bouncing [21]. Typically linear stages with gear motors are used in low-cost moving lens telescopes. Offering greatest cost-efficiency, these devices are suitable for (quasi-)static displacement of the lens, though inducing significant jiggle during movement. More sophisticated systems use linear stages driven by linear-, servo- or galvanometer-motors.

In microscopy and micro-processing technology reimaging and accordingly focusing optics typically weigh less, so an alternative focal position shifting technique is to directly move the focusing optics. Again linear stages could be used for this purpose. Lens holders with attached linear piezo actuators have proven to offer high precision movement over an adequate movement range. Latter technique is commercially available and established in application, especially optical microscopy [22, 23].

In Table 12.6 key parameters for both moving lens telescopes and moving focusing optics are stated.

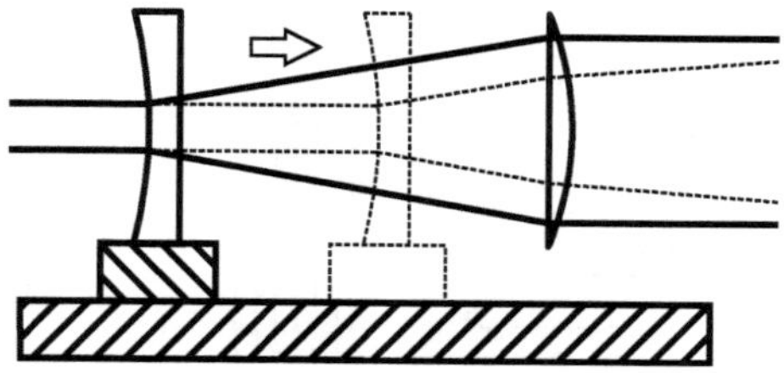

Fig. 12.9 Schematic layout of moving lens focal position shifting device

Table 12.6 Typical key parameters of moving lens telescopes and moving focusing lenses[a]

Moving	Input aperture	Mean focal length	Refraction power range	Repeatability better than	Step response time
Lens in telescope	2 mm… 30 mm	∞…200 mm or −200 mm…−∞	~0.5 dpt	0.1 % of full refraction power range	10 ms @ full range
Microscope objective	Same as focusing optics	Same as focusing optics (typ. < 5 mm)	n.a., but max. $\Delta z = \pm$ 0.25 mm	±10 nm	10 ms @ full range
Focussing lens	Same as focusing optics	Same as focusing optics (typ. < 50 mm)	n.a., but max. $\Delta z \approx \pm$ 25 mm	~0.1 % of Δz range	~100 ms… 1 s @ full range

[a]According to datasheets of Scanlab AG, Arges GmbH, Piezosystem Jena GmbH and Physik Instrumente (PI) GmbH & Co. KG

12.3.2 Deformable Mirrors

Instead of using refractive optical elements to vary the divergence, one can use reflective elements, i.e. deformable mirrors. These types of focal position shifting devices consist of a thin mirror plate, which is deformed by applying force or pressure to either one or more points or surfaces, respectively. The major difference is the resulting deformation of the mirror surface, which leads to the desired variation in divergence. Whereas the deformation was sometimes realized by applying water pressure to the backside of the mirror, today piezo, electrostatic or electromagnetic actuators are commonly used as they exceed in actuation frequency and accuracy.

Single actuator deformable mirrors typically consist of a single piezo actuator, which is attached to the center of the round mirror backside surface by either a ring-shaped or punctual coupling, cp. Fig. 12.10. On actuation the mirror is deformed to either a convex or concave shape. The deformed shape equals a spherical or parabolic mirror for only a portion of the mirror, the so-called clear aperture [24, 25].

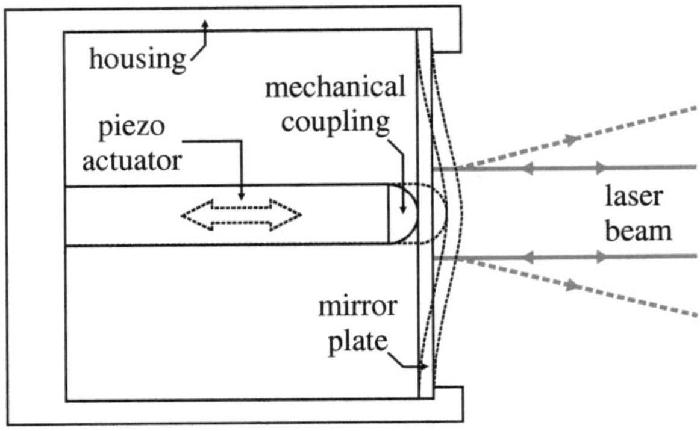

Fig. 12.10 Schematic layout of single piezo actuator deformable mirror

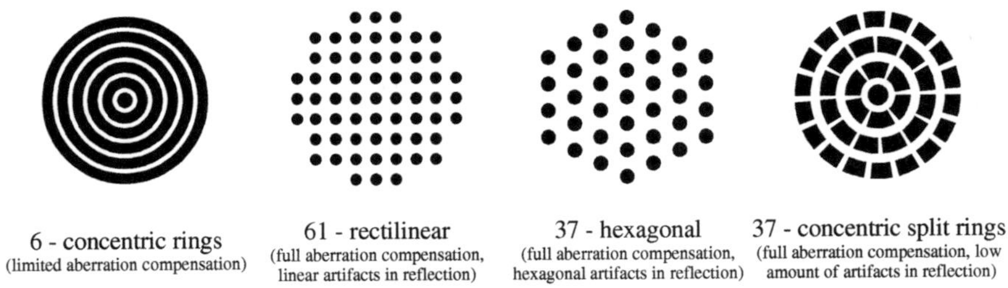

Fig. 12.11 Selection of actuator arrangements of multi-actuator deformable mirrors in *top-view*

Multi actuator deformable mirrors consist of up to $\sim$1000 actuators (piezo, electrostatic or other types), which can be controlled separately. As before, the actuators are attached to the backside of the mirror surface. Typically actuator arrangements include rectangle, ring-shaped, hexagonal and other geometries. Figure 12.11 shows a selection of arrangements used in deformable mirrors, where the black surfaces represent actuator element shapes in top-view (i.e. looking through the frontside of the mirror).

Most multi actuator deformable mirrors are not only capable of changing the divergence of a laser beam but also a selection of wavefront errors. As more actuators are used, more wavefront errors can be corrected or altered. One of the most well-known applications of this technique is the atmospheric aberration compensation in earth-based telescopes [26]. In focal position shifting applications, the multiple actuators are typically not used to compensate wavefront errors, but to both increase actuation frequency and optimize deformed mirror shape [27–29].

As the mirror shape equals a spherical or parabolic mirror on elongation, aberrations would be induced if used off-axis excessively. Multi actuator deformable mirrors are typically capable of compensating only a small portion of these aberrations. Therefore, a small angle of incidence (AOI) has to be assured by the optical setup.

Table 12.7 Typical key parameters of single-actuator and multi-actuator deformable mirrors[a]

Actuators	Input aperture	Mean focal length	Refraction power range	Repeatability better than	Step response time
Single	10 mm… 25 mm[c]	∞ (planar mirror plate)	0.8 dpt	2 % of full refraction power range	4 ms @ full range
Multi[b]	2 mm… > 50 mm[c]	∞ (planar mirror plate)	1.0 dpt	1 % of full refraction power range	4 ms (1 ms) @ full range[d]

[a]According to datasheets of blz GmbH, Agiloptics Inc., Thorlabs GmbH and Edmund Optics Inc
[b]Capable of compensating aberration to some part
[c]Referring to clear (usable) aperture
[d]Shorter time in brackets refers to usage of evacuation window in front of mirror

When the mirror changes its shape, both the mirror mass and adjacent air has to be moved. At high actuation frequencies air induces a high amount of damping and thus limits the achievable stroke. This takes place starting at appr. 200 Hz actuation frequency. To counteract this limitation sealed windows can be placed in front of the mirror plate, enabling evacuation of the volume between mirror plate and window. Low vacuum is sufficient to enhance the applicable actuation frequency to appr. 1000 Hz. The latter limit is caused by mechanical resonance effects [28].

Table 12.7 states the key parameters of deformable mirrors [24]. In comparison to other focal position shifting devices reviewed in this chapter, the deformable mirror is the only reflective type, others are at least to a major part refractive. Thus, no pulse-lengthening group velocity dispersion is induced, which is more important if pulses are shorter (appr. below 200 fs) [10].

12.3.3 Polymer and Liquid Lenses

Within the last decade, polymer and liquid lenses became established as cost-efficient alternatives to moving lens systems especially in miniaturized optical systems needing high focal position shifting capability. There are several approaches to achieve focal length variation using liquids and polymers as lenses. One approach is to embed a liquid into a polymer membrane and deform the shape by applying pressure either directly to the membrane or to a reservoir of liquid connected to the polymer membrane volume. In both cases, the shape and thus focal length is changed accordingly [30, 31]. Another approach is to utilize electrowetting. Electrowetting is the change of wetting properties of typically hydrophobic surfaces when an electrical field is applied. By positioning a liquid interface between both polar and non-polar liquids (e.g. water and oil) in a tapered hole with surfaces capable of electrowetting, the curvature of the liquid interface can be varied by applying an electrical field to the surface, see Fig. 12.12 [32].

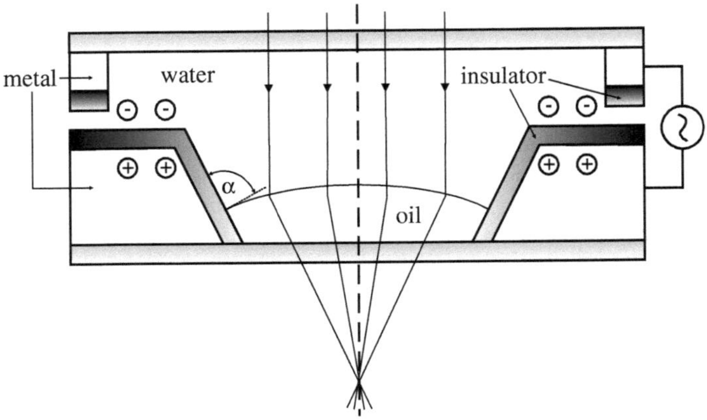

Fig. 12.12 Layout of electrowetting liquid lens with oil/water interface

Although polymer and liquid lenses offer high focal position shifting capability (high refraction power variation) at miniaturized component size, one main disadvantage arises due to the use of polymers and liquids. The typical transmission for wavelengths in the visual range is 90–97 %. As a result, the applicable average power of an incident laser beam has to be limited. Details on damage thresholds are not yet available in full detail, as the main scope of these devices is imaging technology, miniature-cameras in particular. As a coarse guideline the intensity should be lower than appr. 1 kW/cm^2 (cw) and 25 kW/cm^2 (short and ultrashort pulsed) to prevent damage. Also, quantitative information on the amount of induced group velocity dispersion is not available directly. Application in USP laser processing is therefore still very limited today.

Polymer and liquid lenses achieve key parameters as stated in Table 12.8.

12.3.4 Other Methods of Focal Position Shifting

Besides the reviewed devices for focal position shifting, the following selection summarizes the actual state of research.

Focal position shifting can also be achieved by use of liquid crystals in terms of spatial light modulation (SLM) with few electrodes. The liquid crystals are embedded in between two specially designed indium tin oxide (ITO) electrodes. When voltage is applied to the electrodes, the liquid crystals are arranged in such way, that the resulting optical path length distribution equals a lens. Focal length is varied by changing the applied voltage. As birefringence of liquid crystals is utilized, this technique works with linearly polarized light solely. The refraction power variation is appr. 5 dpt [33, 34]. When more than one ITO electrode is used (as done in commercially available spatial light modulators), wavefront correction

Table 12.8 Typical key parameters of polymer and liquid lenses[a]

Type	Input aperture	Mean focal length	Refraction power range	Repeatability better than	Step response time
Liquid in membrane	10 mm… 20 mm	~50 mm	Max. 50 dpt	<1 % of full refraction power range	15 ms @ full range
Electrowetting	3 mm	∞	Max. 20 dpt	<1 % of full refraction power range	20 ms @ full range

[a]According to datasheets of Varioptic SA, Optotune AG and Holochip Corp

comparable to multiple actuator deformable mirrors is feasible. These devices, referred to as SLMs, will be discussed in more detail later in this chapter.

So-called tunable acoustic gradient index (TAG) lenses consist of a piezoelectric ring surrounding a transparent, liquid medium contained by windows. When AC voltage is applied to the piezoelectric ring, sound waves are induced into the liquid, resulting in a standing and rotationally symmetric pressure and density oscillation. As the density corresponds to the index of refraction, the wavefront of an incident laser pulse can be modified corresponding to the instantaneous pattern of the tunable acoustic gradient index lens when the laser pulse is transmitted. In a first demonstration a refractive power in the range of $\pm 6\ \mathrm{m}^{-1}$ could be discretely selected by varying the phase shift between laser pulse incidence and AC driving voltage. Remarkably it is possible to achieve switching times at full refraction power range of as low as 1 µs, as every single pulse is focused by the momentary state of the tunable acoustic gradient index lens. Up to now, tunable acoustic gradient index lenses are in prototype state. Extended use in future is probable as key parameters exceed commercially available technologies [35].

Electro-optical and acousto-optical elements are also studied regarding the suitability to realize high speed focal position shifting (switching times below 10 µs). Up to now, acousto-optical lenses lack an acceptable level of transmission efficiency (typ. < 60 %). As the operating principle is diffraction, transmission efficiencies will likely not increase significantly. Electro-optical lenses are capable of one-dimensional lensing only (cylinder lensing) and are in an early state of research [36, 37].

12.3.5 Summary and Outlook

For easy comparison, Table 12.9 summarizes the key parameters of all commercially available focal position shifting technologies reviewed in this chapter.

In most applications the moving lens telescope is still the standard solution for focal position shifting up to now. Though deformable mirrors exceed in focal position shifting frequency and aperture, the higher investment costs prevent wide use in micro-processing. Further improvements in MEMS technology, which is also

Table 12.9 Typical key parameters of different focal position shifting technologies

Technology	Input aperture	Mean focal length	Refraction power range	Repeatability better than	Step response time
Moving lens in telescope	2 mm…30 mm	∞…200 mm or −200 mm…−∞	~0.5 dpt	0.1 % of refraction power range	10 ms @ full range
Moving microscope objective	Same as focusing optics	Same as focusing optics (typ. < 5 mm)	n.a., but max. $\Delta z \approx \pm$ 0.25 mm	± 10 nm	10 ms @ full range
Moving focussing lens	Same as focusing optics	Same as focusing optics (typ. < 50 mm)	n.a., but max. $\Delta z \approx \pm$ 25 mm	~0.1 % of Δz range	~100 ms… 1 s @ full range
Single actuator deformable mirror	10 mm…25 mm	∞ (planar mirror plate)	0.8 dpt	2 % of full refraction power range	4 ms @ full range
Multi actuator deformable mirror	2 mm… > 50 mm	∞ (planar mirror plate)	1.0 dpt	1 % of full refraction power range	4 ms (1 ms) @ full range
Liquid in membrane lens	10 mm…20 mm	~50 mm	Max. 50 dpt	<1 % of full refraction power range	15 ms @ full range
Electrowetting lens	3 mm	∞	Max. 20 dpt	<1 % of full refraction power range	20 ms @ full range

used in some deformable mirrors, may overcome this limitation in future. Novel, low-cost deformable mirrors could clearly enhance focal position shifting capabilities in application. This especially accounts to very short pulses below 100 fs pulse duration, as deformable mirrors are the only reflective focal position shifting devices reviewed (all others are refractive), thus not inducing any pulse-lengthening group velocity dispersion.

Polymer and liquid lenses offer highest focal position shifting capability but lack a sufficiently high damage threshold especially for high-power USP lasers. Probably this technology will further establish in imaging technology. For application in micro-processing distinct improvements in damage thresholds would be needed, which are not yet observable.

12.4 Systems for Beam Profile Shaping

A Gaussian beam profile, which is typically emitted by USP lasers, enables the generation of small microstructures in micro-processing at a relatively large depth of field (i.e. Rayleigh length). Though, the value of ablation efficiency[7] is

[7]Ablation efficiency is the amount of material ablated by the laser pulse in real process divided by the value of ablated material when the total pulse energy would have been used for ablation.

theoretically limited to appr. 37 % when using a Gaussian beam profile and assuming a fixed and invariant ablation threshold for micro structuring with USP lasers [38, 39]. The additional energy (residing in the edges of the Gaussian profile and center overshoot) is converted into heat to a large part. Furthermore the edge steepness of the ablated structure is limited due to the profile shape. Smooth structures prerequisite high pulse overlap, typically >80 %. These limitations can be solved by converting the Gaussian profile into a top-hat (also called flat-top) or super-Gaussian profile for instance. Theoretically, a top-hat intensity distribution enables an ablation efficiency of 100 % and smooth structures may be generated at lower pulse overlap, though the depth of field is reduced. In reality, an ablation efficiency of appr. 80–90 % can be achieved only at optimum conditions. The difference to the theoretical limit of 100 % is a result of heat conduction, limited edge steepness of the top-hat profile and other factors. Besides micro structuring, so-called homogenizing (i.e. producing top-hat intensity distribution) is also established in photolithography and illumination applications (e.g. UV curing).

Because of the mentioned advantages, some USP lasers are specifically designed to emit top-hat beam profiles or mixed beam profiles (e.g. top-hat in horizontal, Gaussian in vertical direction). This is accomplished by designing the laser crystal rod to work in an unstable regime in one or two directions, both perpendicular to the beam propagation axis. By modifying the intensity distribution or reimaging the collimated output beam of the laser, the desired focal spot shape is achieved [38].

For reshaping a Gaussian intensity distribution to a circular or rectangular top-hat intensity distribution, refractive and diffractive beam shaping elements and systems are available. The specific applicability of the different systems mainly depends on the laser beam source. Especially the mode and diameter of the laser beam, the spatial coherence and the laser power influence the selection of the proper beam shaping method. Other factors are the desired final top-hat dimension, the depth of focus and the usability of the shaper. The latter accounts to the adjustment effort and the stability if the beam properties, such as pointing accuracy or intensity distribution, change over time.

Beside the important field of top-hat generation also splitting and multi spot array generation become increasingly important, because the power of USP laser sources is steadily increasing year by year. Given that the scale up of power does not directly correlate with higher process efficiency and process speed while achieving constant structure quality [1], beam splitting and process parallelising become more interesting.

This section will explain and compare the basics on physical functionality of different beam profile shaping setups. Into date applications mainly diffractive and refractive beam shaping elements are used to a large part when beam homogenizing of ultrashort pulsed laser beams is desired, as key parameters such as homogeneity and transparency exceed other beam integration methods. Additionally the flexible beam shaping with spatial light modulators is of interest for researchers for several years now and is close to break through into industrial market. Therefore these two methods will be described in detail.

According to Dickey [40] beam shaping techniques can be grouped into two types, field mapping and beam integration.[8] The field mapping approach works well for a known laser beam distribution like Gaussian laser beams. The beam is shaped by an aspherical lens, lens system or diffractive beam shaper. Field integrators split and overlap the laser beam. These systems can be used to homogenize arbitrary intensity profiles like multimode laser beams with a low spatial coherence.

12.4.1 Field Mapping Methods—Singlet-Lens Systems

The basic principle of a singlet-lens system for beam profile shaping is the redistribution of the intensity profile by modifying the wavefront with specifically designed aspherical lenses. The so-called Powell lens, which is depicted in Fig. 12.13, is an example for such a system. It reshapes a Gaussian beam into a diverging top-hat intensity distribution.

Dimension and line length of the generated top-hat profile is typically in the range of the entrance beam diameter and can be extended to some 100 mm by magnifying objectives. This principle works well for rotational symmetric laser beams in TEM_{00} mode and a $M^2 < 1.3$. Applications are typically thin layer ablation or thin film photovoltaic industry [2, 41, 42].

Hoffnagle et al. [43] describe a refractive beam shaper which can be used to sort the light into a top-hat distribution using a pair of aspherical lenses which are arranged in a Galilei or Kepler-type telescope setup (Fig. 12.14). This method enables the generation of a collimated, speckle free beam with low divergence. The output of these systems may be reimaged to vary the dimensions of the top-hat distribution. The disadvantages of such systems are—equally to the single lens approach—the strict dependence on the entrance profile and delicate alignment procedure. Alignment errors and fluctuations of the laser beam have a strong influence on the achieved uniformity. The size of the top-hat is in the range of the entrance beam diameter and the beam is propagating over a large distance with a nearly constant homogeneous beam profile.

12.4.2 Diffractive Focal Beam Shaper

The shaping of a laser beam in the focal region of a lens is especially interesting for micro material processing applications. The diameter of the focus is physically

[8]Though, further beam shaping techniques are explained in the last part of this chapter, which do not accord to this categorization.

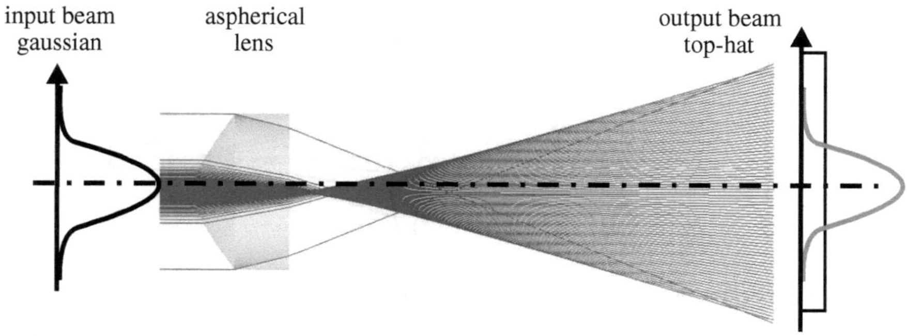

Fig. 12.13 Basic principle of single lens field mapping method (Powell lens)

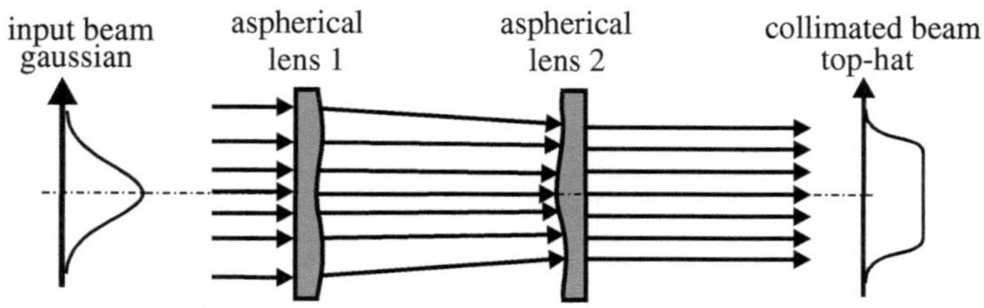

Fig. 12.14 Basic principle of Galilei telescope type: field mapping method for generation of a collimated beam

limited by diffraction. Hence, the size of the shaped top-hat beam is always slightly larger compared to the diffraction limited beam.

Several methods for focal beam shaping are derived directly from Fourier Optics theory, because of the reciprocal relationship between the object plane profile and the focal plane profile. Given that the Fourier transform of a *sinc* function results in a *rect*-function, which represents a 1-dimensional top-hat distribution function, leads to the conclusion that the complex amplitude of the entrance light field has to be shaped to a *sinc* or *bessinc* function in order to generate a square or round top-hat, respectively. This can be achieved by phase and/or amplitude modulation with the help of diffraction gratings [44], phase plates [45] or aperture obscurations [46]. The components must be optimized for a given input and system configuration but in general small deviations of the phase and/or amplitude function will not affect the result significantly. Figure 12.15 shows the principle setup of a focal beam shaper. An optical element generates the transformation function (like *sinc*- or Bessel-function) and a spherical lens is used to Fourier transform this function in the target plane in order to generate a top-hat intensity distribution.

This approach is especially interesting for focal spot shaping. The achievable diameter of the spot in the focus plane is 1.5–2 times the diffraction limited spot size

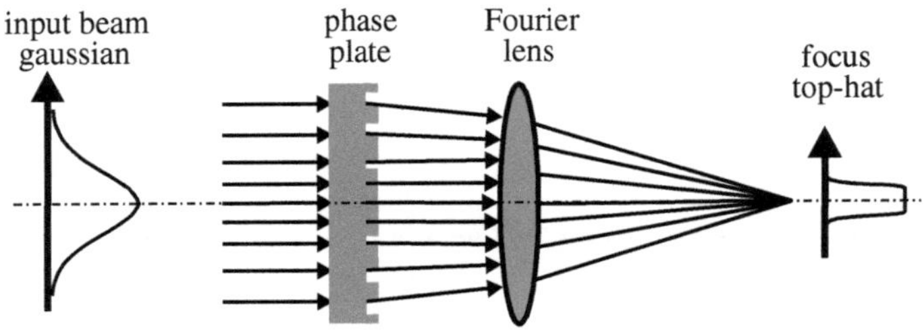

Fig. 12.15 Basic principle of Galilei telescope type field mapping method for focal beam shaping

of the laser beam without the beam shaping element. The focal spot profile may be varied to some limit along the propagation direction (e.g. top-hat, ring-shape, super-Gaussian intensity profiles). While allowing higher variance regarding focal beam shape, this setup is affected by delicate alignment and the prerequisite of a Gaussian input beam shape again. Nevertheless, this device is already established in some applications [39].

12.4.3 Diffractive Far Field Diffusers

Widely used diffractive optical elements for transformation of a laser beam are known as far field diffuser. In general these diffusers are diffractive optical elements, consisting of a periodic binary or multilevel amplitude and/or phase structure, which transforms the entrance distribution into a number of diffraction orders. After the element the beam will propagate with a defined angular spectrum and shaped distribution in the far field. The advantage of far field diffractive diffusers is, that the output profile is relatively independent of parameters like entrance beam diameter, entrance beam profile and overall alignment [40]. The disadvantage of these elements is the manifestation of a speckle pattern, which results in a strong intensity modulation in the target plane.

12.4.4 Beam Integration Methods—Multi Aperture Systems

The basic principle of the field integration method is the multiplication of the light source by special optical elements and superposition of the individual light sources (in one area for instance). Typically, this is achieved by using one of following optical elements.

- Multi aperture element (Fly's eye condenser, facetted mirrors)
- Kaleidoscope (glass rod with a square, rectangle or hexametrical cross section)

- Fiber with a square or rectangular core
- Refractive diffuser

The basic principle of refractive diffusers, e.g. thin plates of fused silica with a stochastically etched surface topology, is well-known (stochastic refraction). Typically these devices are not used in micro-processing as small feature sizes are hardly feasible. Kaleidoscopes and fibers with non-circular core can be understood to mix the beam during passing due to multiple reflections at the core boundary. Typically, the output of these devices is reimaged to achieve the desired geometry. For adequate homogeneity long kaleidoscope rods or fibers have to be used (typ. several meters). The damage threshold of these systems is limited, the group velocity dispersion lengthens the pulse when the initial pulse duration is below appr. 1 ps [10] and the systems are typically not compact. Hence, they are rarely used in USP laser processing.

12.4.5 Homogenizer Based on Lens Arrays

There are two main types of lens array beam homogenizers, non-imaging and imaging homogenizers [40], see Fig. 12.16. Both types use (micro) lens arrays (LA) to split the incident beam into beamlets. These beamlets pass through a spherical lens and overlap at the homogenization plane located at the back focal plane (FP) of the spherical lens. The spherical lens causes parallel bundles of rays to converge in the homogenization plane and is therefore called a Fourier lens (FL). The non-imaging homogenizer consists of a single lens array and a spherical lens. The imaging homogenizer differs in the use of two micro lens arrays before the Fourier lens. Both types of lens array homogenizers are schematically depicted in Fig. 12.16.

Square-type lens apertures of the first micro lens array LA_1 generate a square top-hat intensity distribution in the focal plane. Accordingly, circular or hexagonal micro lenses will generate a circular or hexagonal top-hat distribution. Due to the imaging of the sub-apertures into the target plane, this method is mainly used to generate top-hat dimensions larger than 1 mm.

A regular lens array is a periodic structure showing effects like grating inference and Talbot self-imaging [47, 48]. Light interacting with a periodic structure will always keep traces of this periodicity in its further propagation. This remaining periodicity usually generates an unwanted modulation in the focal plane and limits the degree of uniformity that can be achieved in the top-hat. The number N of modulation peaks in the target plane can be calculated with the Fresnel number FN, which depends on the micro lens properties according to (12.6) and (12.7), where p_{LA} is the pitch of the micro lens array, D_{FT} is the dimension of the top-hat in the

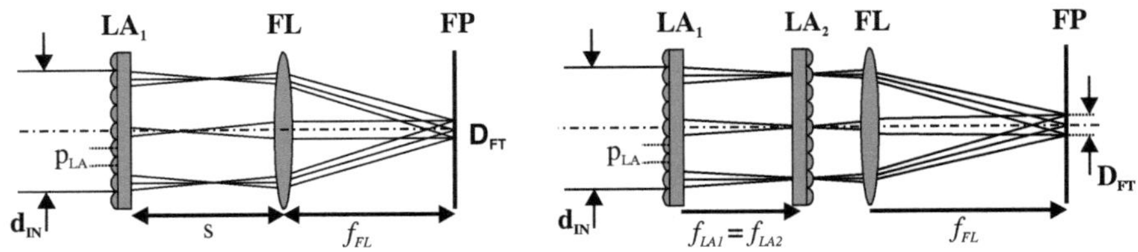

Fig. 12.16 Schematic representation of non-imaging (*left*) and imaging (*right*) homogenizer

target plane, f_{FL} is the focal length of the Fourier lens and f_{LA} is the focal length of the micro lens array.

$$FN \approx \frac{p_{LA} \cdot D_{FT}}{4 \cdot \lambda \cdot f_{FL}} \tag{12.6}$$

$$N = \frac{p_{LA}^2}{\lambda f_{LA}} = 4FN \tag{12.7}$$

The success of using multi aperture lenses for ultrashort pulsed laser applications depends strongly on the temporal coherence length of the laser beam source. The optical path difference OPD of the generated beamlets is essential for the appearance of interference fringes in the target plane and has to be considered before using this method. Generally the OPD lies in the order of some mikrons, considering micro lenses with pitch and radius of curvatures in the milimeter range. I.e. the method is better suited for ps-lasers (coherence length some mm) than for fs-lasers (coherence length some 10 μm) due to the value of OPD. In practice, imaging homogenizers exceed non-imaging homogenizers in uniformity of the top-hat intensity distribution, especially when the width of the beam shape in the focal plane is smaller than appr. 1 mm.

Another way to reduce the intensity modulations is the use of non-periodic elements. Wippermann et al. [49] use a so-called chirped micro lens array for this method.

The capability of generating multiple beam interference with multi aperture elements, such as micro lens arrays, can be used for the multiplication of the laser spot. A fly's eye condenser with small lens pitch will generate an interference pattern of equi-distant spots in the plane of the Fourier lens. The advantage of using a fly's eye condenser with two identical micro lenses is that the envelope of the spot matrix is a top-hat distribution and therefore the intensity in every single spot is approximately constant. Figure 12.17 shows the intensity distribution in the Fourier plane of two identical micro lens arrays in a Fly's eye configuration.

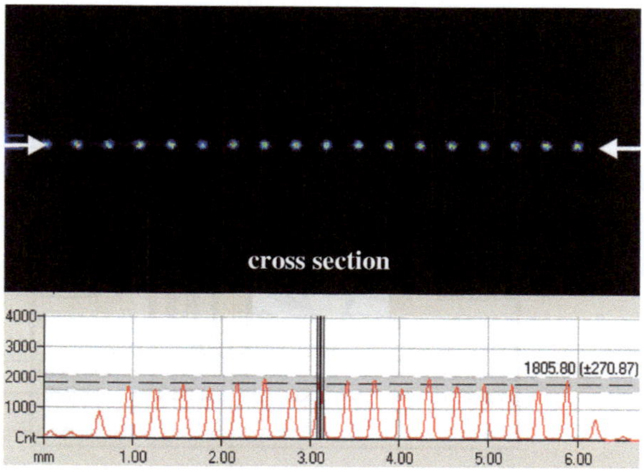

Fig. 12.17 Measured intensity distribution of a multispot generation in the Fourier plane of a lens: p_{LA} = 70 μm, f_{LA} = 220 μm, f_{FL} = 60 mm. Laser system: 532 nm wavelength and 12 ps pulse duration

12.4.6 Flexible Beam Shaping with Spatial Light Modulators

With the aid of liquid crystal light modulators, which are generally known as spatial light modulators (SLM), the phase and/or amplitude of a light wave can be influenced according to the principle of the electrical control of the optical characteristics of a nematic liquid crystal layer. Other technologies for modulation of the phase include: magneto-optic SLMs [50], deformable mirror SLMs [51], optical addressable photorefractive crystals [52] and multiple-quantum-well (MQW) SLMs [53]. With the help of a SLM both field integrating and field mapping methods can be implemented.

12.4.7 Setup of a Liquid Crystal SLM

A schematical setup of a reflective SLM is shown in Fig. 12.18. The system consists of a pair of electrodes with a liquid crystal layer between. The pixelized and separately switchable electrodes enable a local change of the electric field and therefore a change of the liquid crystal orientation. The incident beam is being reflected by the dielectric or metal coating between the pixelated electrodes and the liquid crystal compartment. The birefringent liquid crystal can be defined by two indices of refraction similar to a waveplate. The unique feature of the liquid crystal layer is the extraordinary refractive index, which is dependent on the angle of the molecules in respect to the normal of the entrance and exit layer, typically defined as the z-axis or direction of propagation. The phase retard depends on the layer thickness, the change of refractive index and the wavelength. In general the coating limits the damage threshold of the device to less then 1 GW/cm^2 for an aluminum mirror and 50 GW/cm^2 for a dielectric coating [54].

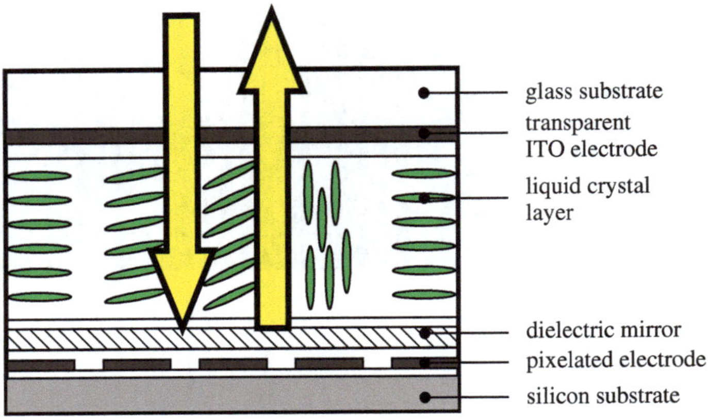

Fig. 12.18 Schematical layout of a liquid crystal display

Table 12.10 Typical key parameters for spatial light modulators[a]

Resolution	Pixel size	Fill factor	Active area	Addressing	Frame rate
Max. HD 1920 × 1080	8–20 μm	>98 %	12 × 16 mm^2	8 bit (256 Grey Level)	60 Hz–1 kHz

[a]According to datasheets of Hamamatsu Photonics K.K., Holoeye Photonics AG and Boulder Nonlinear Systems

State of the art SLMs are optimized to provide a phase shift of approx. 2π in the visible and near infrared spectral range. Typically, the reflectivity is >90 % and the diffraction efficiency exceeds 80 % [54]. In general a graphics board is sending the phase information (phase mask) to the device in terms of a grey level image (via DVI) with a frame rate of 60 Hz. Light, which is not shifted in phase, is visible in the form of a zero order in the Fourier plane. Additionally the image consists of a set of higher diffraction orders. The suppression of the unwanted zero order will be discussed later. In summary, spatial light modulators offer typical parameters as stated in Table 12.10.

12.4.8 Calculation of Phase Mask

The information for the target intensity distribution is stored in matrix form in a so-called phase mask, which can be calculated by means of specific algorithms or with the help of optical simulation software. Well known algorithms for the generation of computer generated holograms (CGHs) already exist and can also be used to calculate the phase masks for SLMs. The most common algorithm is the Gerchberg-Saxton algorithm (GS) [55]. The GS algorithm is an iterative algorithm

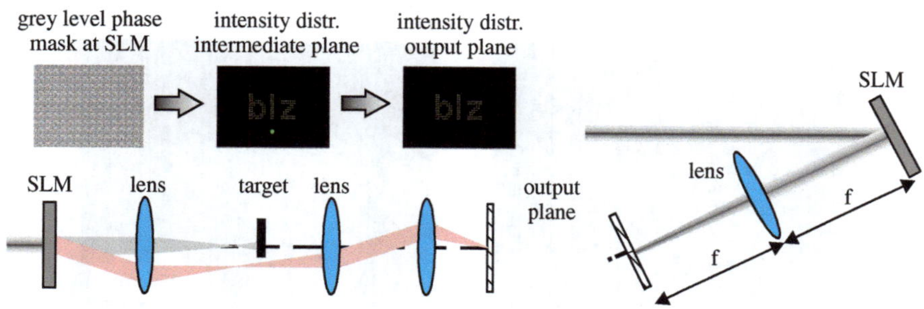

Fig. 12.19 Principle different setups of spatial light modulator

for retrieving the phase of a pair of light distributions related by a propagating function, such as the Fourier transform, if their intensities at their respective optical planes are known. Also non iterative algorithms are established and used for the generation of 3-D intensity distributions [56]. Additionally Fresnel holograms by multiplexing individual Fresnel lenses can be used [57, 58] in order to change the optical power of the system or to shift the focus in the working area.

SLMs are typically used in a 4f- or 2f-setup (Fig. 12.19). In a 4f-setup the zero order can be simply suppressed by assembling a beam block in the intermediate image plane [59]. Several other methods such as phase compression [60], phase conjugation [61] or by working with Fresnel lenses in the Fresnel regime [62] exist for zero order elimination.

12.4.9 SLMs for Material Processing

SLMs with a high damage threshold became available during the last years. The improvements were achieved by technological advances in the production of SLMs, especially the coating technology. The low response time of the liquid crystals allows a quick change of the intensity distribution within milliseconds. Thus, SLMs are used for the flexible beam shaping in short pulse and ultrashort pulsed laser applications, such as in material processing. A SLM offers a wide range of possibilities for flexible laser beam shaping, e.g. the generation of top-hats [63], multi-spot arrays [64] or arbitrary intensity distributions [63]. SLMs can be used in different applications like thin layer ablation [65], parallel processing [66], waveguide writing [67] or 3-D processing [68] to name only a few. Exemplary intensity distributions generated with a SLM are shown in Fig. 12.20.

In Fig. 12.21 the phase function (a), the simulated (b) and the measured (c) intensity distributions of a letter combination in the output plane consisting of spots are shown. Figure 12.21d shows the ablation result in stainless steel (1.4301), which was implemented with the phase mask. The diagrams illustrate the diverse and flexible use of SLMs for laser beam shaping in laser material processing.

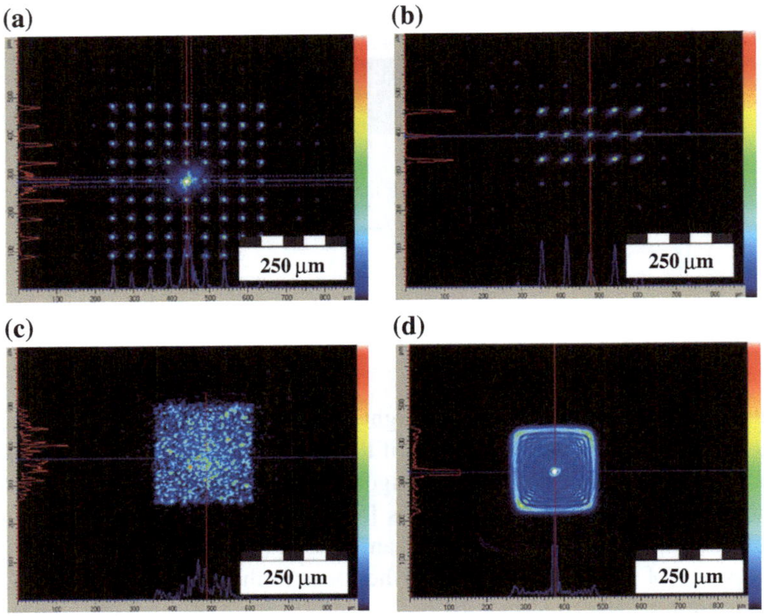

Fig. 12.20 Different intensity distributions generated by spatial light modulator and tested on a ps-laser (12 ps): **a** Spot-Array 9 × 9. **b** Spot-Array 3 × 5. **c** Top-hat diffuser. **d**) Arbitrary spot pattern

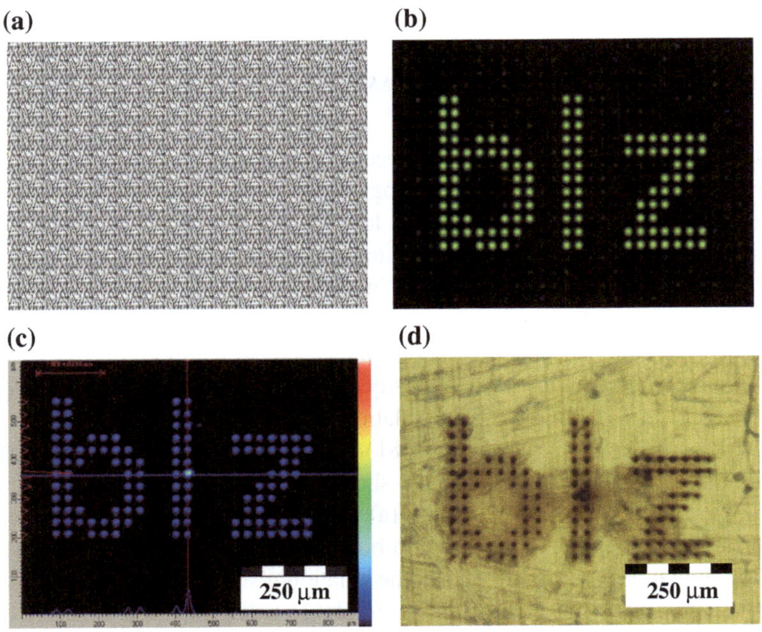

Fig. 12.21 Phase mask (**a**), simulated intensity distribution (**b**) and measured intensity distribution (**c**) of a letter combination, which was structured into stainless steel (**d**) with a picosecond laser. Pulse duration 12 ps, wavelength 532 nm, 100 pulses, pulse energy 10 µJ, repetition rate 200 kHz, focal length 75 mm

12.4.10 Other Methods of Beam Profile Shaping

Besides the use of the optical systems mentioned above, novel methods for the shaping of an USP laser beam are developed. A selection of those, which could be applied in future, is presented in the following.

The so-called two-photon-absorption (TPA) homogenizing utilizes the well-known fact that ultrashort laser pulses of higher intensities induce significant two-photon-absorption. In TPA homogenizing, the (non-uniform) laser pulse is transmitted through an absorber with a high TPA coefficient. Therefore, spikes in the beam profile are attenuated to a lower intensity. This works with USP lasers only, as TPA is a prerequisite. The advantage is that the TPA homogenizing setup is very robust and no exact adjustment is necessary (in contrast to lens systems). Though, laser power is absorbed (i.e. lost) and the achievable beam profile is fixed and dependent on the incident beam profile. Up to now this technique could not be established in application and is purely of scientific interest [69, 70].

Tunable acoustic gradient index (TAG) lenses, see section on 'other methods of focal position shifting', can be used for focal position shifting but also for beam profile shaping to some limit. Up to now a shaping from a Gaussian into a ring-shaped beam profile was demonstrated, where the diameter of the ring could be varied by the frequency of the driving voltage. The diameter can be changed at switching times of appr. 40 μs. As micro-processing with USP lasers mainly uses top-hat or super-Gaussian beam profiles, industrial use of tunable acoustic gradient index lenses for beam profile shaping will probably be limited [71].

12.4.11 Summary Laser Beam Shaping

Table 12.11 summarizes the properties of different beam shaping methods. Many different methods for laser beam shaping of a Gaussian beam profile to a top-hat intensity distribution exist. For the selection of a suitable beam shaping method and element, the beam properties of the laser system and also the desired top-hat dimension have to be considered.

12.5 Outlook: Polarization Shaping

A large part of USP laser applications is done with the use of one possible combination of beam guidance, focal position shifting and intensity profile shaping. Generally the polarization state of the laser beam is constant over the whole beam diameter and in time (e.g. linear or circular polarization). For a range of applications it would be beneficial if the polarization state is dependent on the position in the beam profile or on time. This can be achieved by spatial or temporal polarization shaping.

Table 12.11 Comparison of different laser beam shaping methods

Technology	Beam quality	Top-hat dimensions	Adjustment
Refractive beam shaper	Single mode	min. 1.5x… 2x diffraction limited spot	Critical (beam diameter)
Multi aperture systems	Multimode	1 mm…up to some 100 mm	Not critical
Diffractive diffuser	Single/multimode	1 mm…up to some 100 mm	Not critical
Diffractive beam shaper	Single mode	min. 1.5x…2x diffraction limited spot	Critical (beam diameter)
Spatial light modulator	Single mode	min. 1.5x…2x diffraction limited spot	Critical (beam diameter)
Multi actuator array	Single mode	min. 1.5x…2x diffraction limited spot	Critical

In temporal polarization shaping both, polarization vector angle and polarization state (linear, elliptical and circular), can be varied within one short pulse. Generally the used setups consist of an entrance grating to spatially decompose the laser pulse by frequency (i.e. wavelength), variation of polarization of different frequency spans by a spatial light modulator (liquid crystal array) and final recomposition using another grating. Hence, the polarization shaping takes place in the Fourier domain of the laser pulse. As sufficiently constant (locked) phase over the whole frequency span and broad bandwidth are preconditions for temporal polarization shaping, femtosecond pulses are generally used (typ. pulse duration <100 fs, bandwidth >10 nm). The laser power efficiency of temporal polarization shaping is typ. 15 % or less. In micro-processing this technique is very uncommon not only because of this fact. Up to now no significant improvements are observed in material ablation when temporal polarization shaping is applied. However, in chemistry and biology (studies on chiral molecules and DNA), laser spectroscopy (control of molecular alignment and quantum states), attosecond generation, information encoding/decoding and other fields of research high interest exists in temporal polarization shaping [72–74].

In spatial polarization shaping, the polarization state is a variable dependent on the beam profile position (constant in time). Besides other polarization profiles, the so-called radial and azimuthal polarization modes turned out to yield most benefit for a large part of possible applications. These polarization profiles are schematically depicted in Fig. 12.22 besides linear and circular polarization for comparison. Figure 12.22 also states commonly used abbreviations L, C, R and A for according polarization profiles [75].

Specially designed lasers emit R and A [76], but generally USP laser sources emit L. There exists a range of devices reshaping L to R and A. This includes the interferometric combination of two L polarized laser beams in TEM01 mode,

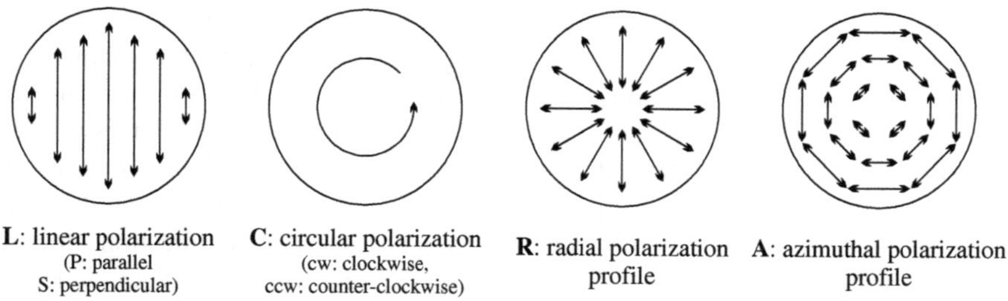

Fig. 12.22 Profiles of a selection of polarization types

spatially varying retarders[9] [77] and the use of liquid crystal arrays in terms of spatial polarization modulation [78]. The latter two devices are commercially available. While spatially varying retarders are used for static change of polarization profile (static spatial polarisation shaping), spatial polarization modulation using liquid crystals typically allows for the change of polarization profile within time-scales of approximately 20 ms (quasi-static spatial polarisation shaping—in comparison to typical pulse-to-pulse delays of 1 ms and shorter).

There are a number of applications where R and A polarization profiles lead to improved results. Especially in such applications, where the angle of incidence of the laser beam is large (above appr. 45°), i.e. laser cutting and drilling at high aspect ratios, R polarization was calculated to increase ultimate depth and cutting speed by a factor of 1.5–2 compared to C polarization [79]. First experimental results partly confirm the calculated ones in terms of an increase of the material removal rate by a factor of 2 in ultrashort pulsed laser cutting of stainless steel [80, 81]. Though, an increase of ultimate depth was not observed in the calculated extent. Despite these recent results, C polarization is still common in USP laser cutting and drilling to a large part.

12.6 Conclusions and Outlook

Up to now beam guidance, focal position shifting, intensity and polarization profile shaping were studied separately to a large part. For each sub-topic viable technologies are at hand and most promising developments are observable. E.g. galvanometer-based and acousto-optical scanners are the best choice for a variety of ultrashort pulsed laser applications and MEMS and KTN scanners will presumably become evenly viable on short- to medium-term timescales (as either more cost-efficient or much faster alternatives). The same accounts to focal shifting

[9]Spatially varying retarders consist of waveplates cut into circle sectors. Every waveplate's optical axis is arranged in such way that when all circle sectors are arranged to a full circle, TEM00 laser beams with L polarization are transformed to R or A polarization.

devices, where moving lenses are the standard to date. Damage thresholds of liquid, polymer and TAG lenses have to be investigated and increased eventually to be viable alternatives. Adaptive optics might be used if additional wavefront correction is necessary, e.g. if high amount of spherical aberration is expected. Otherwise the increased costs and complicated implementation of such devices are typically preventing broader usage. Established technologies for refractive and diffractive intensity profile shaping are most commonly well-suited for application already. Though, both intensity profile shaping and, as a relatively new field of recent research, very promising polarization profile shaping are just at the start to be used in application more broadly.

Although the combination of some or all of these technologies will yield non-trivial challenges, remarkable potential can be achieved. E.g. regarding structuring, cutting and drilling with USP lasers, the polarization and intensity profile may be adapted continuously to the position of the beam on the substrate and the according optimal process conditions in dependence of angle of incidence, structure depth and desired ablation geometry. This would yield a significant increase of structure quality and ablation rate. However, a large amount of degrees of freedom have to be controlled simultaneously and in dependence on each other. As this will likely not be feasible with manual adjustment, automated closed-loop control and adapted optimization algorithms will be needed, comparable to those already developed for temporal pulse shaping in ultrashort pulsed laser glass processing for instance [81]. Furthermore, single devices, which cover some technologies in parallel, such as beam guiding and intensity profile shaping for instance, would be beneficial clearly. A method to do both with a single acousto-optical deflector pair was demonstrated recently [82]. Applicants and researchers will most likely favour such devices over a non-coupled combination of several ones due to typically lower cost, lower complexity and increased flexibility.

References

1. S. Bruening et al., Ultrafast scan technologies for 3D-μm structuring of metal surfaces with high repetitive ps-laser pulses. Phys. Proc. **12**, 105–115 (2011)
2. E. Steiger et. al., Optimization of the structuring processes of CI(G)S thin-film solar cells with an ultrafast picosecond laser and a special beam shaping optics, in *Proceedings of ICALEO 2009* (2009), pp 1292–1297
3. R. Paschotta, *Encyclopedia of laser physics and technology* (Wiley-VCH, Weinheim, 2008)
4. M. Rosete-Aguilar et al., Calculation of temporal spreading of ultrashort pulses propagating through optical glasses. Rev. Mex. Fis. **54**(2), 141–148 (2008)
5. H.A.M. Snelders, J.S.C. Schweigger, His romanticism and his crystal electrical theory of matter. Isis **62**(3), 328–338 (1971)
6. R.P. Aylward, Advanced galvanometer-based optical scanner design. Sens. Rev. **23**(3), 216–222 (2003)
7. Physik Instrumente (PI) GmbH & Co.KG, Piezo tip/tilt mirrors & scanners: fundamentals (2013), http://www.physikinstrumente.com/en/products/prdetail.php?sortnr=300300. Accessed 11 June 2013

8. Brimrose Corp., Introduction to acousto-optics (2013), http://www.brimrose.com/pdfandwordfiles/aointro.pdf. Accessed 11 June 2013
9. Brimrose Corp., Introduction to A-O deflectors/scanners (2013), http://www.brimrose.com/pdfandwordfiles/aointro.pdf. Accessed 11 June 2013
10. I.A. Walmsley et al., The role of dispersion in ultrafast optics. Rev. Sci. Instrum. **72**(1), 1–29 (2001)
11. D. Vučinić, T.J. Sejnowski, A compact multiphoton 3D imaging system for recording fast neuronal activity. PLoS ONE **2**(8), 1–12 (2007)
12. A.E. Wallin et. al., Real-time control of optical tweezers. Proc. SPIE **6644** (2007). doi:10.1117/12.737269
13. H. Miyajima et. al., A MEMS electromagnetic optical scanner for a commercial confocal laser scanning microscope. J. MEMS **12**(3), 243–251 (2003)
14. H. Ra, Two-dimensional MEMS scanner for dual-axes confocal microscopy. J. MEMS **16**(4), 969–976 (2007)
15. D.W. Wine et. al., Performance of a biaxial MEMS-based scanner for microdisplay applications. SPIE **4178**, 186–196 (2000)
16. Fraunhofer IPMS, LinScan—quasi-static micro scanning mirror (2013). http://www.ipms.fraunhofer.de/common/products/MSD/linscan-e.pdf. Accessed 11 June 2013
17. K.T. Gahagan et al., Integrated electro-optic lens/scanner in a $LiTaO_3$ single crystal. Appl. Opt. **38**(7), 1186–1190 (1999)
18. D.A. Scrymgeour et al., Large-angle electro-optic laser scanner on $LiTaO_3$ fabricated by in situ monitoring of ferroelectric-domain micropatterning. Appl. Opt. **40**(34), 6236–6241 (2001)
19. K. Nakamura et.al., Wide-angle, low-voltage electro-optic beam deflection based on space-charge-controlled mode of electrical conduction in KTa1-xNbxO3. App. Phys. Lett. **89**, 131115-1–131115-3 (2006)
20. P. Bechtold, D. Bauer, M. Schmidt, Beam profile deformation of fs-laser pulses during electro-optic scanning with KTN crystals. Phys. Proc. **39**, 683–692 (2012)
21. H. Hügel, T. Graf, *Laser in der Fertigung* (Vieweg & Teubner, Wiesbaden, 2009)
22. J.Y. Lin, R.P. Huang, P.S. Tsai, C.H. Lee, Wide-field super-resolution optical sectioning microscopy using a single spatial light modulator. J. Opt. A: Pure Appl. Opt. **11**(015301), 1–6 (2009)
23. L. Brigo et. al., Water slip and friction at a solid surface. J. Phys.: Condens. Matter **20**, 354016, 1–5 (2008)
24. U. Urmoneit, M. Dirscherl, M. Guggenmoos, Scannerspiegel trifft Adaptive Optik—Neue Systemtechnik zur schnellen 3D-Strahlablenkung, in Laser in der Elektronikproduktion und Feinwerktechnik - LEF 2006, ed. by M. Geiger, S. Polster. Meisenbach, Bamberg (2006)
25. U. Urmoneit, M. Dirscherl, 3D-Strahlablenkung mit Adaptivspiegelscanner. Laser **2**, 38–40 (2006)
26. S.M. Ammons et. al., Laboratory demonstrations of multi-object adaptive optics in the visible on a 10 meter telescope. Proc. SPIE **7015**, 5 (2008)
27. T. Bifano et. al., Micromachined deformable mirrors for adaptive optics. Proc. SPIE **4825**, 10–13 (2002)
28. K. Bush, A. Marrs, M. Schoen, Electrostatic membrane deformable mirror characterization and applications. Proc. SPIE **5894**, 58940E.1–58940E.15 (2005)
29. K. Bush et. al., Electrostatic membrane deformable mirror wavefront control systems: design and analysis. Proc. SPIE **5553**, 28–38 (2004)
30. H. Ren et. al., Tunable-focus liquid lens controlled using a servo motor. Opt. Express **14**(18), 8031–8036 (2006)
31. H. Ren, S.T. Wu, Variable-focus liquid lens by changing aperture. Appl. Phys. Lett. **86** (211107), 1–3 (2005)
32. C. Gabay et. al., Dynamic study of a Varioptic variable focal lens. Proc. SPIE **4767**, 159–165 (2002)
33. A.F. Naumov et. al., Liquid-crystal adaptive lenses with modal control. Opt. Lett. **23**(13), 992–994 (1998)

34. M. Ye et. al., Properties of variable-focus liquid crystal lens and its application in focusing system. Opt. Rev. **14**(4), 173–175 (2007)
35. A. Mermillod-Blondin, E. McLeod, C.B. Arnold, High-speed varifocal imaging with a tunable acoustic gradient index of refraction lens. Opt. Lett. **33**(18), 2146–2148 (2008)
36. M. Krishnamurthi et. al., Two dimensional dynamic focusing of laser light by ferroelectric domain based electro-optic lenses. Appl. Phys. Lett. **90**(201106), 1–3 (2007)
37. A. Kaplan, N. Friedman, N. Davidson, Acousto-optic lens with very fast focus scanning. Opt. Lett. **26**(14), 1078–1080 (2001)
38. D. Keming, Thin layer ablation with lasers of different beam profiles: energy efficiency and over filling factor. Proc. SPIE **7202**, 72020Q–72020Q-9 (2010)
39. A. Laskin, V. Laskin, Advanced refractive beam shaping optics for advanced laser technologies, in *Proceedings of PICALO 2010 Paper 1005* (2010), pp. 1–6
40. F.M. Dickey, C.H. Scott, *Laser Beam Shaping: Theory and Techniques* (Marcel Dekker, New York, 2000)
41. S. Haas et al., High speed laser processing for monolithical series connection of silicon thin-film modules. Prog. Photovolt: Res. Appl. **16**, 195–203 (2008)
42. S. Haas et al., Analysis of the laser ablation processes for thin-film silicon solar cells. Appl. Phys. A **92**(4), 755–759 (2008)
43. J.A. Hoffnagle, C.M. Jefferson, Design and performance of a refractive optical system that converts a gaussian to a flattop beam. Appl. Opt. **39**, 5488–5499 (2000)
44. W.B. Veldkamp, Laser beam profile shaping with interlaced binary diffraction gratings. Appl. Opt. **21**(17), 3209–3212 (1982)
45. E. Jäger et al., Vorrichtung zur Strahlformung eines Laserstrahls. Offenlegungsschrift DE102010005774A1 (2011)
46. I. Gur, D. Mendlovic, Diffraction limited domain top-hat generator. Opt. Comm. **145**, 237–248 (1998)
47. I. Harder, et. al., Homogenization and beam shaping with micro lens arrays. Proc. SPIE **5456**, 99–107 (2004)
48. B. Besold, N. Lindlein, Fractional Talbot effect for periodic micro lens arrays. Opt. Eng. **36** (44), 1099–1105 (1997)
49. F. Wippermann, et. al., Beam homogenizers based on chirped micro lens arrays. Opt. Expr. **15**, 6218-6231 (2007)
50. K. Iwasaki et al., Practical magneto-optic spatial light modulator with single magnetic domain pixels. IEEE Trans. Magn. **44**(11), 3296–3299 (2008)
51. R.M. Boysel, A 1920 X 1080 element deformable or device for high definition displays. IEEE Trans. Electron Devices **38**(12), 2715 (1991)
52. Hong, H. John et al., Photovoltaic spatial light modulator. J. Appl. Phys. **69**, 2835–2840 (1991)
53. T.L. Worchesky et al., Large arrays of spatial light modulators hybridized to silicon integrated circuits. App. Opt. **35**(8), 1180–1186 (1996)
54. Hamamatsu datasheet and user guideline for X 10496
55. R.W. Gerchberg, W.O. Saxton, A practical algorithm for the determination of the phase from image and diffraction plane pictures. Optik **35**, 237 (1972)
56. J. Liesener et al., Multi-functional optical tweezers using computer-generated holograms. Opt. Comm. **185**, 77–82 (2000)
57. S. Hasegawa et al., Holographic femtosecond laser processing with multiplexed phase Fresnel lenses. Opt. Lett. **31**(11), 1705–1707 (2006)
58. S. Hasegawa, Y. Hayasaki, Holographic femtosecond laser processing with multiplexed phase Fresnel lenses displayed on a liquid crystal spatial light modulator. Opt. Rev. **14**(4), 208–213 (2007)
59. Z. Kuang et al., Fast parallel diffractive multi-beam femtosecond laser surface micro-structuring. App. Surf. Sci. **255**(13–14), 6582–6588 (2009)
60. J. Liang et al., Suppression of the zero-order diffracted beam from a pixelated spatial light modulator by phase compression. Appl. Opt. **51**(1616), 3294–3304 (2012)

61. L. Ma et al., Partition calculation for zero-order and conjugate image removal in digital in-line holography. Opt. Exp. **20**(2), 1805–1815 (2012)
62. A. Jesacher et al., Diffractive optical tweezers in the Fresnel regime. Opt. Exp. **12**(10), 2243–2250 (2004)
63. D. Liu et al., Picosecond laser beam shaping using a spatial light modulator. Proc. of ICALEO **2012**, 735–738 (2012)
64. N. Matsumoto et al., High-quality generation of a multispot pattern using a spatial light modulator with adaptive feedback. Opt. Lett. **37**(15), 3135–3137 (2012)
65. Z. Kuang et al., Ultrashort pulse laser patterning of indium tin oxide thin films on glass by uniform diffractive beam patterns. Appl. Surf. Sci. **258**, 7601–7606 (2012)
66. Z. Kuang et al., Diffractive multi-beam surface micro-processing using 10 ps laser pulses. Appl. Surf. Sci. **255**, 9040–9044 (2009)
67. M. Sakakura et al., Fabrication of three-dimensional 1 × 4 splitter waveguides inside a glass substrate with spatially phase modulated laser beam. Opt. Exp. **18**(12), 12136–12143 (2010)
68. A. Jesacher, M.J. Booth, Parallel direct laser writing in three dimensions with spatially dependent aberration correction. Opt. Exp. **18**(20), 21090–21099 (2010)
69. O.E. Martinez, Stabilization, attenuation and shaping of picosecond laser pulses by two-photon absorption in dyes. Opt. Qant. Electr. **11**, 223–228 (1979)
70. P. Simon, J.H. Klein-Wiele, Vorrichtung und Verfahren zur Strahlhomogenisierung von Ultrakurzpulslasern. Patent No. DE10249532, Deutsches Patent- und Markenamt (2004)
71. A. Mermillod-Blondin, E. McLeod, C.B. Arnold, Dynamic pulsed-beam shaping using a TAG lens in the near UV. Appl. Phys. A **93**, 231–234 (2008)
72. T. Brixner, G. Gerger, Femtosecond polarization shaping. Opt. Lett. **26**(8), 557–559 (2001)
73. M. Plewicki et. al., Independent control over the amplitude, phase and polarization of femtosecond pulses. Appl. Phys. B **86**, 259–263 (2007)
74. T. Brixner et al., Adaptive shaping of femtosecond polarization profiles. J. Opt. Soc. Am. B **20** (5), 878–881 (2003)
75. V.G. Niziev, A.V. Nesterov, Influence of beam polarization on laser cutting efficiency. A. Phys. D **32**, 1455–1461 (1999)
76. R. Oron et. al., The formation of laser beams with pure azimuthal or radial polarization. Appl. Phys. Lett. **77**(21), 3322–3324 (2000)
77. G. Machavariani et. al., Efficient extracavity generation of radially and azimuthally polarized beams. Opt. Lett. **32**(11), 1468–1470 (2007)
78. M. Stalder, M. Schadt, Linearly polarized light with axial symmetry generated by liquid-crystal polarization converters. Opt. Lett. **21**(23), 1948–1950 (1996)
79. O.J. Allegre et al., Laser microprocessing of steel with radially and azimuthally polarized femtosecond vortex pulses. J. Opt. **14**(8), 085601 (2012)
80. O.J. Allegre et al., Ultra-short pulse laser micro-machining of metals with radial and azimuthal polarization. Proceedings of ICALEO **2011**, 917–925 (2011)
81. A. Mermillod-Blondin, Size correction in ultrafast laser processing of fused silica by temporal pulse shaping. Appl. Phys. Lett. **93**(2), 021921 (2008)
82. P. Bechtold, R. Hohenstein , M. Schmidt, Beam shaping and high-speed, cylinder-lens-free beam guiding using acousto-optical deflectors without additional compensation optics. Opt. Exp. **21**(12), 14627–14635 (2013)

Part III
Medical and Technical Applications

Chapter 13
Femtosecond Lentotomy: A Prospect for a Treatment to Regain the Accommodation Ability

Silvia Schumacher and Uwe Oberheide

Abstract Presbyopia is the age-related loss of the accommodation of the lens of the eye which affects every person in the fifth decade of life. When presbyopia occurs, continuous growth of the lens fibers results in sclerosis of the lens tissue which is accompanied by a decrease in flexibility. Initially, this impairs the dynamic adaptation from far- to short-sightedness, until ultimately it fails completely. Currently, the conventional approach to compensate for the loss of accommodation is the use of reading glasses for short-sightedness. Although new surgical treatment methods have been developed, so far none of them allow a dynamic accommodation. An alternative approach is the restoration of the flexibility of the lens using a procedure based on the non-linear interaction of ultrafast laser pulses and tissue. The non-linearity of the photodisruption effect can be used to create micro-incisions inside the lens without opening the eye globe. These defined gliding planes thereby restore the lost flexibility. This treatment method, known as fs-lentotomy, enables regeneration of real dynamic accommodation. The fs-lentotomy treatment technique as well as the effect of laser irradiation on the tissue was evaluated. For the first time, various 3-D structures for gliding planes were successfully generated in experiments with human donor lenses of different ages. An average increase in anterior-posterior lens thickness of 100 μm accompanied by a decrease of equatorial lens diameter was observed as a direct consequence of fs-lentotomy. This is attributed to increased flexibility, as the force of the capsule bag deforms the lens tissue to a higher degree. Using the Fisher's spinning test, a 16 % average flexibility increase was ascertained in human donor lenses. The control of the position of the gliding planes was found to be extremely important for safe and successful surgery. In addition to the experiments, calculations of the biomechanics during accommodation were carried out using the finite element method. This indicated that the achievable increase in flexibility of the lens depends on the applied cutting pattern. In vivo experiments with the lab prototype surgical instrument showed that laser

S. Schumacher (✉)
Laserzentrum Hannover e.V., Hollerithallee 8, Hannover 30419, Germany
e-mail: silvia.schumacher@web.de

U. Oberheide
Augenklinik Am Neumarkt, Schildergasse 107-109, 50667 Cologne, Germany

S. Nolte et al. (eds.), *Ultrashort Pulse Laser Technology*,
Springer Series in Optical Sciences 195, DOI 10.1007/978-3-319-17659-8_13

incisions inside a rabbit eye lens caused no growing opacification (cataract) over a 6 month follow-up period. However, the incisions were still detectable using Scheimpflug imaging and histopathological techniques, although the visibility of the incisions was declining. No distinctive features were observed upon evaluating thermal exposure of the rabbit retina during fs-lentotomy. It is expected that no damage will occur in the human retina, as exposure of the human retina is lower than exposure of the rabbit retina, due to the larger human eye bulb. The basic scientific investigations of fs-lentotomy show that it is possible to recover the flexibility of ex vivo human donor lenses. Consequently, the requirements for regaining a dynamic accommodation exist. Furthermore, no side effects were observed during the wound healing process and during a 6 months follow-up period. Based on the presented findings, it can be concluded that fs-lentotomy has the potential to become a well suited procedure for the treatment of presbyopia.

13.1 Introduction

The average life-span in western industry states is increasing due to constant well-balanced diet and ongoing medical progress. Thus the amount of elderly people is growing and their needs cannot be neglected. Beside their health also their life quality is very important, and thus the need to improve age related physical restrictions is increased. One physical restriction which affects every elderly human being is the loss of near vision and accommodation ability (presbyopia).

According to the nowadays accepted Helmholtz' theory of accommodation, the optical power change of the eye is initiated by the contraction of the ciliary muscle. The ciliary body moves inwards and releases the tension of the zonular fibers. This loss of tension allows the elastic lens capsule to mold the lens into the more accommodated spherical shape, increasing the optical power of the lens [1–3].

Although the development of presbyopia is noticed in the fifth decade of life the complete loss of accommodation occurs within the next 15 years. The reasons of developing presbyopia are still discussed; one well accepted cause for the failure of accommodation is the progressive sclerosis of the crystalline lens [4, 5]. The flexibility of the lens decreases and the contraction force of the lens capsule fails to change the lens geometry to steeper curvatures and an increased lens thickness. Despite the loss of deformation ability of the whole lens, the other structures incorporated in the accommodation process like the ciliary muscle, zonular fibers and lens capsule are mainly unaffected by age and stay active the whole life [6–8].

The sclerosis process of the lens is caused by the steady growth of the lens. The loss in flexibility thus can be measured from early childhood years on [9]. The continued growth of new collagen fibers at the lens endothelium results in an increase in lens thickness and a compression of the existing fibers. Hence the nucleus and also the cortex of the lens become stiffer and less flexible. Over the years, the restoring force of the lens capsule fails to mold the hard lens tissue and the accommodation amplitude finally drops down to zero, irrespective of ciliary muscle contraction.

So far the conventional treatment method is to wear reading glasses. Nevertheless the glasses are regarded as a sign of age and are limited to mainly two, or three distance of good vision (near, (intermediate) and far) and thus a demand of a permanent and covering all distances (dynamic focusing) treatment exists, although there is no medical reason for a treatment. First therapy methods which are supposed to substitute the reading glasses are available (e.g. monovision, intraocular lenses, corneal inlays or scleral expansion bands) [10–15], but none of these procedures offers a dynamic accommodation ability.

Advances in modern laser technology offer a growing area of application in ophthalmology. Especially the introduction of the femtosecond laser established a new era of treatment possibilities. Since 2002 ultrashort pulse lasers have become well established in refractive surgery [16–18]. The outstanding precision and the advantage of scanning the femtosecond laser pulses inside transparent material offers the possibility to cut arbitrary three dimensional structures inside the cornea. Thus, the use of ultrashort pulse lasers has progressed from fs-LASIK to other intrastromal cutting applications, such as lamellar and penetrating keratoplasty [19, 20] and femtosecond assisted cataract surgery [21–23]. The cutting inside the tissue is generated by the effect of photodisruption, which is based on nonlinear absorption. Since the ocular media in the anterior segment are transparent to near infrared laser wavelengths, ultrashort pulse laser photodisruption is not limited to the cornea, but can also be focused deeper into the eye.

Myers and Krueger first of all described the clinical prospect of deeper delivery inside the eye, and suggested treatment of the crystalline lens with ultrashort laser pulses to overcome presbyopia in 1998 [24]. They showed that by performing shaped intralenticular incisions with a nanosecond laser, the lens tissue softens and its flexibility is potentially re-established [25]. One potential drawback of focusing ns laser pulses inside the lens, are the strong side effects (strong disruption forces and large residual gas bubbles which cause tissue damage) linked with the treatment. These side effects, however, can be dramatically reduced by the use of ultrashort (femtosecond) laser pulses, due to the lower pulse energy need to induce the laser induced optical breakdown [26, 27].

In early investigations of the treatment of enucleated porcine lenses, Ripken et al. showed that cutting inside the lens tissue with femtosecond laser pulses is feasible without apparent side effects. At the same time, the flexibility of the lens was increased depending on the particular cutting pattern performed [28].

This femtosecond laser treatment of the crystalline lens with the goal to enhance the lens's flexibility is called "fs-lentotomy". The aim of the presented investigations is to subsequently show that the flexibility of human presbyopic cadaver lenses can be improved as well and that the fs-lentotomy treatment has no further side effects to living tissue and causes no damage to the eye.

13.2 Proof of Concept of Fs-Lentotomy

To proof the concept of fs-lentotomy experimental investigations on human donor lenses as well as finite element simulation of a human lens during accommodation with and without induced fs-laser cuts were performed.

13.2.1 Experimental Investigations

13.2.1.1 Materials and Methods

Tissue Samples

For the current study, 41 human donor lenses from autopsy eyes for corneal transplants were available. The sample collection followed the tenets of the Declaration of Helsinki. The average age was 55 ± 12 years (ranging from 20 to 67 years). The postmortem time until laser treatment varied from 32 to 71 h (average: 49 ± 10 h) The lenses were taken from the eye bulb after corneal dissection and were stored in average 22.7 ± 0.8 h at room temperature in physiologic saline solution (0.9 %). To avoid dehydration during the experiments and the laser treatment the lenses were kept moist with frequently dripping saline solution onto them.

Gliding Planes

A simple pattern of gliding planes, which was verified by Ripken et al. [28] to improve flexibility, was used for the experiments. It consists of three main geometrical parts: two annular rings, two cylinders and a set of 12 planes (see Fig. 13.1).

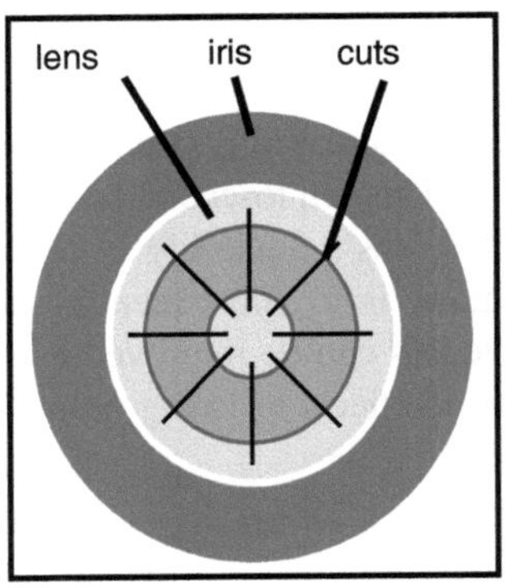

Fig. 13.1 "Steering-Wheel" cutting pattern

Two cylinders in depth direction of the lens with annular rings as top and bottom part. This pattern is then divided into segments by 12 radial planes. In principle, all geometric parameters of the pattern (e.g. radii of cylinders, depth of structures) are variable. However in this study the geometry of the pattern was kept fixed to ensure a comparability of all treated lenses despite of the fact that the relative size of the pattern compared to the lens size might also have an effect. The outer and the inner diameter of the pattern was set to $r_{out} = 2.5$ mm and $r_{in} = 1.0$ mm, respectively. The height of the pattern was $d = 1$ mm and it was applied 1 mm below the anterior pole.

Laser Treatment

For the application of the fs-laser induced gliding planes the Ti:Sa laser system Bright from Thales (Paris, France) was used. Its ultrashort laser pulses have a minimum duration of $\tau = 125$ fs at a repetition rate of $\nu_{rep} = 5$ kHz and $\lambda_c = 780$ nm center wavelength. The pulse duration is controlled by a single shot autocorrelator. The applied pulse energies varied between $E_{pulse} = 0.5$ and 1.5 μJ.

The laser beam was three dimensionally scanned by a galvanometer-scanner unit (GSI Lumonics, USA) and a mechanical translation stage (M126-DG, Physical Instruments, Germany) which moves the fixation unit of the crystalline lens relative to the focusing lens. A f-theta focusing optics (f = 75 mm) offered a spatial positioning resolution below one micron, a spot size of 5 μm and an operating range of 9 mm in diameter. The fixation unit consists of a special mount which centers the lens with respect to the focusing optics and a glass plate that applanates the lens surface up to 200–400 μm. Figure 13.2 shows the setup of the application unit including the scanner and the lens fixation.

The scanning parameters were chosen in accordance to the best parameters evaluated by Ripken et al. in porcine lenses [28]. The lateral spot separation was 6 μm and the axial one 40 μm, respectively. Furthermore the pattern was scanned twice with double spot separation and shifting the pattern by one distance of the spot separation during the second scan as recommended by Heisterkamp et al. to achieve best cutting smoothness [29].

Flexibility Measurement

The quantitative deformation ability of the crystalline lens was measured by applying rotational forces on the lenses as suggested by Fisher [30]. The centrifugal force during rotation simulates the outwards directed tension of the zonular fibers when the ciliary muscle relaxes in the unaccommodated eye [31]. Of course this method reflects not the true forces which act on the lens during accommodation, but it is a method to measure a quantitative change in deformation ability. The used rotation platform consists of a ring whose inner diameter is slightly smaller than the equatorial diameter of the crystalline lens. This offers the possibility of observing almost the whole shape of the lens with a CCD-camera during rotation. Images are

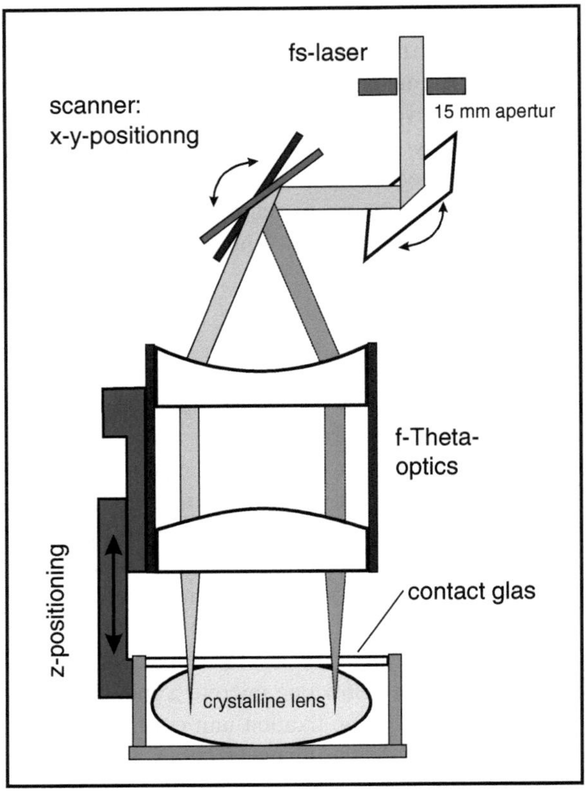

Fig. 13.2 Principle of the three-dimensional scanning unit consisting of two galvanometer scanning mirrors and a linear translation stag

taken at different rotational speeds and the geometry change during rotation can be analyzed by processing the digital images. The whole setup is shown in Fig. 13.3. According to data from Rosen [32], a 55 year old (average age in this study) human crystalline lens has a weight of approximately 244 mg and an equatorial radius of 4.73 mm. Thus a first approximation of the applied centrifugal force $F_c = m\upsilon r$ (m lens mass, υ angular velocity, r median radius) at a rotational speed of 1620 rpm (maximum speed used in this study) is 17 mN.

Fisher's spinning test is performed directly before and after laser treatment. Between the two spinning tests, the laser treatment is applied to all lenses in the same manner. The only parameter which is changed during the laser treatment is the pulse energy. Depending on the yellow coloring and developing cataract of the old human donor lenses, the pulse energy has to be adjusted between $E_{pulse} = 0.5$ and $E_{pulse} = 1.5$ μJ. The specific energy for each lens was subjected to the coloring of the lens and was chosen empirically from experience of former experiments.

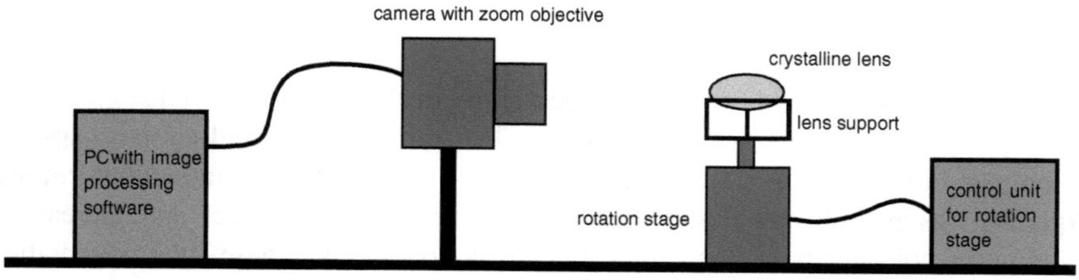

Fig. 13.3 Setup of Fisher spinning test

Analysis

The images of the lens taken during the spinning test were analyzed with an image viewing software. The anterior to posterior (a-p) thickness, equatorial diameter and radii of curvature of the central area (r = 2.8 mm anterior, r = 2.0 mm posterior) were measured and scaled to a mm-scale included in the images. The image resolution of the camera and the blurring due to the rotation of the lens lead to an observational accuracy of the lens thickness and equatorial diameter of about 30 μm. The accuracy of the fitted radii of curvature is about 0.5 mm

The different lens sizes require a standardization of the size for comparison. Therefore, the gauge for the change of a lens are its normalized geometric values, e.g. its normalized lens thickness $\eta = d_{rot}/d_0$ at a given rotational speed which is dimensionless. The crystalline lens' thicknesses at rotation and in non-rotating, stationary position are d_{rot} and d_0 respectively.

A comparison of the lens a-p thickness and equatorial diameter at static (0 rpm) and the normalized lens thickness η and normalized radii of curvature ξ at a rotational speed of 1620 rpm before and after laser treatment were performed. An increased lens thickness in non-rotating, static (0 rpm) position and a decrease in normalized lens thickness and an increase in normalized radii of curvature, respectively, represent an increase in deformation ability and therefore in flexibility.

The accommodation process is not simply characterized by the change of lens thickness; the actual change in optical power is mainly due to the change in anterior and posterior curvature and can be calculated in a first approximation using the conventional thick lens formula. The change of optical power (OP) due to the Fisher test can be obtained by comparing the unrotated case, which corresponds to the full accommodated state for near sight, to the 1620 rpm rotated case, which corresponds to a state which is close to the unaccommodated state for far sight. The normalized optical power due to the spinning test is $\Theta = OP_{rot}/OP_0$

All data was tested for statistically significance by using the student-t-test. Furthermore a control group of 10 lenses were investigated which underwent the same procedure as the treated lenses ones, just without laser pulses applied.

13.2.1.2 Results

The gliding planes could be applied successfully in all human donor lenses included in the following graphs and tables without damaging the surrounding lens capsule. Compared to a previous study on enucleated pig lenses [28], a homogenous cutting quality was not always possible due to the developing cataract and increasing yellowish coloring with age of the human donor lenses. These inhomogeneities necessitate a constant adjustment of the applied pulse energy. Unfortunately, at this point there is no method of ascertaining the right pulse energy for an optimal cut depending on the coloring of the lens. This results in an inhomogeneous quality of the cutting pattern of the human donor lenses. A typically obtained cutting pattern is shown in Fig. 13.4. These eyes with partly cutted patterns were common among all treated human donor eyes and as a consequence the enhancement potential of the cutting pattern therefore is reduced compared to a complete pattern. Nevertheless all eyes were included in the study [33].

Change in Deformation Ability

To overcome presbyopia, it is important that the forces of the lens capsule mold the lens to the more spherical accommodated shape after the laser treatment. In this case, the geometry of the lens should change directly after the treatment due to the restored deformability of the lens if the cuts increase the lens flexibility. The first indication of a more spherical shape of the lens is an increased lens thickness and a

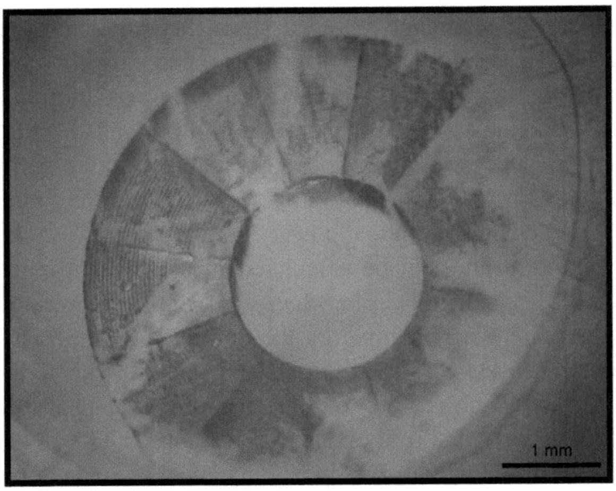

Fig. 13.4 A typical quality (about 60 % successful) of an obtained cutting pattern inside a human donor lens. A 61 year old lens; pulse energy 1.0 μJ. In the areas where no small gas bubbles are visible the tissue had inhomogeneities and the pulse energy was not high enough to induce an optical breakdown. Therefore no cut is created there. The circular edge around the pattern is the applanation edge, the lens is still applanated

decreased diameter, respectively. Figure 13.5 shows the relationship of the anterior–posterior lens thickness and equatorial diameter before and after laser treatment with no rotation applied (0 rpm) for the Fisher spinning test. The majority of the 41 treated lenses exhibit an enlargement of their thickness. The average increase of thickness of the 41 lenses is 97 ± 14 μm ($p < 0.001$, statistically significant change), whereas the change varies between −260 and +440 μm. On the other hand the lens diameter is decreasing in most of the lenses after the laser treatment. The average decrease is −120 ± 18 μm ($p < 0.001$, statistically significant). The change varies between +310 and −550 μm. The control group of 10 lenses showed in average a decrease of −154 ± 74 μm ($p > 0.05$) in lens thickness and a decrease in diameter of −141 ± 71 μm ($p > 0.05$) which are both not statistically significant at a 0.05 level and therefore no change occurred as anticipated.

Figure 13.6 shows the difference of normalized lens thickness η of 26 lenses at a rotational speed of 1620 rpm before and after laser treatment. (This measurement was performed only successfully on 33 of the available 41 lenses; additionally on seven lenses of the 33 measured lenses it was not possible to obtain the required data from the images showing the lens during rotation). As can be seen, the normalized lens thickness decreases on average by 0.013 ± 0.006 ($p < 0.05$, statistically significant) from 0.918 ± 0.007 to 0.905 ± 0.008, that means the lens flattens more than before. Thus the deformation ability of the lenses in total thickness increases

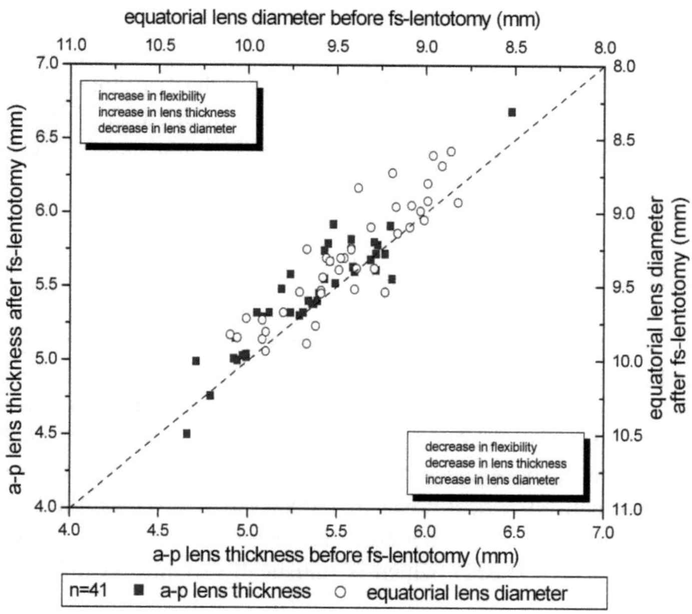

Fig. 13.5 Change of the anterior–posterior lens thickness (*left*) and diameter (*right*) of 41 lenses after laser treatment compared to before the treatment at 0 rpm. The *straight line* indicates no change. Above the line an increase in thickness and diameter occurred, below the line a decrease. In average the lens thickness increased and the diameter decreased due to the laser treatment

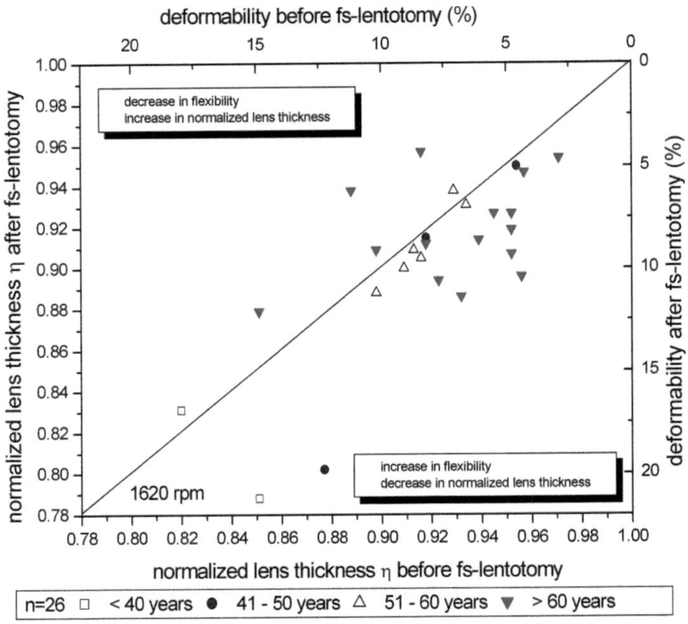

Fig. 13.6 Normalized lens thickness η of 26 lenses at rotational speed of 1620 rpm after laser treatment compared to the normalized thickness before the treatment. The *straight line* indicates no change

from 8.2 % before the laser treatment to 9.5 % after the laser treatment; that is an increase in deformation ability of about 16 %. The control group of untreated lenses showed in average an increase of 0.02 ± 0.02 ($p > 0.05$, not statistically significant) in normalized lens thickness [33].

Improvement in Optical Power Change

In Fig. 13.7 the change of normalized optical power (calculated from a-p thickness, anterior and posterior curvature) due to the laser treatment is shown for 26 lenses. The normalized optical power Θ due to a rotational speed of 1620 rpm before the treatment ranged from Θ = 0.61 to Θ = 1.03 with an average of Θ = 0.78 ± 0.02. After the treatment the normalized optical power decreased in all cases except one. The average normalized optical power is reduced to Θ = 0.65 (±0.02) and ranges from Θ = 0.42 to Θ = 0.87. Therefore the average change in optical power is increased by 59 % ($p < 0.001$), which is statistically significant [33].

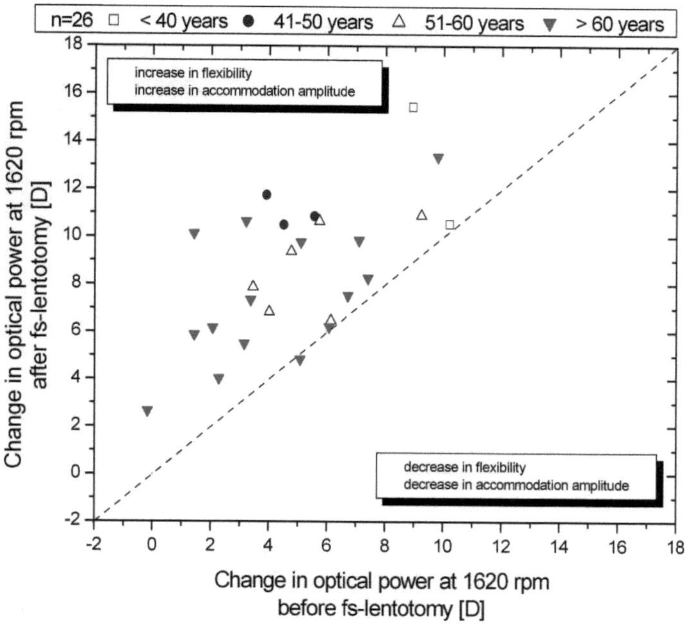

Fig. 13.7 First approximation of the optical power at rotational speed of 1620 rpm after laser treatment compared to before the treatment. The *straight line* indicates no change

13.2.2 Finite Element Model of the Human Lens

In order to understand the effect of the induced gliding planes on the biomechanical behavior of the lens tissue in more detail, a finite element model of the human eyes was developed.

Within the last 15 years several FEM models of the accommodation process which differ by complexity were published by several authors [34–42]. The model used for the fs-lentotomy simulation is adapted from the well established eye model of Burd et al. [37] and is realized in the software environment of ANSYS (Version 10.0, ANSYS INC., Canonburg, USA). The model differentiates the lens tissue volume according to anatomy in three zones: lens nucleus, lens cortex and lens capsule (see Fig. 13.8a). Each zone has its specific dimensions and material properties and is regarded in this model as homogenous, linear-elastic and isotropic. The chosen Young's moduli and poisson ratio for each zone are summarized in Table 13.1.

In order to simulate accommodation, the zonular fibers, which transmit the power of the ciliary muscle to the lens, are divided into three bundles (anterior, posterior and equatorial) with a different number of single fibers. The bundles have a ratio of 6:3:1 which results in an unequal stiffness distribution (anterior: $66 \cdot 10^{-3}$ N/mm, posterior: $33 \cdot 10^{-3}$ N/mm, equatorial $11 \cdot 10^{-3}$ N/mm) [37].

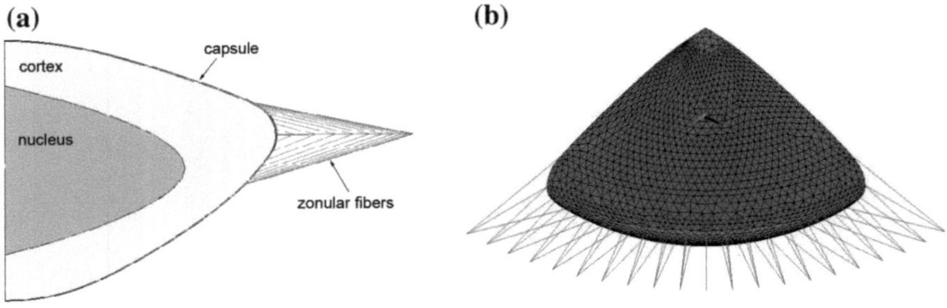

Fig. 13.8 **a** Cross-section of the lens tissue. **b** 3D model of the lens with 21zonular fiber bundles

Table 13.1 Young's modulus and poisson ratio of the three lens volumes [31, 43, 44]

	Young's modulus (N/mm^2)	Poisson number	Citations
Cortex	$3.4170 * 10^{-3}$	0.49	[31]
Nucleus	$0.5474 * 10^{-3}$	0.49	[31]
Capsule	1.27	0.47	[43, 44]

Despite the fact that the lens itself is rotational symmetric the lens model is created 3-dimensional in order to provide the potential to implement non rotational symmetric gliding plane patterns like the steering-wheel pattern used in the experiments. In order to downsize the number of elements and therefore calculation time, the model was restricted to a quarter of the lens. 21 bundles of zonular fibers were attached with equal spacing to the lens quarter. The model is shown in Fig. 13.8b.

In the beginning of the simulation the lens is in the fully accommodated state where no force is applied by the zonular fibers onto the lens capsule. During the simulation the loose ends of the zonular fibers, which would be attached to the ciliary muscle, are displaced radial outwards by 0.363 mm. This causes a traction force on the lens capsule and results in a flattening of the lens contour. The lens diameter is roughly stretched by 5 % with a maximum force of 0.1 N on the lens capsule.

13.2.2.1 Implementation of the Gliding Planes

Within the performed investigation three different gliding plane patterns were implemented into the lens model, and the resulting deformation during the accommodation process was evaluated.

One pattern is the steering wheel pattern (SW) used in the experiments. The dimensions were kept in accordance to the experiments as well. The outer and the inner diameter are 5 and 2 mm, respectively. The total height is 2.25 mm and it is placed 1.25 mm above and 1 mm below the lens equator. The three patterns inside

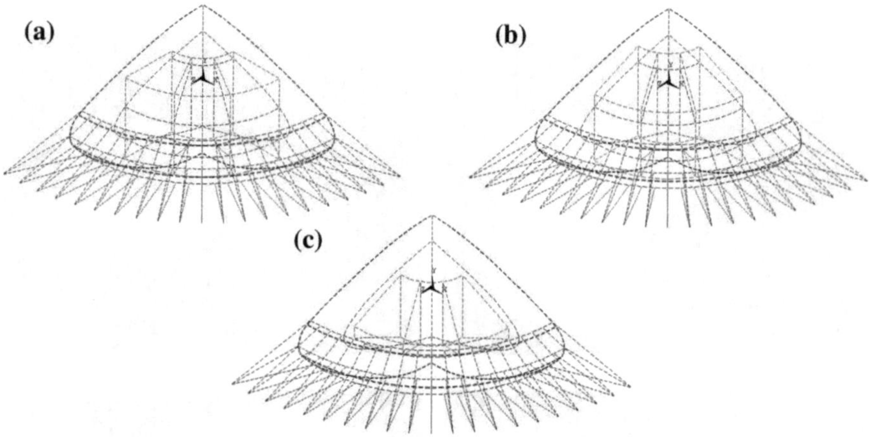

Fig. 13.9 Different cutting patterns which are applied to the lens model, **a** steering wheel (SW), **b** CSW cutting pattern, **c** DCSW cutting pattern

the lens model are shown in Fig. 13.9. The second pattern, conic steering wheel, (CSW) is similar to the steering wheel pattern, but the front and back surface have an angle of 25° so that the plane is parallel to the direction of tension caused by the zonular fibers. The last pattern, double conic steering wheel (DCSW) is more different. The inner and outer diameter of the pattern is kept, but the height is chosen that way that the pattern is placed only inside the lens nucleus. This results in a height of the inner and the outer cylinder of 1.8 and 0.34 mm, respectively, and an angle of the front and back plane of 22.8°.

The cutting patterns have a very small Young's modulus of $1 \cdot 10^{-6}$ Nmm^{-2} and a poisson ratio of 0.499 (very close to 0.5 for an incompressible material). The cuts have a thickness of 5 μm.

13.2.2.2 Theoretical Affirmation of Thickness Change

The resulting deformation of the lens induced by a displacement of the zonular fiber end points by 0.363 mm for a native lens and the three implemented cutting patterns is shown in Fig. 13.10. All four cases show a flattening due to the pull of the zonular fibers, but the flattening has a different occurrence. Table 13.2 summarizes the anterior-posterior lens thickness, the lens diameter and radii of curvature of all cases. The lenses with cutting pattern applied show a stronger deformability compared to the native lens. The steering wheel (SW) and the DCSW pattern have a 14 % higher deformability compared to the native lens, the CSW had only an increase of 7 % respectively.

Most important for the visual outcome is the change in radii of curvature. For each state the optical power of the lens can be calculated by using the thick lens equation. Table 13.3 summarizes the optical power of the lenses with and without

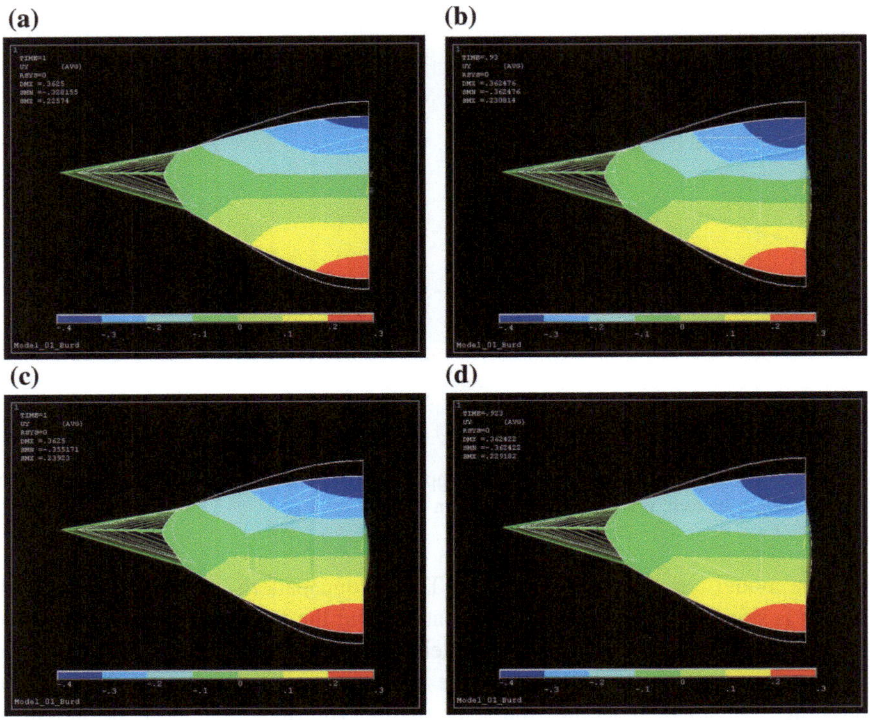

Fig. 13.10 Change of geometry of all four models at maximum displacement of 0.363 mm. **a** Native lens, **b** steering wheel (SW) pattern, **c** CSW pattern, **d** DCSW pattern

Table 13.2 Change in geometry of all four simulations at maximum displacement of 0.363 mm of the zonular fibre compared to the case without any displacement

Pattern	Change of a-p lens thickness d (mm)	Change of equatorial lens diameter L (mm)	Anterior radius r_a (mm)	Posterior radius r_p (mm)	Increase in deformability (%)
None	0.554	0.386	9.75	4.25	–
SW	0.631	0.404	10.75	4.30	13.9
CSW	0.594	0.410	10.00	4.25	7.2
DCSW	0.633	0.362	10.50	4.30	14.3

displacement and, therefore, the accommodation amplitude. Comparing the native amplitude to the ones with gliding planes an increase of 3.2 % can be found with the DK pattern and an increase of 9.0 % for the SW pattern. The optical power for the CSW pattern remains the same.

Thus the FEM simulation shows that induced gliding planes have the ability to change the accommodation behavior of the lenses. It was also shown that

Table 13.3 Change in optical power at maximum displacement

Pattern	Optical power full accommodated (D)	Optical power full deaccommodated (D)	Accommodation amplitude (D)	Increase (%)
None	38.48	31.23	7.25	–
SW	38.48	30.58	7.9	+9.0
CSW	38.48	31.26	7.22	−0.4
DCSW	38.48	31.00	7.48	+3.2

depending on the geometry of the patterns the deformation change is different, it can result in an increase in accommodation ability but it can also kept unchanged.

13.2.3 Discussion

The experiments demonstrate the possibility of creating defined cutting patterns inside the human crystalline lens tissue. The previous results on porcine lenses showing an increased deformation ability after laser treatment [28] could be confirmed on partly presbyopic human donor lenses. Whereas reproducing the cutting pattern at constant quality in all young clear porcine lenses was possible with optimized laser parameters [28], this task was more challenging on human donor lenses. The tissue characteristic of each human lens was different due to age, postmortem time and individual variety. Each lens had a different yellowish coloring and most of the lenses had a partially developed cataract resulting in an inhomogeneous tissue which increased scattering. At the moment there is no method of determining the exact pulse energy needed to induce optical breakdown at every position in the lens tissue. Thus the cutting quality of the patterns was not perfect in many of the treated lenses. Overcoming this limitation will be future work.

Nevertheless the experiments for enhancing the deformation ability of the crystalline lens show an average increase of deformation ability due to the Fisher's spinning test of 16 %. This value is lower than the value obtained on porcine lenses by Ripken et al. [28]. This may have two reasons. First, the deformation ability and geometry of young porcine lenses compared to presbyopic human lenses is different from the beginning and reacts in a slightly different way to the treatment. Second, due to the aforementioned challenges the cutting pattern was often not generated of perfect quality, which might result in a reduced gain in deformation ability. Due to this second fact, the averaged flexibility does not reflect the true possibility of regaining large quantities of deformation ability in the aged lenses.

In some lenses the deformation ability could be doubled, which was not previously achieved in porcine lenses [28]. This large gain in flexibility is due to the fact that human donor lenses suffer from the progressing sclerosis whereas the young porcine lenses are very flexible. This fact also explains the thickness increase of the human donor lenses directly after laser treatment which was never noticed on young porcine lenses before [28]. The restoring force of the lens capsule is able to mold the lens back to a shape more spherical compared to the untreated lens. The fact, that this was not observed on porcine lenses, eliminates the assumption that this increase in thickness is caused by the remaining small residual bubbles which disappear within 2 h after the treatment. Nevertheless, maintaining a control by measuring the lenses thickness again after some time is inappropriate, because the water intake of the lenses distorts the measured thickness values [45, 46]. The water intake also influences the initial thickness of the human lenses which reach our lab up to 3 days postmortem. Therefore all measured absolute values in this study have to be used with caution. Since the measurements were performed within 12 min the relative values comparing the lens before and after laser treatment are both affected in the same manner of water intake and therefore this effect can be neglected. This is supported by the fact that the relative obtained values show no relevant dependence on the ratio of diameter and thickness.

The effect of the water intake influences also the measurement of the radii of curvatures. The lenses are thicker and have altered radii of curvatures compared to in vivo measured values published in literature [7, 32, 47, 48, 49]. This alters the calculation of the optical power and therefore the accommodation amplitude. Thus the absolute values do not reflect the reality, but the relative values obtained before and after the laser treatment still show the potential of the treatment method. The values presented of an average increase of 59 % might overestimate the real success on living tissue, because the average change in optical power shown by in vitro tissue during the Fisher's spinning test is also higher compared to in vivo measurements of the accommodation amplitude [47, 50].

However the data clearly shows that the treatment method presented succeeds in increasing the change in optical power. Therefore it seems that the induced micro cuts inside the crystalline lens tissue have the potential to relieve the effects of presbyopia and offer an improvement in accommodation ability. The applanation of the lenses during the laser treatment could also have an impact on the mechanical behavior of the lenses during the experiments. However the performed study on the control group of 10 lenses showed no statistically significant changes of the lenses after the applanation. Therefore despite the larger deformation due to the applanation no conspicuous mechanical behavior is caused.

The next step in developing a successful clinical treatment method resulting from these promising experiments is optimizing the geometry of the cutting pattern to a maximum gain in accommodation amplitude and investigating possible undesired side effects. The optimum pattern is difficult to investigate experimentally, because every lens acts slightly differently during accommodation. Therefore theoretical work has to been done in addition.

Comparing the experimental results to the results of the performed FEM simulations is difficult. The main result of increased deformation ability by the induced gliding planes could be shown with both methods. The comparison of the values obtained is more or less not practicable. The FEM model is based on an idealized lens and there is definitive a difference between the simulated forces and the force which occurs during the Fisher spinning test. Thus a comparison of absolute or even relative values is not appropriate in the opinion of the authors. However, the FEM simulation showed very successful that the geometry and the placement of the pattern have a significant influence on the outcome in terms of increased accommodation amplitude. Higher deformation ability implies not automatically in higher accommodation amplitude. The change in radii of curvature is important to increase the accommodation amplitude and a wrong geometry might increase deformation but not accommodation. The steering-wheel pattern showed in this evaluation the best results, but further and more detailed investigations might show an even more suitable pattern, which might be even depending on lens size and initial geometry.

13.3 Safety Study of Fs-Lentotomy

The effectiveness of the treatment on porcine and human cadaver lenses was shown in the last section. In order to verify the results on living humans another aspect has to be addressed first: safety.

Besides the desired cutting effect of the tissue two side effects may occur in the eye. First the deposition of light energy into the crystalline lens causes an optical breakdown and thus a change of the lens tissue in the focal region. The high laser light intensities might also alter the lens tissue along the optical path of the light due to nonlinear effects. This could perhaps cause implications to the highly sensitive crystalline lens's metabolism and might result in cataract formation. Secondly, the light energy is not consumed completely at the focal volume and propagates further into the eye to the retina. The strongly divergent laser light will have not enough intensity to cause a laser induced optical breakdown (LIOB), but each laser pulse might cause a very small thermal heating. A single pulse should do no harm to the retina, but the accumulation of the several thousand pulses of a treatment might.

A preliminary in vivo investigation with a 5 kHz femtosecond laser system on the lenses of six living rabbits performed by Kruger et al. pointed out that the crystalline lens of a rabbit eye stays clear for at least 3 months after treatment.

In TEM analysis it was shown that the zone of changed tissue around the cut is about 0.5 μm and the surrounding hexagonal lens fibers appear normal and undisturbed. However, one rabbit did show visible evidence of cataract formation in both the treated eye and the control eye, and because of the bilaterality of occurrence, was believed to be unrelated to the laser treatment [51, 52].

Nevertheless both effects, retina damage and the induction of cataract, have to be investigated further.

In order to have practical treatment times (old system for experiment had 6 min) and a mobile unit, a new laser unit was developed which offered a treatment time of 25 s.

13.3.1 Methods and Material

The new mobile femtosecond-lentotomy system has in principle the same optical setup like the system described in the experimental part. The main difference is the new laser unit. The FCPA μJewel D-400 femtosecond laser system (IMRA America Inc, Ann Arbor, Michigan, USA) emits at a central wavelength of $\lambda_c = 1041$ nm with a maximum pulse energy of $E_{pulse} = 3.6$ μJ. The pulse duration is $\tau = 306$ fs at a repetition rate of $\nu_{rep} = 100$ kHz. Also the scanning device is a new adapted faster module. The eye globe of the rabbit is basically fixed to the three-dimensional scanning in the same way than the human donor lenses before. A suction ring fixates the eye, while the cornea is applanated by a flat glass contact [53].

30 Chinchilla Bastard rabbits at 10 weeks of age were enrolled and cared for in accordance with the recommendations of the German national guidelines for animal studies (No. 33.9-42502-04-07/1356). One eye of each animal was subjected to the laser surgery while the other served as an untreated control. The animals were sedated intramuscularly using ketamine 10 % and medetomidin hydrochloride 0.1 % (Domitor; Orion Pharma, Espoo, Finland). Prior to the laser treatment, a topical anesthetic drop of oxybuprocaine hydrochloride 0.4 % (Conjuncain EDO; Dr Gerhard Mann GmbH, Stulln, Germany) was administered.

Preoperative evaluation of the rabbit eyes included slit-lamp optical coherence tomography (OCT) (SLOCT; Heidelberg Engineering GmbH, Heidelberg, Germany) and Scheimpflug imaging (Topcon SL-45; Topcon Optical Instruments, Tokyo, Japan). To provide better visualization and exposure of the lens during examination and treatment, the rabbit pupils were dilated with tropicamide (Mydriaticum Stulln; Pharma Stulln GmbH, Stulln, Germany). The eyes were also kept moist with saline solution before and after the procedure. The intralenticular laser cutting pattern was delivered as a combination of radial and annular layers as described in the previous section with an inner diameter of 1 mm, an outer diameter of 4.0 mm, a depth of 1.5 mm, and 12 planes applied to the left eye of each rabbit. The pulse energy varied between $E_{pulse} = 1.2$ and $E_{pulse} = 1.6$ μJ, and the pulses were placed with a spot separation of 6–7 μm in the x-y plane and 50 μm along the optical axis into the lens tissue. With the fast repetition rate of $\nu_{rep} = 100$ kHz, the laser treatment was finished within 25 s.

After removal of the suction ring, the location and intensity of the generated effect was investigated with a surgical microscope (Zeiss AG, Jena, Germany) equipped with a charge coupled device camera. Subsequently, slit-lamp OCT and Scheimpflug images were taken. Reversal of the animal's sedation was accomplished by atipamezol hydrochloride 0.5 % (Antisedan, Orion Pharma). Further evaluation of the rabbit eyes was performed using the same techniques (surgical microscope, slit-lamp OCT, and Scheimpflug camera) 14 days, 1, 3 and 6 months postoperatively. Histopathological sections of the retina and the lens tissue were prepared in order to investigate the surrounding tissue in more detail. The sections were taken from 5 rabbits on day zero, on 10 rabbits on day 14, on 10 rabbits on month 3 and on 5 rabbits on month 6.

13.3.2 Results

The femtosecond-lentotomy treatment was successfully applied to the left lens of each rabbit. However, due to the limited cooperation of the sedated rabbits, the eye globe was not perfectly aligned to the suction unit in all cases, resulting in potentially decentered cutting patterns. Figure 13.11 shows exemplarily images of the micro incisions inside the rabbit's lens immediately after laser treatment, one month post, three months post and 6 months post surgery. Directly post surgery the cutting pattern is clearly visible with all three imaging techniques (see Fig. 13.11a). The residual gas bubbles of the photodisruptive effect, which disappear within an hour after treatment, are dominantly visible in the images. In the case of the Scheimpflug images the bubbles interfere with the exposure.

However, the tissue changes are bleached and no gas bubbles are observed at 1 month. This fading of the cutting pattern was also observed in the OCT and Scheimpflug images. As time is proceeding the visibility of the micro incisions is even more fading. Especially with the OCT imaging technique the micro incision are hardly detectable after 1 month of follow up (see exemplarily Fig. 13.11IIc–d). In the higher resolution cross section images of the Scheimpflug camera (see Fig. 13.11III) and in the surgical microscope images (see Fig. 13.11I) the micro incisions are still visible after 6 months however the visibility is fading progressively over time. More important is that the scattering of the incisions detectable in the Scheimpflug images are not spreading in the adjacent tissue, like it would be expected in the case of cataract formation. This is supported by the findings in the histopathological sections of the treated lens tissue. Directly after the fs-lentotomy

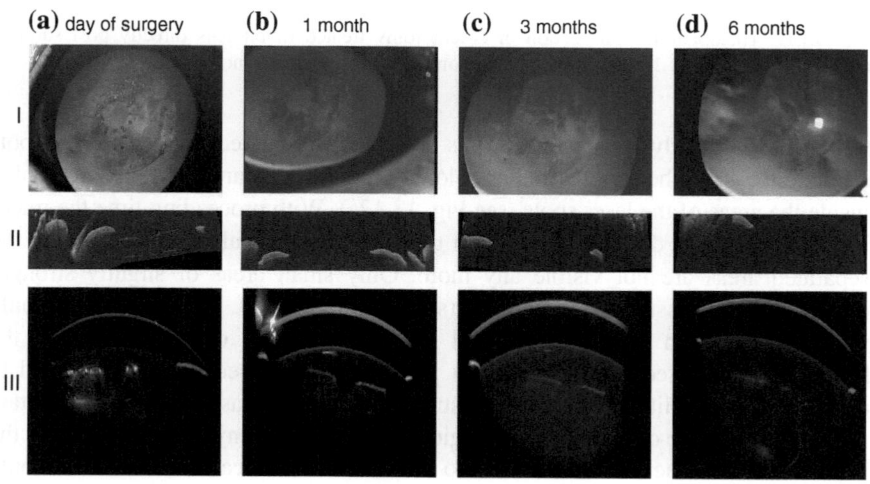

Fig. 13.11 *I* Surgical microscope images, *II* OCT images and *III* Scheimpflug images (pulse energy 1.4 μJ, spot separation x = 7 μm. **a** Directly post surgery, **b** 1 month post surgery, **c** 3 months post surgery, **d** 6 month post surgery.)

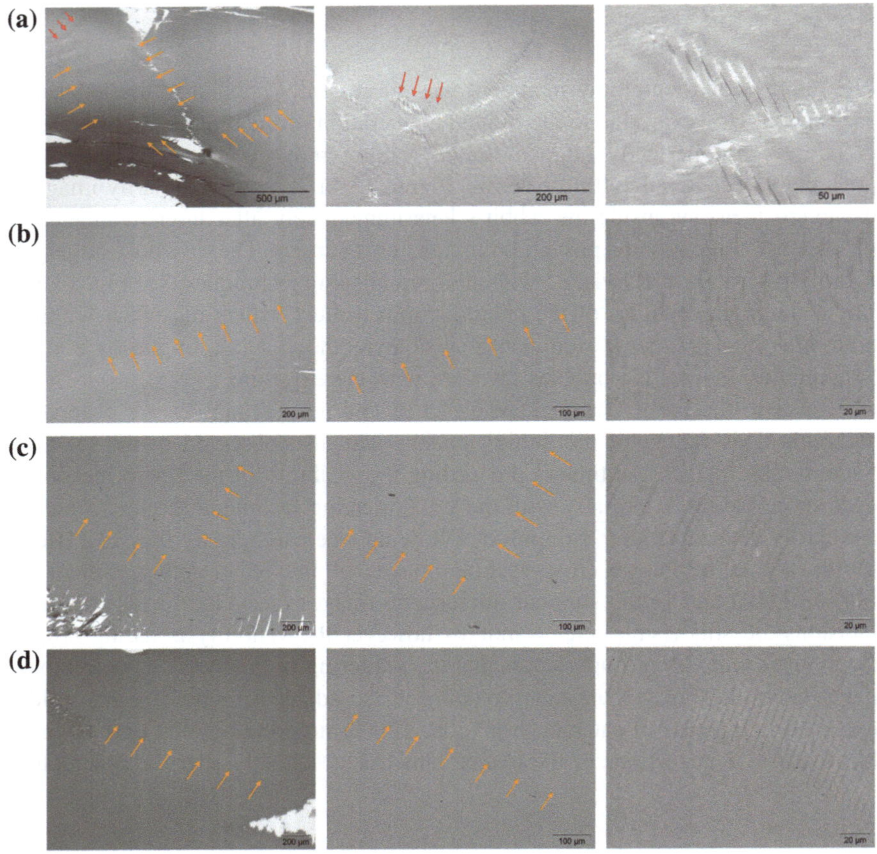

Fig. 13.12 Histopathological section of fs-lentotomy treated rabbit lens directly post surgery, **a** day of surgery, **b** at 14 day post-op, **c** 3 months post-op and **d** 6 months post-op

treatment the clearly separated tissue is detectable in the sections. The laser spots separate the lens fibers and stronger colored areas of denatured proteins are visible inside the areas of the laser spots (see Fig. 13.12a). With proceeding time the micro incisions are more difficult to detect in the histopathological sections, because the separated areas are not visible any more. Only small areas of slightly stronger coloring of the tissue are remaining (see Fig. 13.12b–d). These areas are only visible under high magnification and digital processing of the images of the histopathological section. The contrast is digitally increased in Fig. 13.12b–d to enhance the visibility of the slightly stronger colored areas. The most important finding is that none of the histopathological section shows any changed tissue in the adjacent areas of the micro incisions. So no indication of a cataract formation could be detected in any of the rabbit lenses.

The second aim of the study was the exclusion of a possible thermal damage of the retina due to transmitted laser irradiation. Therefore the rabbits' retinas were examined if any thermal induced change occurred. The histopathological section of the irradiated and control eyes looked the same in all cases. No thermal induced changes of the retinal pigment epithel or photoreceptor layer could be detected that would show changes due to thermal heating. In some cases the retina was detached from the choroid. This appears in both treated and untreated eyes and is therefore regarded as an artifact. This is supported by the fact that no hypertrophy of the retinal pigment epithelial (RPE) cells was observable which should also occur if the detachment of the retina was due to a thermal heating.

13.3.3 Discussion of Results

The in vivo delivery and localization of femtosecond laser pulses in rabbit lenses verifies the possibility of performing specific cutting patterns as the treatment of femtosecond-lentotomy, without any immediate hazardous side effects to the cornea and lens capsule. Photodisruption inside the lens tissue leads to small gas-filled bubbles, which remain as small, faint opacities after the bubbles disappear. The gas bubbles vanish after some minutes because the gaseous content of the bubbles dissolves. The greatest challenge within these in vivo rabbit experiments was the fixation and alignment of the eyes beneath the scanner system. As a consequence, some eyes were not centrally treated, but rather, decentered towards the lens equator. Thus, the laser was partly shielded by the iris, so that an asymmetric intralenticular cutting effect took place in these eyes. This shortcoming can be clinically prevented by coaxial fixation of the conscious patient just prior to the application of the suction ring. The OCT images taken prior to the laser treatment provide essential information for targeting the appropriate depth at which to place the cutting pattern. Optical coherence tomography images taken immediately after the laser surgery show that the cut was placed, as intended, well below the lens capsule to avoid capsule rupture. Optical coherence tomography localization therefore plays an integral and essential part in the feedback of intralenticular laser treatment, which is well suited to this system. Slit-lamp analysis after treatment indicates no frank opacity or other cataractous change of the lens, as might be anticipated from such a treatment. Furthermore, Scheimpflug imaging verifies the location and intensity of any opacity or other abnormalities and confirms the absence of cataractous changes following bubble resolution. The extent of the cutting pattern is well visualized due to the scattering of the light on the remaining gas bubbles. The larger the gas bubbles created inside the lens, the greater the intensity of light scattering by Scheimpflug imaging. At 14 days postoperative, each of the above imaging techniques showed a reduction and fading of the cutting pattern. Also, the cutting pattern appeared better localized and less intense in the OCT and Scheimpflug images. Actual cataract formation, which would be visible as a progressive opacity, increasing the light scattering and spreading into the surrounding tissue [54] was not observed.

In summary, the study showed that it is feasible to deliver in vivo fs laser induced microincisions into rabbit eyes. During the first 6 months no indication of cataract formation and retinal damage is found. Nevertheless, more in vivo studies in regard to long term cataract formation and increase of accommodation amplitude have to be performed in order to transfer this treatment concept to a successful clinical method for restoring accommodation.

13.4 Conclusions

In conclusion the concept of fs-lentotomy for improving presbyopic human crystalline lens flexibility is promising. The performed experimental investigations and theoretical simulations showed six noteworthy effects. First, the lenses anterior–posterior thickness is enhanced by an average of 97 μm accompanied with a decrease in equatorial lens diameter and steeper radii of curvatures due to the treatment which is associated with the fact that the lens capsule is now able to mold the lens tissue to a more spherical shape. Thus this results in an increase in optical power. This observation shows the successful aim of the fs-lentotomy. Second, the dynamic change of lens geometry due to a simulated accommodation process can be increased by the laser treatment. This shows that the flexibility of the whole lens can be increased and a significant larger change in optical power is feasible. Thirdly, compared to former studies on clear porcine lenses [28] it is more challenging to create smooth and accurate cutting patterns inside the inhomogeneous tissue of older human donor lenses. The optimization of this task will be future work.

Fourthly, the in vivo application on rabbits showed that no immediate hazardous side effects to the cornea or the lens capsule occurred. During the first 6 months no indication of cataract formation in the crystalline lens was found.

Fifthly, histologically, no changes on RPE, granular and photoreceptor layers seem to arise due to fs-lentotomy of the crystalline lens. Due to larger distances between lens and retina in humans, light intensity at the retina will be even smaller in humans giving an upper estimate in this study.

Sixthly, the OCT imaging is a direct and well suited tool to plan and monitor the position of the induced cutting pattern.

The investigation has demonstrated a promising increase in deformation ability of the lens and the safety of fs-lentotomy. Nevertheless the evidence that the results can be applied directly to in vivo human lenses still has to be adduced. Therefore the next step in developing a future clinical treatment method will be to transfer the application to humans.

References

1. H. Helmholtz, in *Helmholtz's Treatise on Physiological Optics*, ed. by J. Southall. Mechanism of Accommodation (Dover, New York, 1909), pp. 143–173
2. A. Glasser, P.L. Kaufman, The mechanism of accommodation in primates. Ophthalmology **106**, 863–872 (1999)

3. E. Fincham, The mechanism of accommodation. Br. J. Ophthalmol. **8**, 7–80 (1937)
4. K.R. Heys, S.L. Cram, R.J.W. Truscott, Massive increase in the stiffness of the human lens nucleus with age: the basis for presbyopia? Mol. Vis. **10**, 956–963 (2004)
5. H.A. Weeber, G. Eckert, W. Pechhold, R.G.L. van der Heijde, Stiffness gradient in the crystalline lens. Graefes Arch. Clin. Exp. Ophthalmol. **245**, 1357–1366 (2007)
6. D.A. Atchison, Accommodation and presbyopia. Ophthalmic Physiol. Opt. **15**, 255–272 (1995)
7. A. Glasser, M.C. Campbell, Biometric, optical and physical changes in the isolated human crystalline lens with age in relation to presbyopia. Vision. Res. **39**, 1991–2015 (1999)
8. S.A. Strenk, L.M. Strenk, S. Guo, Magnetic resonance imaging of aging, accommodating, phakic, and pseudophakic ciliary muscle diameters. J. Cataract Refract. Surg. **32**, 1792–1798 (2006)
9. A. Glasser, M.C. Campbell, Presbyopia and the optical changes in the human crystalline lens with age. Vision. Res. **38**, 209–229 (1998)
10. R.A. Schachar, Cause and treatment of presbyopia with a method for increasing the amplitude of accommodation. Ann. Ophthalmol. **24**(445–7), 452 (1992)
11. G.U. Auffarth, H.B. Dick, Multifocal intraocular lenses: a review. Ophthalmologe **98**, 127–137 (2001)
12. D. Azar, M. Chang, C. Kloek, S. Zafar, K. Sippel, S. Jain, in *Hyperopia and Presbyopia*, eds. by K. Tsubota, B.B. Wachler, D. Azar, D. Koch. Monovision Refractive Surgery for Presbyopia (Marcel Dekker Inc., New York, 2003), pp. 189–208
13. M. Küchle, B. Seitz, A. Langenbucher, G.C. Gusek-Schneider, P. Martus, N.X. Nguyen, Group TEAILS. Comparison of 6-month results of implantation of the 1CU accommodative intraocular lens with conventional intraocular lenses. Ophthalmology **111**, 318–324 (2004)
14. G. Baïkoff, G. Matach, A. Fontaine, C. Ferraz, C. Spera, Correction of presbyopia with refractive multifocal phakic intraocular lenses. J. Cataract Refract. Surg. **30**, 1454–1460 (2004)
15. AcuFocus. AcuFocus ACI 7000 (2008), http://www.acufocus.com. Accessed 31 July 2008
16. H. Lubatschowski, G. Maatz, A. Heisterkamp, U. Hetzel, W. Drommer, H. Welling, W. Ertmer, Application of ultrashort laser pulses for intrastromal refractive surgery. Graefes Arch. Clin. Exp. Ophthalmol. **238**, 33–39 (2000)
17. P.S. Binder, One thousand consecutive IntraLase laser in situ keratomileusis flaps. J. Cataract Refract. Surg. **32**, 962–969 (2006)
18. K. Stonecipher, T. Ignacio, M. Stonecipher, Advances in refractive surgery: microkeratome and femtosecond laser fl ap creation in relation to safety, effi cacy, predictability, and biomechanical stability. Curr. Opin. Ophthalmol. **17**, 368–372 (2006)
19. B. Seitz, A. Langenbucher, C. Hofmann-Rummelt, U. Schlötzer-Schrehardt, G. Naumann, Nonmechanical posterior lamellar keratoplasty using the femtosecond laser (femto-plak) for corneal endothelial decomposition. Am. J. Ophthalmol. **136**, 769–772 (2003)
20. M. Holzer, T. Rabsilber, G. Auffarth, Penetrating keratoplasty using femtosecond laser. Am. J. Ophthalmol. **143**, 524–526 (2007)
21. D.V. Palanker, M.S. Blumenkranz, D. Andersen, M. Wiltberger, G. Marcellino, P. Gooding, D. Angeley, G. Schuele, B. Woodley, M. Simoneau, N.J. Friedman, B. Seibel, J. Batlle, R. Feliz, J. Talamo, W. Culbertson, Femtosecond laser-assisted cataract surgery with integrated optical coherence tomography. Sci. Transl. Med. **2**(58), 58–85 (2010). doi: 10.1126/scitranslmed.3001305
22. N.J. Friedman, D.V. Palanker, G. Schuele, D. Andersen, G. Marcellino, B.S. Seibel, J. Batlle, R. Feliz, J.H. Talamo, M.S. Blumenkranz, W.W. Culbertson, Femtosecond laser capsulotomy. J. Cataract Refract. Surg. **37**(7), 1189–1198 (2011). doi: 10.1016/j.jcrs.2011.04.022
23. Z. Nagy, A. Takacs, T. Filkorn, M. Sarayba, Initial clinical evaluation of an intraocular femtosecond laser in cataract surgery. J. Refract. Surg. **25**(12), 1053–1060 (2009). doi: 10.3928/1081597X-20091117-04
24. R.I. Myers, R.R. Krueger, Novel approaches to correction of presbyopia with laser modification of the crystalline lens. J. Refract. Surg. **14**, 136–139 (1998)

25. R.R. Krueger, X.K. Sun, J. Stroh, R. Myers, Experimental increase in accommodative potential after neodymium: yttrium-aluminum-garnet laser photodisruption of paired cadaver lenses. Ophthalmology **108**, 2122–2129 (2001)
26. A. Heisterkamp, T. Ripken, T. Mamon, W. Dommer, H. Welling, W. Ertmer, H. Lubatschowski, Nonlinear side effects of fs pulses inside corneal tissue during photodisruption. Appl. Phys. B Lasers O. **74**, 419–425 (2002)
27. A. Vogel, J. Noack, G. Hüttman, G. Paltauf, Mechanisms of femtosecond laser nanosurgery of cells and tissues. Appl. Phys. B Lasers O. **81**, 1015–1047 (2005)
28. T. Ripken, U. Oberheide, M. Fromm, S. Schumacher, G. Gerten, H. Lubatschowski, fs-Laser induced elasticity changes to improve presbyopic lens accommodation. Graefes Arch. Clin. Exp. Ophthalmol. **246**, 897–906 (2008)
29. A. Heisterkamp, T. Mamom, O. Kermani, W. Drommer, H. Welling, W. Ertmer, H. Lubatschowski, Intrastromal refractive surgery with ultrashort laser pulses: in vivo study on the rabbit eye. Graefes Arch. Clin. Exp. Ophthalmol. **241**, 511–517 (2003)
30. R.F. Fisher, The force of contraction of the human ciliary muscle during accommodation. J. Physiol. **270**, 51–74 (1977)
31. R.F. Fisher, The elastic constants of the human lens. J. Physiol. **212**, 147–180 (1971)
32. A.M. Rosen, D.B. Denham, V. Fernandez, D. Borja, A. Ho, F. Manns, J.M. Parel, R.C. Augusteyn, In vitro dimensions and curvatures of human lenses. Vision. Res. **46**, 1002–1009 (2006)
33. S. Schumacher, U. Oberheide, M. Fromm, T. Ripken, W. Ertmer, G. Gerten, A. Wegener, H. Lubatschowski, Femtosecond laser induced flexibility change of human donor lenses. Vision. Res. **49**, 1853–1859 (2009)
34. R.A. Schachar, T. Huang, X. Huang, Mathematic proof of Schachar's hypothesis of accommodation. Ann. Ophthalmol. **25**, 5–9 (1993)
35. H.J. Burd, S.J. Judge, M.J. Flavell, Mechanics of accommodation of the human eye. Vision. Res. **39**, 1591–1595 (1999)
36. R.A. Schachar, A.J. Bax, Mechanism of human accommodation as analyzed by nonlinear finite element analysis. Compr. Ther. **27**, 122–132 (2001)
37. H.J. Burd, S.J. Judge, J.A. Cross, Numerical modelling of the accommodating lens. Vision. Res. **42**, 2235–2251 (2002)
38. E.A. Hermans, M. Dubbelman, G.L. van der Heijde, R.M. Heethaar, Estimating the external force acting on the human eye lens during accommodation by finite element modelling. Vision. Res. **46**, 3642–3650 (2006)
39. Z. Liu, B.Wang, X. Xu, Y. Ju, J. Xie, C. Bao, in *Engineering in Medicine and Biology Society IEEE-EMBS 2005*. Finite Element modeling and simulating of accommodating human crystalline lens (2006) pp. 11–14
40. E. Hermans, M. Dubbelman, R. van der Heijde, R. Heethaar, The shape of the human lens nucleus with accommodation. J. Vis. **7**, 1601–1610 (2007)
41. H.A. Weeber, R.G.L. van der Heijde, On the relationship between lens stiffness and accommodative amplitude. Exp. Eye Res. **85**, 602–607 (2007)
42. E.A. Hermans, M. Dubbelman, G.L. van der Heijde, R.M. Heethaar, Change in the accommodative force on the lens of the human eye with age. Vision Res. **48**, 119–126 (2008)
43. R.F. Fisher, Elastic constants of the human lens capsule. J. Physiol. **201**, 1–19 (1969)
44. S. Krag, T. Olsen, T.T. Andreassen, Biomechanical characteristics of the human anterior lens capsule in relation to age. Invest. Ophthalmol. Vis. Sci. **38**, 357–363 (1997)
45. R.C. Augusteyn, M.A. Cake, Post-mortem water uptake by sheep lenses left in situ. Mol. Vis. **11**, 749–751 (2005)
46. R.C. Augusteyn, A.M. Rosen, D. Borja, N.M. Ziebarth, J.M. Parel, Biometry of primate lenses during immersion in preservation media. Mol. Vis. **12**, 740–747 (2006)
47. M. Dubbelman, G.L. Van der Heijde, H.A. Weeber, Change in shape of the aging human crystalline lens with accommodation. Vision Res. **45**, 117–132 (2005)

48. F. Manns, V. Fernandez, S. Zipper, S. Sandadi, M. Hamaoui, A. Ho, J.M. Parel, Radius of curvature and asphericity of the anterior and posterior surface of human cadaver crystalline lenses. Exp. Eye Res. **78**, 39–51 (2004)
49. P. Rosales, M. Dubbelman, S. Marcos, R. van der Heijde, Crystalline lens radii of curvature from Purkinje and Scheimpflug imaging. J. Vis. **6**, 1057–1067 (2006)
50. A. Duane, Studies in monocular and binocular accommodation, with their clinical application. Trans. Am. Ophthalmol. Soc. **20**, 132–157 (1922)
51. R.R. Krueger, J. Kuszak, H. Lubatschowski, R.I. Myers, T. Ripken, A. Heisterkamp, First safety study of femtosecond laser photodisruption in animal lenses: tissue morphology and cataractogenesis. J. Cataract Refract. Surg. **31**, 2386–2394 (2005)
52. G. Gerten, T. Ripken, P. Breitenfeld, R.R. Krueger, O. Kermani, H. Lubatschowski, U. Oberheide, In-vitro- und In-vivo-Untersuchungen zur Presbyopiebehandlung mit Femtosekundenlasern. Ophthalmologe **104**, 40–46 (2007)
53. S. Schumacher, M. Fromm, U. Oberheide, G. Gerten, A. Wegener, H. Lubatschowski, In vivo application and imaging of intralenticular femtosecond laser pulses for the restoration of accommodation. J. Refract. Surg. **24**, 991–995 (2008)
54. A. Gwon, F. Fankhauser, C. Puliafito, L. Gruber, M. Berns, Focal laser photoablation of normal and cataractous lenses in rabbits: preliminary report. J. Cataract Refract. Surg. **21**, 282–286 (1995)

Chapter 14
Towards Industrial Inspection with THz Systems

M. Stecher, C. Jördens, N. Krumbholz, C. Jansen, M. Scheller, R. Wilk, O. Peters, B. Scherger, B. Ewers and M. Koch

Abstract Due to their high spatial resolution, THz imaging systems (based on time domain spectrometers) have a broad variety of applications in scientific as well as industrial applications in the future—especially through the upcoming of reliable titanium sapphire lasers and fiber amplified femtosecond lasers. With such THz scanning highly sensitive measurements of materials as well as physical phenomena and simultaneously real-time imaging with real-time data extraction can be realized which can be used for production monitoring and/or more accurate and novel inline control systems. The article gives an overview on recent state of the art THz systems based on femtosecond lasers and their application scenarios, like detection

M. Stecher
Carl Zeiss SMT GmbH, Productmanagement, Rudolf-Eber-Straße 2, 73447 Oberkochen, Germany

C. Jördens
Metrology Department, Volkswagen AG, Berliner Ring 2, 38440 Wolfsburg, Germany

N. Krumbholz
Technische Entwicklung der Volkswagen AG, Wolfsburg, Germany

C. Jansen
Philipps-Universität Marburg, Experimentelle Halbleiterphysik, 35032 Marburg, Ehemals, Germany

M. Scheller
Optical Sciences, University of Arizona, Tucson, Arizona, USA

R. Wilk · O. Peters
Menlo Systems GmbH, Am Klopferspitz 19A, Martinsried, Germany

M. Koch (✉)
Phillips-Universität Marburg, Experimentelle Halbleiterphysik, Renthof 5, 35032 Marburg, Germany
e-mail: martin.koch@physik.uni-marburg.de

B. Scherger
Active Safety, Delphi Deutschland GmbH, Delphi Platz 1, 42119 Wuppertal, Germany

B. Ewers
Deutsches Zentrum für Luft-und Raumfahrt e.V., Institut für Aerodynamik und Strömungstechnik, Bunsenstr. 10, 37073 Göttingen, Germany

S. Nolte et al. (eds.), *Ultrashort Pulse Laser Technology*,
Springer Series in Optical Sciences 195, DOI 10.1007/978-3-319-17659-8_14

of foreign bodies in foods, inline inspection of paper layer thickness and area mass as well as non-destroying impact or tensile tests or fiber orientation checks in Compound-Polymers.

14.1 Introduction

The frequency band of the electromagnetic spectrum between the microwaves and the optical spectrum gained a growing amount of interest in the last decade. This so called terahertz (THz) region has been in the "dark" and rather unexploited until the end of the last century due to the lack of compact, cost-effective and efficient generation and detection schemes in this frequency range. Since then the potential applications for THz technology are investigated rapidly in the scientific as well as in the industrial community. Like microwaves, THz radiation can penetrate through clothing, plastics, paper, wood, and leafs, while metals and polar liquids like water are basically opaque. Due to the fact that THz wavelengths are shorter than millimeter waves, they offer a higher spatial resolution, ranging below some hundreds of microns for high frequency components. This brings a great advantage for THz imaging in a variety of applications, such as security scans at airports (body scans, luggage and mail inspection; [1–7]), industrial inline control and inspection (e.g. polymer, paper, pharmaceutical or food industry; [8–12]), and plant physiology (water status or dry stress influences; [13–15]). A bit further down the road there might be THz telecommunications [16–19], mainly for high speed, short-range indoor links and hot spots. These systems operating at a few hundreds of GHz could be the future generation of wireless technology in ten to fifteen years time.

Up to now the number of developed or rediscovered methods of generation has increased a lot, ranging from classical microwave technology based on electrical diode sources to ultra short pulsed optical generation on photoconductive semiconductors or in nonlinear crystals. The same holds true for the detection schemes, which until recently were mainly based on thermal methods (Bolometers, Golay cells and pyroelectric sensors) and were suffering from slow speed and low sensitivity. In addition, these detection schemes give only integrated power measurements not sufficient for spectral analysis. Yet, recently there has been a lot of progress in the development of THz detector arrays and cameras [20–23] which might change the game in the near future. Besides, cost-effective passive devices to guide and manipulate THz waves have been developed which might become important ingredients to all sorts of THz systems. This includes lenses [24–26], gratings [27], mirrors [28], waveguides [29, 30], and waveplates [31]. Furthermore, it turned out that 3D printing is an interesting way to fabricate such devices [32, 33].

Recent THz generation and detection schemes are based on using either optical difference frequency generation [34–43], or employing femtosecond laser pulses to to emit pulsed THz radiation from photoconductive switches or antennas [44–48].

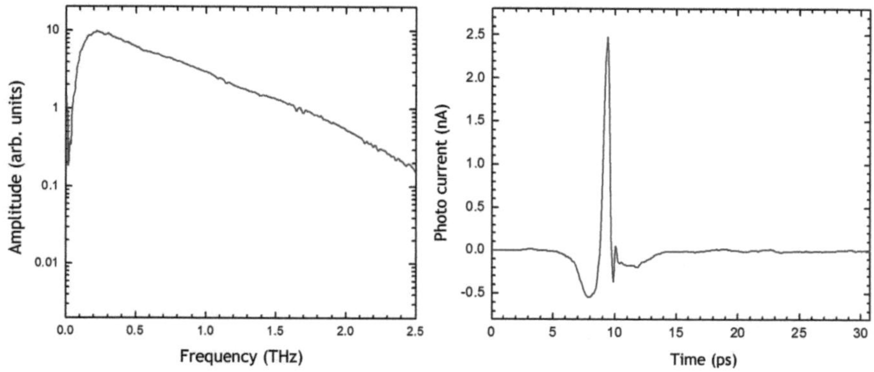

Fig. 14.1 Time domain waveform (*left*) and frequency domain spectra (*right*) of a THz pulse recorded in a conventional THz TDS system

Where continuous wave (cw) THz systems offer a very good spectral resolution but only at a narrow frequency band, pulsed laser sources have the advantage of providing information across a wide spectrum with only one single pulse measurement (cf. Fig. 14.1). Therefore, we are only focusing on pulsed THz systems and spectrometers, commonly referred to as THz time domain spectrometers (TDS), in this overview. A far more detailed overview on THz technology can be found in several overview articles [49–52]. There is also a cost effective approach in between, the recently introduced so called quasi-time-domain spectroscopy [53, 54]. In this approach a continuous wave multimode semiconductor laser diode worth a few dollars replaces the cost-intensive femtosecond laser. A more comprehensive overview on this promising novel technique is given in the last section of this chapter.

Imaging systems based on pulsed THz sources were first demonstrated in the mid 1990 [55] and the principle behind this technique was described in several publications [56–58]. The cornerstone of a coherent pulsed THz system is a femtosecond (fs) laser, usually a titanium sapphire or erbium doped fiber laser.

A schematic of a coherent THz setup is shown in Fig. 14.2. When the femtosecond laser pulse hits the biased photoconductive gap of the emitter antenna, a short electrical pulse is triggered. Abiding to Maxwell's law, any accelerated carriers emit electromagnetic radiation and the emitted field is approximated by the derivative of the current between anode and cathode of the antenna [46, 47]. If the rise time of the pulse is fast enough the emitted radiation exhibits strong components in the THz region. A silicon lens on the other side of the antenna structure precollimates the generated THz radiation. By using polymer lenses or off-axis parabolic mirrors the beam is collimated and may be guided to an intermediate focus on a sample of interest. The THz electromagnetic field is detected after refocusing on the receiver antenna, like the emitter a photoconductive switch triggered by an optical pulse. The semiconductor material has a very short electron-hole lifetime in order to achieve the best signal recording. The detection

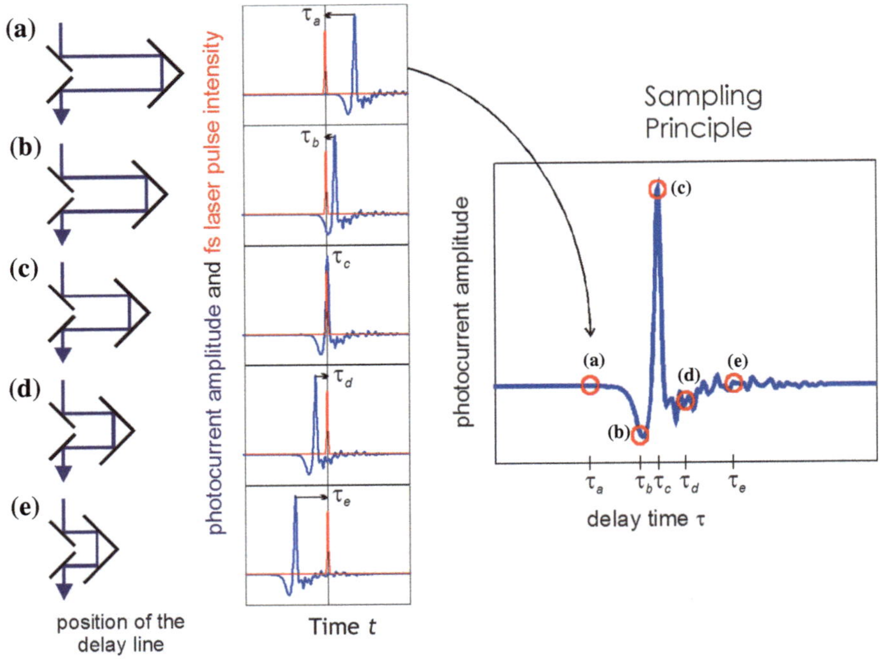

Fig. 14.2 Schematic on coherent sampling of THz pulses (Courtesy of F. Rutz)

process is slightly different to the emitting one, rather than having an applied electrical field accelerating the carriers, the detection scheme is based on the inverse effect. Since it is not possible to record the pulses of sub-picosecond lengths in real time, the detector antenna is gated for the detection. When the active region of the detector antenna is illuminated by the laser pulse the incoming THz radiation leads to a net current within the antenna structure in the order of a nanoampere. This current can be amplified and detected as a measure of the incoming amplitude of the electrical field. Since the laser pulse duration is shorter than the THz signal by several orders, it can be used to switch the antenna. Only the part of the terahertz pulse correlating with the gating time will contribute to the current signal, normally averaged over a large number of pulses. Delaying the detector pulse gradually in time allows for "scanning" through the THz pulse and allows for obtaining the full waveform as a function of time (see Fig. 14.2).

As a clear economical potential has been identified for THz technology, the systems have to become more compact, faster and more cost-effective. In the following, we will give an overview on recent state of the art THz systems based on femtosecond lasers and their application scenarios. We will discuss avenues for these systems to enable broad applicability for THz technology in the future.

14.2 Application Scenarios

A large step for THz systems was made by the upcoming of reliable titanium sapphire lasers and fiber amplified femtosecond lasers. Their ultra short pulses can be used for generating and detecting THz pulses in various ways. The coherent detection of phase-locked THz pulses and the rejection of incoherent background noise enable highly sensitive measurements of materials and physical phenomena. As many applications of THz sensing are done in transmission mode, it should also be noted that reflection geometries can be realized as well. While transmission is the approach of choice for inline inspection of polymer production or compounding processes, the reflection method allows one to evaluate intransparent or highly absorbing samples, like water, other polar liquids or thick samples add (see e.g. [59–62].

In this section a closer look on a few industrial THz systems will be offered, analyzing detection schemes and their specific application and purpose. The method of operation will be shown on an inline process control for the polymer extrusion. A system is presented which allows for the detection of additives and the purity of the main compound. A second possible scenario for process control is imaging of food products with low water content during fabrication. On the fly scanning for unwanted particles within opaque substances, e.g. in chocolate, is a great advantage over visible inspection schemes. Impurities which are of the same water or fat density as the surroundings can also be detected using THz waves, which even might not be picked up by X-ray scanners.

14.2.1 Inline THz Polymeric Compounding Monitoring

The first industrial application scenario we have a closer look at, is inline monitoring with a THz TDS system. As such a system needs to be robust, transportable and flexible to meet the needs in a highly specialized production line; one of the main goals is to take the existing technology for lab setups a step further. The biggest issue for a reliable and functioning laser based THz system is the insensitivity towards vibrations. The way around this problem is a fully fiber based guiding of the optical beams to the photoconductive antennae. A big portion on the way to realize such a system was the further development of fiber coupled antennae. A cleaved fiber end is positioned for maximum THz signal above the gap of the antenna. Fixing the fiber in respective to the gap is achieved by index-matching glue to simultaneously reducing the reflection losses as well as to protect the antenna and the end face of the fiber from external environmental influences and stresses [63, 64]. This approach guarantees an optimal THz output and enhances the systems robustness and versatility tremendously (cf. Fig. 14.3). In addition, the system has all laser sensitive parts enclosed with minimum laser safety standards, allowing it to be operated in any industrial area.

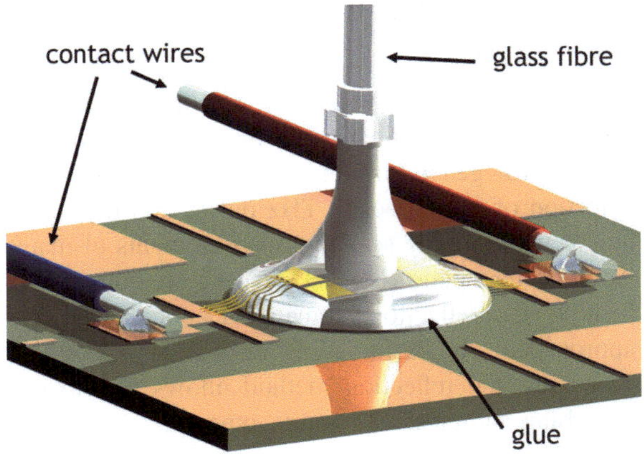

Fig. 14.3 Glued fiber-coupled antenna chip

The sensor head for the polymeric compound monitoring had to be specially made to guide the THz beam [64]. Two fused silica plates are used as windows to access the extruder mass with the THz radiation. These glass windows are nearly transparent for frequencies up to 1.0 THz. In between the plates the polymer die is piped through a narrow slit (in the experimental setup adjustable width of 0.5–10 mm). The sensor head was designed for working at temperatures up to 260 °C and melt pressure levels of up to 26 bar. Existing control sensors provided information on melt temperature and pressure during the extrusion process.

While the THz radiation has to penetrate not only the composited die but also the silica windows, analyzing and calibrating the system is essential to gain live value of the compound added into the process. With varying the temperature of the fused silica windows no changes for the frequency range from 0.1 to 0.4 THz could be detected. The melt on the other hand is rather sensitive to temperature changes. A plot of the measured refractive index over temperature is shown in Fig. 14.4.

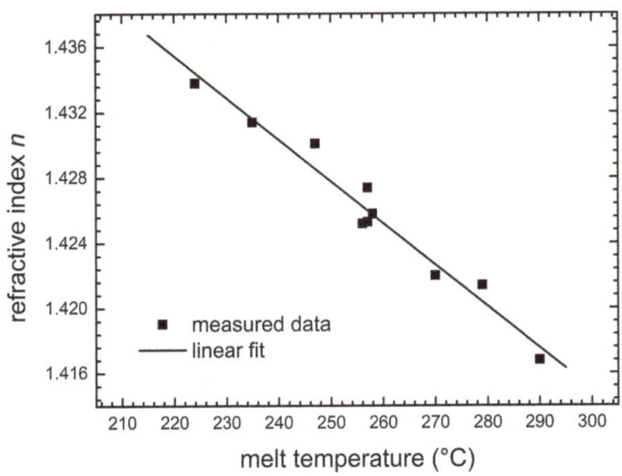

Fig. 14.4 Refractive index over melt temperature of molten polypropylene (PP)—averaged over the frequency interval from 200 to 350 GHz [64]

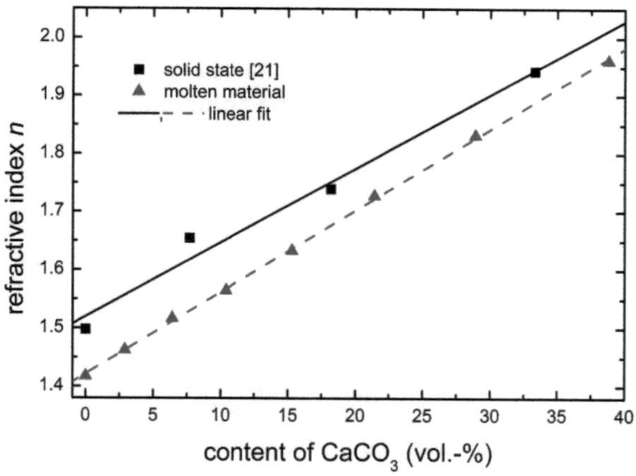

Fig. 14.5 Refractive index of molten (*circles*) and solid (*diamonds*) polypropylene–$CaCO_3$ compound as a function of the volumetric additive content—averaged over the frequency interval from 200 to 350 GHz [64]

On a lab based system the functionality of the system was tested. Therefore different polypropylene melts with an increasing concentration of calcium carbonate ($CaCO_3$) were measured. Due to the higher refractive index of $CaCO_3$ the combined refractive index is scaling with the additive content. Figure 14.5 shows a linear dependency of the refractive index of solid and molten compound over the volume percentage of $CaCO_3$.

A fully functional inline sensor system has to provide sufficient information in real-time. Though kilohertz scanning spectrometers are technically possible [65, 66], high speed scanning of full terahertz pulses leads to sophisticated and expensive setups. As in this application the main interest is in changes of the refractive index and not the index itself, the setup can be simplified. The change in refractive index leads to a change in travel time for the pulse, and hence one is only interested in the time delay of the pulse. In order to avoid time consuming screening of the full THz pulse, one locks at the front slope of the pulse and calibrates the amount of change in the THz amplitude to a certain phase delay and therefore a change of refractive index. Figure 14.6 demonstrates the principle of the fixed measurement point. The steeper the slope the more sensitive the system is to changes in the additive percentage. Yet, a long slope increases the detectable range of additive concentrations, so there is a trade-off to be considered.

Real time measurements taken during the compounding process of polypropylene with calcium carbonate ($CaCO_3$) were taken under real environmental conditions. The filler quantity was stepwise varied over time. As expected, the terahertz measurement shows a delay of 1–2 min for the changed filler content to arrive at the extruder outlet. Additionally, a real time sensing of the wash out of the filler can be monitored (cf. Fig. 14.6). This is due to the travel time of the die through the extruder, where the mixing of the additives takes place. The exponential change in

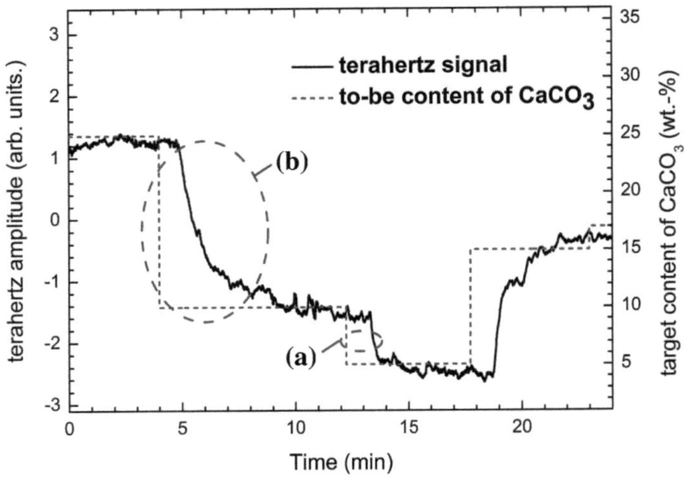

Fig. 14.6 Real-time THz signal (*solid*) measured on polypropylene mixed with $CaCO_3$ (*dashed*) the concentration of which is varied during the course of the measurement. **a** Residence time of the additive inside the extruder, **b** wash out time [64]

the refractive index and therefore within the melt can be explained by a gradual wash-out Recently, similar results have been obtained for rubber [67].

In conclusion, this example shows that THz technology is well suited for determining additive values in polymers. Real time operation for a nondestructive and contact free industrial sensor setup has been demonstrated. A linear dependency of the additive under inspection and the refractive index in the THz range could be established. But THz systems may be used in more sophisticated imaging applications like for security or food inspection. In the following inline foreign body detection in chocolate is described as an example for fast and reliable inline imaging feasibility.

14.2.2 Detection of Foreign Bodies in Foods (Chocolate)

In the food industry the detection of contaminations and unwanted foreign bodies plays a vital role. These inclusions can state a serious health risk to consumers. Existing testing and detection systems, like metal detectors or X-ray scans, are already in use. As metallic defilements are easily discovered with metal detectors, the focus here is on non nonmetallic foreign bodies like glass, stone or plastic particles. X-rays are not suited to detect plastic pieces in fat-based foods, like chocolate, because both consist mainly of the same elements, hydrogen and carbon. Similar detection problems exist for glass contaminations, which are prevented by holding regular checks of glass inventory within the fabrication environment. This demonstrates that there is a great need for a reliable and sophisticated capability of

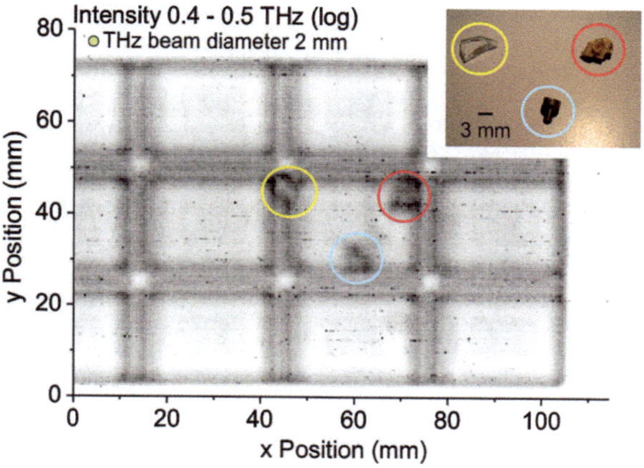

Fig. 14.7 Terahertz image of a whole milk chocolate bar containing three different contaminations: a piece of glass, a stone, and a metal screw; *Inset* photograph of the foreign bodies [68]

detecting nonmetallic particles. THz inline monitoring is a suitable solution for detecting these contaminations, because THz waves are not only providing amplitude information (absorption spectra), but at the same time reveal the phase information, too. This is attributed to the coherent detection scheme embedded with the same "triggering" optical pulses on both photoconductive antenna (emitter and detector).

The setup used for this measurement scheme is a standard pulsed THz spectrometer with photoconductive LT-GaAs antennas. It was driven by a femtosecond titanium sapphire laser operating at 800 nm generating pulsed THz radiation which was guided by off-axis parabolic mirrors. In order to maintain a high spatial resolution and large enough signal at the same time, the THz radiation was focused to a spot size of about 2 mm. To get a full image across the whole chocolate bar a raster scanning method was applied for first images. Yet, for a future real time inline measurement setup, one needs to have several antenna pairs or scanning across the sample under inspection in a rapid pace.

Inclusions and contaminations were placed within a commercially acquired bar of milk and hazelnut chocolate. Foreign bodies were carefully hidden in the chocolate by localized heating so it was impossible to see these inclusions with the naked eye. THz intensity image from frequencies between 0.4 and 0.5 THz revealed the hidden metal screw and a stone, as well as the glass splinter (cf. Fig. 14.7) in between the structure of the chocolate [68].

The discrimination between these insertions is done by looking at the transmission intensity image as well as the phase delay in order to guarantee the highest detection rate for all contaminations. Through the pure chocolate bar, the measured time domain signal has typically one single pulse peak. But obstructing the THz beam with any different material having a different refractive index results in a double peak structure (Fig. 14.8). This is due to the fact that parts of the THz pulse travel around the obstacle and parts travel through it—mostly at a slower pace.

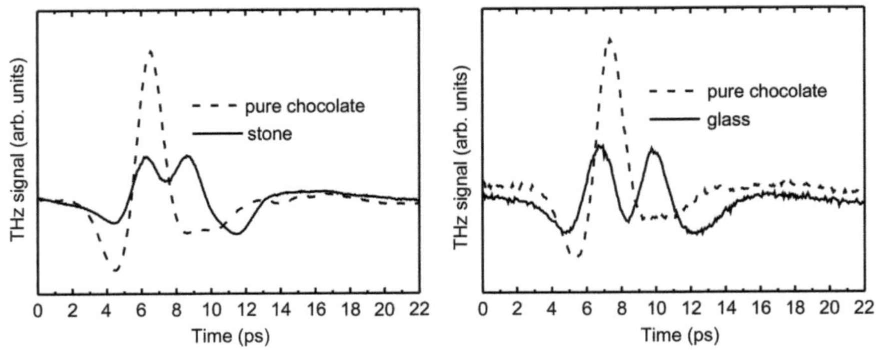

Fig. 14.8 Terahertz waveforms through pure chocolate compared with waveforms through contaminated chocolate: stone (*left*), glass splinter (*right*) [68]

To increase the accuracy of the detection imaging algorithm the height profile of the chocolate bar is also taken into account [68]. In that laboratory setup an existing scanning unit was available (Atos HR scanner) for simulating inline scanning. Yet, the scan was rather time consuming, typically taking up to one or even more hours. However, this sensor fusion system offers a clear reduction of the false alarm rate. THz technologies proofed perfectly suitable to detect foreign bodies in foods in a noninvasive, contact-free manner. A faster scanning approach is presented later in this chapter.

14.2.3 Inspection of Paper Layer Thickness and Area Mass

Another highly desired field of operation for THz technology is in the paper production industry. There is a great need for an accurate and real time thickness measurement across the whole layer of paper. This task needs a high technological effort as the paper layers are running on the production line at speeds of 15–30 m/s. Pulsed THz spectrometers are ideal to distinguish the paper layer thickness and density at the same time.

Paper usually consists of more than one layer and an additional finishing layer of varnish. During the fabrication process the individual layers are jointed together and as a last step the finishing layer is deposited on top of the running paper sheets. To have minimal costs it is of highest interest to deposit a minimum of varnish. For exact inline determination of this layer one has to know the thickness of the paper layers. To obtain the paper thickness from the THz time-domain data the so called quasi space algorithm can be used (see [69] for details). The right part of Fig. 14.9 shows data obtained for paper. The correct thickness is given by the position of the minimum of the curve. As the single paper layers can be measured before they are jointed together, a software tool can extrapolate the thickness of the varnish, potentially by making use of the multiple echo pulses from the multi layer systems.

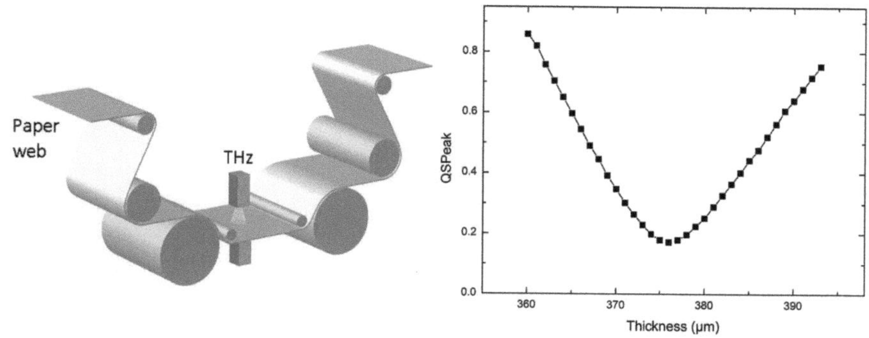

Fig. 14.9 Schematic of an inline THz thickness determination (*left*) and extracted layer thickness (*right*)

Active research is done on optimizing the averaging time to have a shorter response time on variations in the finishing layer thicknesses and area mass of the paper layers. A schematic of the inline setup can be seen in the left of Fig. 14.9.

THz technology can also be used for further points of interest within the paper production. To provide a high quality it would be desirable to have inline monitoring of the water content in the paper layers, as well as the homogeneity of the layers in respect to each other. In addition, it would be handy to monitor desired additives, next to eliminate undesired inclusions in the final product.

14.2.4 Fiber Orientation in Compound-Polymers

As a last industrial application we take a closer look on fiber-reinforced materials. These compounds exhibit very special mechanical characteristics, like high tensile strength and a low specific density, and are therefore perfectly suited for the aviation and automobile industry. The specific properties rely heavily on the amount of added fibers as well as their orientation and it is crucial for safety tests to have accurate knowledge on these parameters. Up to now, a 100 % inspection ratio is only possible by using X-ray scanners. All other methods like bending, impact or tensile tests are destroying or damaging the parts under investigation, rendering them useless for further usage. Here, THz spectroscopy can substitute the hazardous ionizing X-rays and offer even more information on the samples properties, due to the fact that many polymers are transparent for THz radiation.

In 2009 Jördens et al introduced a method to determine the orientation of birefringent materials [70]. This method assumes that fibres and fibrils have an orientation degree of hundred percent. Yet, this is typically not fulfilled in practical samples. Therefore, Jördens et al extended their method to overcome this restriction. They reported on a method to determine the preferential fibre orientation and the fraction of orientated fibres in reinforced plastics [71]. Due to the rod-like shape

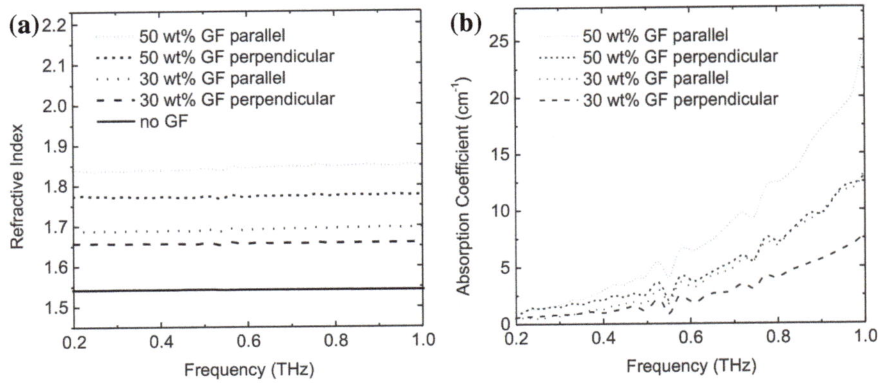

Fig. 14.10 Refractive index of HDPE with different filling levels of glass fibers (**a**) and the corresponding absorption coefficient for different fiber orientations (**b**) [71]

of the glass fibers, a birefringence effect can be detected and leads to different optical properties of the samples depending on the polarization of the THz wave. Calculated and measured values for different fiber orientations are displayed in Fig. 14.10 for a frequency of 420 GHz. Different additive percentages induce a birefringence which is uniformly strong over the lower THz range (0.2–1 THz) but lead to increased absorption for higher frequencies (Fig. 14.10b).

Measured and calculated values differ due to non-ideal alignment of all fibers within the compound. In order to obtain the main fiber orientation of any sample a THz measurement in three different orientation angles is necessary. In principle, these angles can be chosen arbitrarily. Theory shows that the ordinary and extraordinary refractive indexes of the compound span an ellipse [70]. By using the three measurements values and the corresponding formulas for the refractive index, one can calculate the ordinary and extraordinary refractive index.

Solving these formulas provide the angle of fiber orientation. Measuring at different positions across the sample allows for monitoring the uniformity of fiber distribution and arrangement [71] (see Fig. 14.11).

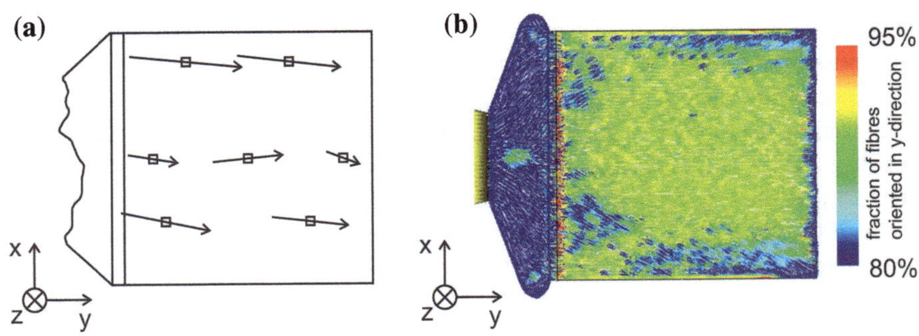

Fig. 14.11 Fiber orientation across a reinforced glass fiber polymer sample (preferential orientation at different positions (**a**) and amount of fibers oriented in y-direction (**b**)) [71]

Besides the fiber orientation THz time-domain measurements also offers information on sample thickness at the same time. By lowering the initial costs for femtosecond lasers and the system components THz TDS spectrometers are becoming a valuable alternative in non-destructive industrial testing systems.

14.3 Commercially Available Time Domain Systems

In the last few years THz technology evolved from a pure scientific field of research to a technology that now yields commercially available consumer products with a variety of applications.

Leading companies for optoelectronic devices in the THz regime are Picometrix, Zomega Terahertz Corp., TeraView, Advantest, Menlo Systems and Toptica for example. Here we will take a closer look at a product from Picometrix and its application, namely their time-domain terahertz (TD-THz) system T-Ray 4000®.

This portable handheld device possesses two flexible fiber-coupled terahertz sensors which can be operated in reflection and transmission mode and allows for the support of multiple THz channels. Two different scan ranges with a 320 ps temporal delay and an acquisition rate of 100 Hz or a 80 ps time delay with an acquisition rate of 1 kHz at a bandwidth of 0.2–3.5 THz are available. The sensor antennas can be mounted in a casing on wheels, ensuring a constant measurement distance and allowing for inspection of a large area.

Using this device it was possible to non-destructively examine hidden drawings covered under other layers of paint [72]. It can be also used for dating wood to its age or reading ancient texts on papyrus rolls without opening them. By gathering this information not only a basic comprehension of antique cultures is possible but also a preservation of cultural heritage [73].

However, one of the biggest and important branches for THz technology lies in security and military inspection applications, where non-invasive scans for the detection of hidden objects or liquid explosives can be performed. So the T-Ray 4000® was not only employed for safety inspection of the outer shell of the space shuttle, but also to inspect protection shells of radar domes. It is also intended to be used by the US air force for inspecting the outer shell of stealth fighters as the F-35.

As THz technology advances further and further, the market for commercial products and applications will grow steadily over the next years. But towards mass production and the prospect of having THz inspection in numerous situations it is necessary to make the technology faster, cheaper and more robust technology. In the following we have a closer look on how to continue the research path towards these goals.

14.3.1 Avenues Towards More Robust THz Systems

THz systems driven by femtosecond lasers are highly sensitive to external influences. Especially at high temperatures and in rough industrial environments, like at the end of an extruder for polymeric compounding or food production, accurate and reliable measurements are hard to guarantee. This demands a reliable system, which has fully concealed antenna structures, an encased delay line and a robust control electronic. A full fiber based system would offer all these key features and reduce the laser safety restrictions for the system.

To reduce these side effects to a minimum an all-fiber coupled system would be desirable, particularly for the very sensitive optically triggered semiconductor antennas (Fig. 14.12). Here, instead of having a free space beam alignment onto the antenna gap, a fixed mechanical setup guides the beam to its right position. As both the emitter and the receiver unit consist of a semiconductor-based photoconductive antenna structure, a perfect alignment is essential for optimal THz output. One possibility is to adjust a long fiber pigtail with a FC-APC connector on top of the photoconductive gap for maximum photo current. An index matching UV-curing glue holds the bare fiber end in position (cf. Fig. 14.3). This approach offers not only a robust antenna setup but also long term stability of the THz output power. One should also mind that emitting and detecting antenna can be combined and a transceiver setup can be realized by using separate, in time delayed pulses for generation and detection [74, 75].

Embedding the optical patch along a silica fiber allows for shock resistant operation without alignment issues. These advances are essential for employing THz imaging or scanning systems in an industrial environment. In order to avoid any free space parts at all it is also necessary to have an all-fiber delay line within the setup [76]. In general fiber stretchers alternating elongate the optical paths to the antennas. This mechanical stretching is achieved by either a motor based motion of the cylinders or by piezo agitated half cylinders which the fibers are coiled around (cf. Fig. 14.13). Under a medium strain the fibers are glued on the cylinders for a robust and precise induced delay. By simultaneous stretching and releasing the two

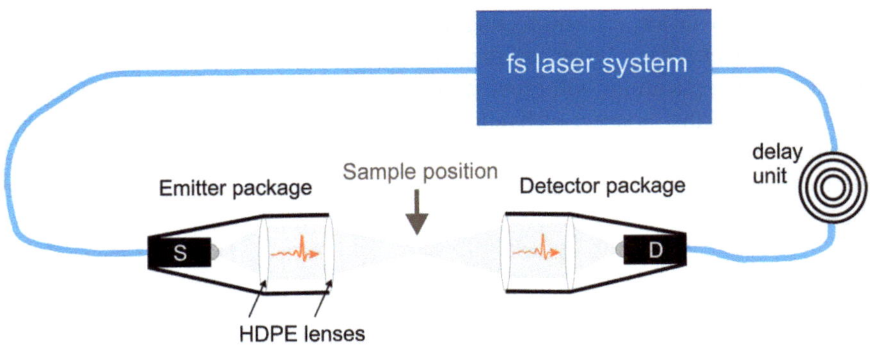

Fig. 14.12 Schematic of an all-fiber THz TDs system

Fig. 14.13 3D schematic of a mechanical fiber stretcher (Courtesy of TEM Messtechnik, see [76] for details)

optical patches one achieves a time delay of up to 280 ps with the mechanical motor. The full waveform can be sampled up to 50 Hz with the mechanical motor. Piezo driven fiber stretchers can work up to 1 kHz, but have a reduced delay and, therefore, can only be used for thin samples with a low refractive index.

14.3.2 Avenues Towards Faster THz Imaging

One of the biggest obstacles for achieving THz imaging in real time is the relatively slow sampling of the THz waveform. Resolving a THz pulse requires between 4.000 and 40.000 data points depending on the desired frequency resolution. Standard mechanical delay lines, which are usually used in laboratory THz time domain setups, focus on high position accuracy rather than on lateral velocity. Nevertheless one need both for achieving detailed real time imaging. Realizing a fast mechanical scanning speed can be achieved by different ways. Possible solutions would be a piezo actuator based fiber stretcher as alluded in the last section or a repetition rate tuning with an intra-cavity piezo driven mirror [77]. Another approach is a rotating disc with embedded retro reflectors [78]. By bouncing the beam off a fixed mirror back to the retro, the beam is folded and the path length change is doubled, and can be scanned with up to 1 kHz.

An alternative extremely cost efficient method is embodied by rapidly rotating a THz transparent dielectric cube [79]. Using this scheme allows for sampling a time

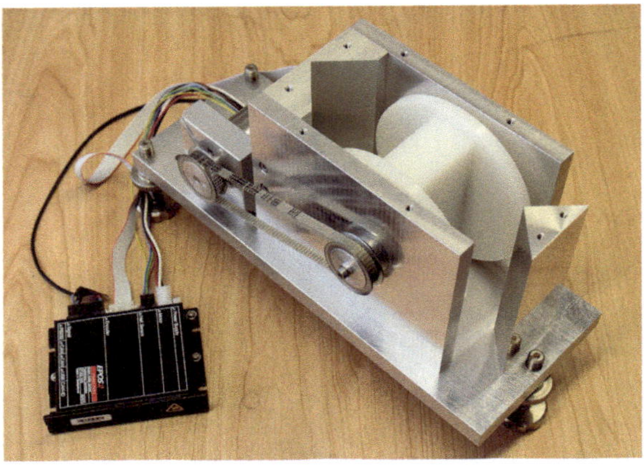

Fig. 14.14 Photograph of an optical delay line relying on a THz transparent dielectric cube (see [79] for details)

window of 40 ps with a scan rate of hundreds of Hertz (see Fig. 14.14 for a photo). But even with thousand image points per second, the area under inspection is still limited or the resolution has to be quite low. So the item to be inspected needs to be scanned in a more rapid pace. Two approaches come to mind—parallel/multifocal imaging or mechanical scanning of the focal point.

Employing a system with more than one pair of antennas enables measurements of more pixels at the same time [80, 81]. On the contrary, the costs increase linearly with every antenna pair, as well as the potentially additional costs for a laser which delivers enough power to drive all antennas simultaneously (Fig. 14.15a). In order to avoid a cost intensive multi-focus approach, it would be preferable to reduce the amount of emitter/receiver antennas. Splitting up the THz radiation generated by one emitter onto several detectors would reduce the cost by nearly 50 % (Fig. 14.15b). Mechanical scanning of the focal point reduces the number of antennas down to two (Fig. 14.15c). The limiting factor is now the mechanical scanning as well as the time to acquire each pulse form.

Scanning the THz focus along a line can be realised by an f-theta lens system. A rotating set of planar mirrors on each side of the lens setup guides the beam at different angles. The special characteristics of the lens setup leads always to a focus

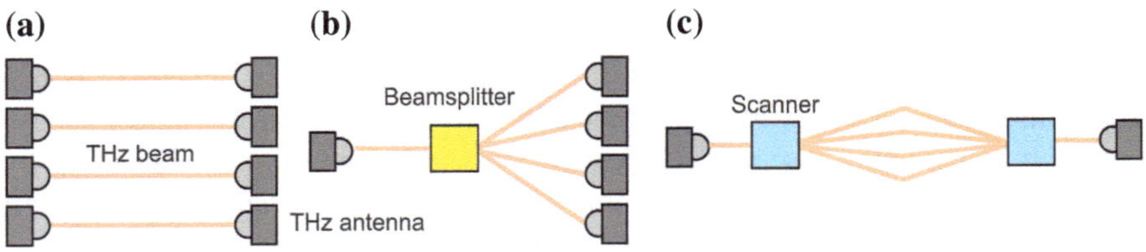

Fig. 14.15 Schematics of a multi-channel system (**a**), a *N* detector power splitting system (**b**), and a scanning system

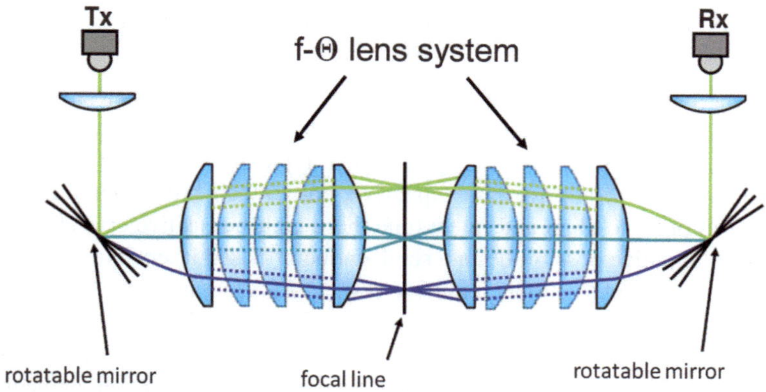

Fig. 14.16 Schematic of a f-theta THz scanning system

between the lenses. Only the position of the focus is dependent on the incident angle of the THz radiation (cf. Fig. 14.16). Rotating the mirrors with several tens of Hertz allows for an inline inspection of goods on a conveyor belt.

So far a first demonstration of such a f-theta system for inline monitoring was successful. A sample of nougat chocolate was prepared with several inclusions and scanned by the system with nearly 20 Hz acquisition rate. For this preliminary study we turned the mirrors by 15 rounds per minute and the chocolate bar moved at a constant speed of only 0.3 mm/s and led to an image of the full bar in approx. 5 min (see Fig. 14.17). Note, that this concept also works in a reflection geometry [82].

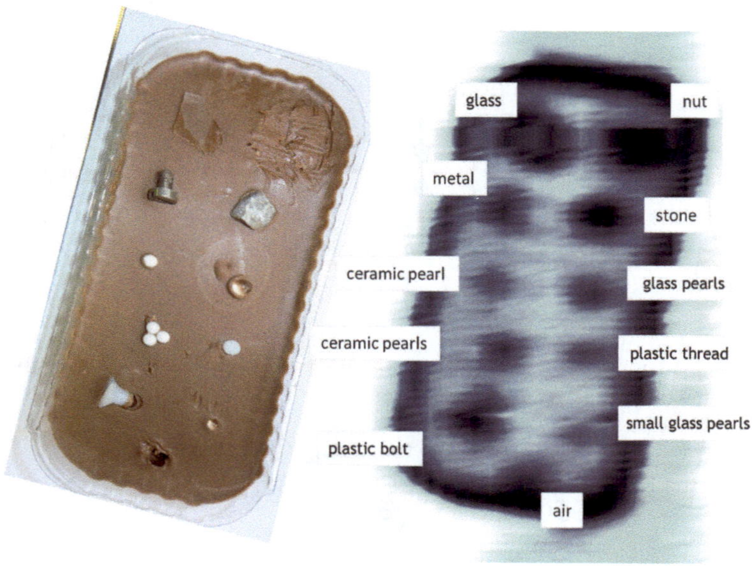

Fig. 14.17 Photograph (*left*) and THz image (*right*) obtained with a f-theta scanning system

14.3.3 Avenues Towards Cost Efficient THz Systems

One of the biggest disadvantages of THz time domain spectrometers has been the high price for femtosecond lasers. Ti:Sapphire based laser systems, mainly employed for scientific research applications, typically cost more than 100.000$ and are not likely to become less expensive, due to their high amount of technical expertise in order to align and finely tune these highly-sensitive crystal based laser cavities. Alternative possibilities are semiconductor-based lasers like diode lasers or vertical-external-cavity surface emitting laser (VECSEL) in which the active gain structure is based on quantum wells grown on a Bragg reflector. These laser diode setups consist of an external free space cavity, typically with an additional pulse compressing mechanism. A THz spectrometer based on an all-semiconductor femtosecond laser (600 fs optical pulses) has been achieved by Jördens et al. in 2008 [83]. The quantum wells of the VECSEL structure are pumped by a laser diode operating at a lower wavelength. In order to emit femtosecond pulses a semiconductor saturable absorber mirror (SESAM) has to be included in the cavity, ensuring a passively mode-locked laser [84–86]. Optimized quantum well structures as well as specially designed SESAMs using the fast optical Stark-effect have led to pulse durations of well under 100 fs in VECSELs [84].

Interestingly, VECSELs allow for modelocking also without a SESAM which makes this laser even less expensive [87, 88]. This scheme is called self-mode locking. Figure 14.18 shows a schematic of a VECSEL cavity operating in the self-mode locked scheme (see [87] for details).

Yet, the above mentioned semiconductor lasers comprise a free space optical resonator which is very sensitive to external influences like vibrations and temperature. Due to the custom alignment for each of these lasers, the price is lower

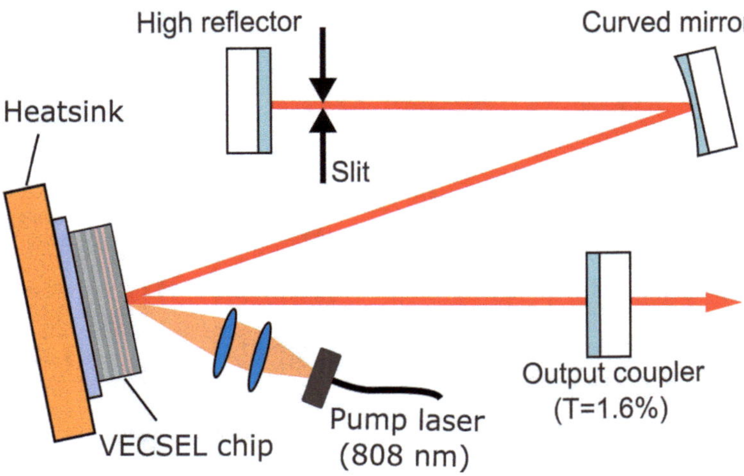

Fig. 14.18 Schematic of a VECSEL cavity operating in the self-mode locked scheme [87]

than a Ti:Sapphire laser but won't reduce much further below several tens of thousands of dollars.

For a cheap and reliable femtosecond source one would like to have products with high selling numbers. Changing from very sensitive free space alignments towards fiber laser, which are used in the telecommunication industry, offer a compact and less expensive alternative. These erbium doped fiber amplified lasers (EDFA) are capable of producing femtosecond pulses below 100 fs and have high average powers as well (up to several watts), if needed. Their price has reduced over the last decade significantly and the low-power models (approx. 100 mW) can be acquired for around 20.000$. The change in wavelength from 800 or 1060 nm up to 1550 nm makes it necessary to switch to a different semiconductor substrate (InGaAs) in order to accommodate the lower photon energy. First time domain spectrometers at 1550 nm were presented in 2007 [89]. Shortly after, Sartorius et al. at the Frauenhofer Institute in Berlin presented an all-fiber THz spectrometer at 1550 nm [90]. Since then also the antenna material and structure was optimized [91, 92].

Several groups and nowadays even commercial companies are offering of the shelve antenna chips and complete fiber-coupled antenna packages for a few thousand dollars. This technology advances enable very reliable all-fiber time-domain spectrometers, now ready to be employed in all kinds of industrial, real world applications. Possible applications in the near future are mobile handheld scanners of liquids at security checks, quality inspection in rough industrial environments and biological or medical sensitive applications.

Yet, these systems still have initial costs of a few ten thousand dollars; a rather low cost alternative approach was presented in 2009 [54]. The so called quasi time-domain spectrometer is based on mixing equidistant modes of a low cost multi-mode laser diode. These laser diodes can be purchased for less than 10$ with sufficient optical output power. Mixing between different modes leads to a comb like frequency spectrum, stretching over several hundreds of GHz (cf. Fig. 14.19).

First demonstration of this principle was presented in 1997 by Matsuura et al. [93]. The advantage of cheap continuous wave sources and broad frequency information is now combined in a coherent detection scheme. A signal trace

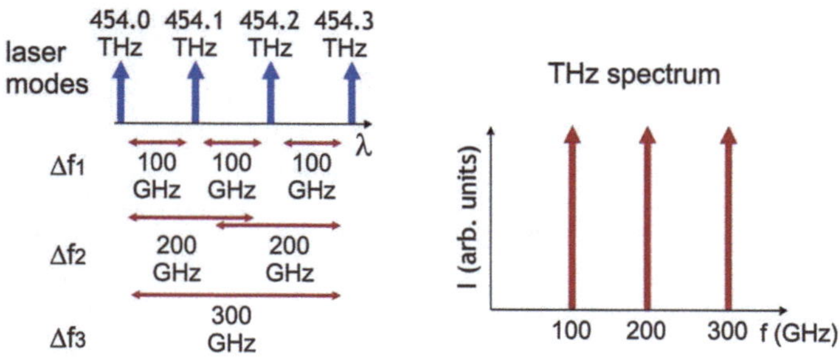

Fig. 14.19 Mixing principle of a quasi time-domain spectrometer [54]

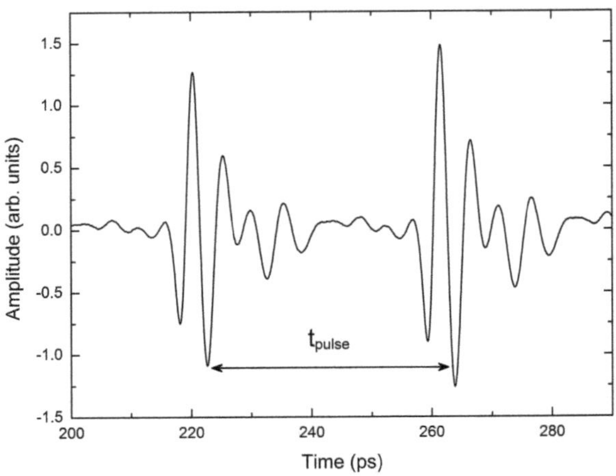

Fig. 14.20 Measured THz time trace of a QTDS system

measured with a coherent detection scheme employing two photoconductive antennas is shown in Fig. 14.20.

The resulting time domain signal comprises a periodic pulse like shape with the time spacing t_{pulse}, determined by the inverse cavity mode spacing. The measured signal can now be interpreted as a regular time-domain spectrometer trace and allows for determination of absorption and refractive index at the same time with a given thickness or specialized analyzing tools [54]. This novel technique enables cheap imaging and other sensing applications. Yet, the trade off is the power distribution in all present optical modes leading to a limited spectral brightness. QTDS also allows to be used for THz TDS imaging and inspection applications [94]. A photograph of a plastic airbag cover with a predetermined breaking point and its corresponding QTDS image are shown in Fig. 14.21.

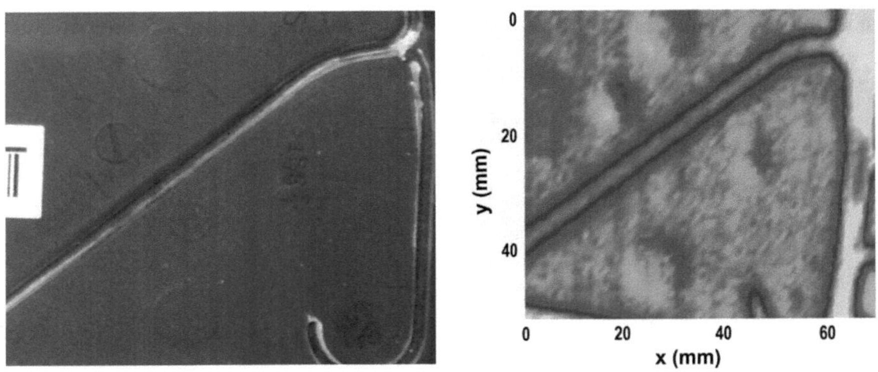

Fig. 14.21 Photograph (*left*) and QTDS THz image of the airbag cover (*right*) [93]

14.4 Conclusion

THz imaging systems based on time domain spectrometers have a great variety of applications in industrial applications. Commercial products covering those demands will more and more emerge in the following years. Their advancement into industrial environment will expedite within the next decade. In several branches THz scanning will enable new and better quality in production monitoring and/or more accurate and novel inline control systems. For THz living up to its full potential and for benefiting a plethora of industrial applications, the systems have to become faster, cheaper and suitable for rough industrial applications.

In this chapter the advances and technological improvements in THz systems have been discussed. First different detection schemes in industrial inspection applications have been presented. THz systems for real-time imaging with real-time data extraction have been developed in the last years and THz technology stepped out of the laboratory environment into real world systems. Yet, before establishing THz technology as a standard technique, operators will still have to validate the results of THz systems by conventional methods to gain confidence in this new non-destructive testing method. The required steps to a fully automated controlled THz inspection system are outlined in the last part of the chapter.

References

1. J.E. Bjarnason et al., Millimeter-wave, terahertz, and mid-infrared transmission through common clothing. Appl. Phys. Lett. **85**(4), 519–521 (2004)
2. K. Yamamoto et al., Noninvasive inspection of C-4 explosive in mails by terahertz time-domain spectroscopy. Jpn. J. Appl. Phys. **43**(3B), L414
3. D. Zimdars et al., Time domain terahertz detection of concealed threats in luggage and personnel. in *Terahertz for Military and Security Applications IV* (SPIE, Orlando (Kissimmee), FL, USA, 2006)
4. M. Lu et al., Detection and identification of illicit drugs using terahertz imaging. J. Appl. Phys. **100**(10), 103104–103105 (2006)
5. N. Krumbholz et al., Handheld terahertz spectrometer for the detection of liquid explosives, in *Millimetre Wave and Terahertz Sensors and Technology II* (SPIE, Berlin, Germany, 2009)
6. Y.C. Shen et al., Detection and identification of explosives using terahertz pulsed spectroscopic imaging. Appl. Phys. Lett. **86**(24), 241116–241123 (2005)
7. K. Kawase, Terahertz imaging for drug detection and large-scale integrated circuit inspection. Opt. Photon. News **15**(10), 34–39 (2004)
8. M. Reid, R. Fedosejevs, Terahertz birefringence and attenuation properties of wood and paper. Appl. Opt. **45**(12), 2766–2772 (2006)
9. P.F. Taday, Applications of terahertz spectroscopy to pharmaceutical sciences. Philos. Trans. Roy. Soc. Lond. Ser. A: Math. Phys. Eng. Sci. **362**(1815), 351–364 (2004)
10. S. Krishnamurthy et al.: Characterization of thin polymer films using terahertz time-domain interferometry. Appl. Phys. Lett. **79**(6), 875–877 (2001)
11. T. Hattori, H. Kumon, H. Tamazumi, Terahertz spectroscopic characterization of paper. in *35th International Conference on Infrared Millimeter and Terahertz Waves (IRMMW-THz)*, 2010

12. A. Sengupta et al., Characterisation of olefin copolymers using terahertz spectroscopy. Electron. Lett. **42**(25), 1477–1479 (2006)
13. N. Born, D. Behringer, S. Liepelt, S. Beyer, M. Schwerdtfeger, B. Ziegenhagen et al., Monitoring plant drought stress response using terahertz time-domain spectroscopy. Plant Physiol. **164**, 1571–1577 (2014)
14. E. Castro-Camus, M. Palomar, A.A. Covarrubias, Leaf water dynamics of *Arabidopsis thaliana* monitored in-vivo using terahertz time-domain spectroscopy. Sci. Rep. **3**, 2910 (2013)
15. C. Jördens et al., Evaluation of leaf water status by means of permittivity at terahertz frequencies. J. Biol. Phys. **35**(3), 255–264 (2009)
16. R. Piesiewicz et al., Performance analysis of future multigigabit wireless communication systems at THz frequencies with highly directive antennas in realistic indoor environments. IEEE J. Sel. Top. Quantum Electron. **14**(2), 421–430 (2008)
17. T. Kleine-Ostmann, T. Nagatsuma, A review on terahertz communications research. J. Infrared Millimeter Terahertz Waves **32**(2), 143–171 (2011)
18. C. Jastrow et al., 300 GHz transmission system. Electron. Lett. **44**(3), 213–214 (2008)
19. S. Priebe, T. Kürner, Towards THz communications – status in research, standardization and regulation. J. Infrared Milli. Terah. Waves **35**(1), 53–62 (2013)
20. D. Dufour, Marchese et al., Review of terahertz technology development at INO. J. Infrared Milli. Terah. Waves, 1–25 (2015)
21. F. Simoens, J. Meilhan J.A. Nicolas, Terahertz real-time imaging uncooled arrays based on antenna-coupled bolometers or FET developed at CEA-Leti. J. of Infrared Mill. Terah. Waves, 1–25 (2015)
22. N. Oda, S. Kurashina et al., Microbolometer terahertz focal plane array and camera with improved sensitivity in the sub-terahertz region. J. of Infrared Milli. Terah. Waves, 1–14 (2015)
23. J. Grzyb, U.Pfeiffer, THz direct detector and heterodyne receiver arrays in silicon nanoscale technologies. J. Infrared Milli. Terah. Waves, 1–35 (2015)
24. B. Scherger, C. Jördens, M. Koch, Variable-focus terahertz lens. Opt. Exp. **19**, 4528 (2011)
25. B. Scherger, M. Scheller, C. Jansen, M. Koch, K. Wiesauer, THz lenses made by compression molding of micro-powders. Appl. Opt. **50**, 2256 (2011)
26. A. Siemion, A. Siemion, M. Makowski, J. Suszek, J. Bomba, A. Czerwiński, F. Garet, J.L. Coutaz, M. Sypek, Diffractive paper lens for terahertz optics. Opt. Lett. **37**, 4320–4322 (2012)
27. B. Scherger, N. Born, C. Jansen, S. Schumann, M. Koch, K. Wiesauer, Compression molded terahertz transmission blaze-grating. IEEE Trans. THz Sci. Techn. **2**, 556 (2012)
28. C. Jansen, S, Wietzke, V. Astley, D. Mittleman, M. Koch, Mechanically flexible polymeric compound 1D photonic crystals for terahertz frequencies. Appl. Phys. Lett. **96**, 111108 (2010)
29. H. Bao, K. Nielsen, O. Bang, P.U. Jepsen, Dielectric tube waveguides with absorptive cladding for broadband, low-dispersion and low loss THz guiding. Sci. Rep. **5**, 7620 (2015)
30. M. Navarro-Cía, M. Vitiello, C. Bledt, J. Melzer, J. Harrington, O. Mitrofanov, Terahertz wave transmission in flexible polystyrene-lined hollow metallic waveguides for the 2.5–5 THz band. Opt. Exp. **21**, 23748–23755 (2013)
31. B. Scherger, M. Scheller, N. Vieweg, S.T. Cundiff, M. Koch, Paper terahertz wave plate. Opt. Exp. **19**, 24884 (2011)
32. S.F. Busch, M. Weidenbach, M. Fey, F. Schäfer, T. Probst, M. Koch, Optical properties of 3D printable plastics in the THz regime and their application for 3D printed THz optics. J. Infrared Milli. Terah. Waves **35**, 993 (2014)
33. A.D. Squires, E. Constable, R.A. Lewis, 3D printed terahertz diffraction gratings and lenses. J. Infrared Milli. Terah. Waves **36**(1), 72–80 (2015)
34. Y.R. Shen, *Principles of Nonlinear Optics* 1984. Medium: X Size, p. 575
35. E.R. Brown et al., Photomixing up to 3.8 THz in low-temperature-grown GaAs. Appl. Phys. Lett. **66**(3), 285–287 (1995)
36. S. Matsuura, M. Tani, K. Sakai, Generation of coherent terahertz radiation by photomixing in dipole photoconductive antennas. Appl. Phys. Lett. **70**(5), 559–561 (1997)

37. R. Wilk et al., Continuous wave terahertz spectrometer as a noncontact thickness measuring device. Appl. Opt. **47**(16), 3023–3026 (2008)
38. K.J. Siebert et al., Continuous-wave all-optoelectronic terahertz imaging. Appl. Phys. Lett. **80** (16), 3003–3005 (2002)
39. M. Scheller, K. Baaske, M. Koch, Multifrequency continuous wave terahertz spectroscopy for absolute thickness determination. Appl. Phys. Lett. **96**(15), 151112–151113
40. M. Scheller et al., Room temperature continuous wave milliwatt terahertz source. Opt. Express. **18**(26), 27112–27117
41. D. Stanze et al., Compact cw terahertz spectrometer pumped at 1.5 μm wavelength. J. Infrared Milli. Terah. Waves **32**(2), 225–232 (2011)
42. K.-I. Maki et al., Characteristics of the beam-steerable difference-frequency generation of terahertz radiation. J. Infrared Milli. Terah. Waves **32**(5), 603–617 (2011)
43. S. Preu, A unified derivation of the terahertz spectra generated by photoconductors and diodes. J. Infrared Milli. Terah. Waves **35**(12), 998–1010
44. D.H. Auston, K.P. Cheung, P.R. Smith, Picosecond photoconducting hertzian dipoles. Appl. Phys. Lett. **45**(3), 284–286 (1984)
45. N. Katzenellenbogen, D. Grischkowsky, Efficient generation of 380 fs pulses of THz radiation by ultrafast laser pulse excitation of a biased metal-semiconductor interface. Appl. Phys. Lett. **58**(3), 222–224 (1991)
46. K. Ezdi, B. Heinen, C. Jördens, N. Vieweg, N. Krumbholz, R. Wilk, M. Mikulics, M. Koch, A hybrid time-domain model for pulsed terahertz dipole antennas. J. Eur. Opt. Soc. Rapid Publi. **4**, 09001 (2009)
47. P.U. Jepsen, R.H. Jacobsen, S.R. Keiding, Generation and detection of terahertz pulses from biased semiconductor antennas. J. Opt. Soc. Am. B **13**(11), 2424–2436 (1996)
48. N. Vieweg et al., Terahertz-time domain spectrometer with 90 dB peak dynamic range. J. Infrared Milli. Terah. Waves **35**(10), 823–832 (2014)
49. P.H. Siegel, Terahertz technology. IEEE Trans. Microwave Theory Tech. **50**(3), 910–928 (2002)
50. M. Tonouchi, Cutting-edge terahertz technology. Nat. Photon. **1**(2), 97–105 (2007)
51. B. Ferguson, X.-C. Zhang, Materials for terahertz science and technology. Nat. Mater. **1**(1), 26–33 (2002)
52. P. Jepsen, D. Cooke, M. Koch, Terahertz spectroscopy and imaging—modern techniques and applications. Laser Photon. Rev. **5**, 124–166 (2010)
53. O. Morikawa, M. Tonouchi, M. Hangyo, Sub-THz spectroscopic system using a multimode laser diode and photoconductive antenna. Appl. Phys. Lett. **75**(24), 3772–3774 (1999)
54. M. Scheller, M. Koch, Terahertz quasi time domain spectroscopy. Opt. Exp. **17**(20), 17723–17733 (2009)
55. B.B. Hu, M.C. Nuss, Imaging with terahertz waves. Opt. Lett. **20**(16), 1716–1718 (1995)
56. W.L. Chan, J. Deibel, D.M. Mittleman, Imaging with terahertz radiation. Rep. Prog. Phys. **70**, 1325 (2007)
57. D. Mittleman, *Sensing with Terahertz Radiation*, vol. 85. (Springer, Berlin, Allemagne, 2003), XIII, 337p.
58. S. Hunsche et al., New dimensions in T-Ray imaging. IEICE Trans. Electron. **E81-C**(2), 269–276 (1998)
59. L. Thrane, R.H. Jacobsen, P. Uhd Jepsen, S.R. Keiding, THz reflection spectroscopy of liquid water. Chem. Phys. Lett. **240**(4), 330–333 (1995)
60. M. Nagai, H. Yada, T. Arikawa, K. Tanaka, Terahertz time-domain attenuated total reflection spectroscopy in water and biological solution. Int. J. Infrared Milli. Waves **27**(4), 505–515 (2006)
61. T. Arikawa, M. Nagai, K. Tanaka, Characterizing hydration state in solution using terahertz time-domain attenuated total reflection spectroscopy. Chem. Phys. Lett. **457**(1–3), 12–17 (2008)

62. A. Soltani, T. Probst, S. Busch, M. Schwerdtfeger, E. Castro-Camus, M. Koch, Error from delay jitter in terahertz attenuated total reflection spectroscopy. J. Infrared Milli. Terah. Waves **35**, 468 (2014)
63. N. Vieweg et al., Fiber-coupled THz spectroscopy for monitoring polymeric compounding processes. in *Optical Measurement Systems for Industrial Inspection V* (SPIE, Munich, Germany, 2007)
64. N. Krumbholz et al., Monitoring polymeric compounding processes inline with THz time-domain spectroscopy. Poly. Test, **28**(1), 30–35 (2009)
65. T. Yasui, E. Saneyoshi, T. Araki, Asynchronous optical sampling terahertz time-domain spectroscopy for ultrahigh spectral resolution and rapid data acquisition. Appl. Phys. Lett. **87**, 061101 (2005)
66. A. Bartels, F. Hudert, C. Janke, T. Dekorsy, K. Köhler, Femtosecond time-resolved optical pump-probe spectroscopy at kilohertz-scan-rates over nanosecond-time-delays without mechanical delay line. Appl. Phys. Lett. **88**(4), 041117 (2006)
67. O. Peters, M. Schwerdtfeger, S. Wietzke, S. Sostmann, R. Scheunemann, R. Wilk, R. Holzwarth, M. Koch, B.M. Fischer, Terahertz spectroscopy for rubber production testing. Poly. Test. **32**, 932 (2013)
68. C. Jördens, M. Koch, Detection of foreign bodies in chocolate with pulsed THz spectroscopy. Opt. Eng. **47**, 037003 (2008)
69. M. Scheller, C. Jansen, M. Koch, Analyzing sub-100-μm samples with transmission terahertz time domain spectroscopy. Opt. Commun. **282**, 1304–1306 (2007)
70. C. Jördens, M. Scheller, M. Wichmann, M. Mikulics, K. Wiesauer, M. Koch, Terahertz birefringence for orientation analysis. Appl. Opt **48**, 2037–2044 (2009)
71. C. Jördens et al., Terahertz spectroscopy to study the orientation of glass fibres in reinforced plastics. Compos. Sci. Tech. **70**(3), 472–477 (2010)
72. J.B. Jackson et al., Terahertz imaging for non-destructive evaluation of mural paintings. Opt. Commun. **281**(4), 527–532 (2008)
73. J.B. Jackson et al., A survey of terahertz applications in cultural heritage conservation science. IEEE Trans. Terah. Sci. Technol. **1**(1), 220–231 (2011)
74. S. Busch, T. Probst, M. Schwerdtfeger, R. Dietz, J. Palací, M. Koch, Terahertz transceiver concept. Opt. Exp. **22**(14), 16841–16846 (2014)
75. C. Jördens et al., Fibre-coupled terahertz transceiver head. Electron. Lett. **44**(25), 1473–1475 (2008)
76. N. Krumbholz et al., Handheld terahertz spectrometer for the detection of liquid explosives. In *SPIE Europe Security+Defence*. (International Society for Optics and Photonics 2009, September) (pp. 748504–748504)
77. T. Hochrein et al., Optical sampling by laser cavity tuning. Opt. Exp. **18**(2), 1613–1617
78. US Patent: US8390910 B2
79. T. Probst, A. Rehn, S.F. Busch, S. Chatterjee, M. Koch, M. Scheller, Cost-efficient delay generator for fast terahertz imaging. Opt. Lett. **39**(16), 4863-4866. (2014)
80. C. Jördens et al., Micro-mirrors for a multifocus terahertz imaging system. Proc. Eur. Microwave Assoc. **2**, 300–304 (2006)
81. B. Pradarutti et al., Terahertz line detection by a microlens array coupled photoconductive antenna array. Opt. Exp. **16**(22), 18443–18450 (2008)
82. S. Katletz et al., Efficient terahertz en-face imaging. Opt. Exp. **19**(23), 23042–23053 (2011)
83. C. Jördens et al., All-semiconductor laser driven terahertz time-domain spectrometer. Appl. Phys. B: Lasers Opt. **93**(2), 515–520 (2008)
84. A.H. Quarterman et al., A passively mode-locked external-cavity semiconductor laser emitting 60-fs pul. Nat. Photon. **3**(12), 729–731 (2009)
85. M. Scheller, T.L. Wang, B. Kunert, W. Stolz, S.W. Koch, J.V. Moloney. Passively modelocked VECSEL emitting 682 fs pulses with 5.1 W of average output power. Electron. lett. **48**(10), 588–589 (2012)

86. K.G. Wilcox, F. Rutz, R. Wilk, H.D. Foreman, J.S. Roberts, J. Sigmund, A.C. Tropper, Terahertz imaging system based on LT-GaAsSb antenna driven by all-semiconductor femtosecond source. Electron. Lett. **42**(20), 1159–1161 (2006)
87. M. Gaafar et al., Harmonic self-mode-locking of optically pumped semiconductor disc laser. Electron. Lett. **50**(7), 542–543 (2014)
88. L Kornaszewski, G. Maker, G.P.A. Malcolm, M. Butkus, E.U. Rafailov, C.J. Hamilton, SESAM-free mode-locked semiconductor disk laser. Laser Photon. Rev. **6**(6), L20–L23 (2012)
89. R. Wilk et al., *THz Time-Domain Spectrometer Based on LT-InGaAs Photoconductive Antennas Exited by a 1.55 μm Fibre Laser*. (Optical Society of America, 2007)
90. B. Sartorius et al., All-fiber terahertz time-domain spectrometer operating at 1.5 μm telecom wavelengths. Opt. Express **16**(13), 9565–9570 (2008)
91. R.J. Dietz, M. Gerhard, D. Stanze, M. Koch, B. Sartorius, M. Schell, THz generation at 1.55 μm excitation: six-fold increase in THz conversion efficiency by separated photoconductive and trapping regions. Opt. Exp. **19**(27), 25911–25917 (2011)
92. R.J. Dietz, B. Globisch, M. Gerhard, A. Velauthapillai, D. Stanze, H. Roehle, M. Schell et al., 64 μW pulsed terahertz emission from growth optimized InGaAs/InAlAs heterostructures with separated photoconductive and trapping regions. App. Phys. Lett. **103**(6), 061103 (2013)
93. M. Tani et al., Multiple-frequency generation of sub-terahertz radiation by multimode LD excitation of photoconductive antenna. IEEE Microwave Guided Wave Lett. **7**(9), 282–284 (1997)
94. M. Scheller, S.F. Dürrschmidt, M. Stecher, M. Koch, Terahertz quasi-time-domain spectroscopy imaging. App. Opt. **50**(13):1884–1888 (2011)

Chapter 15
Laser-Triggered Electron Source for X-Ray Applications

J. Wueppen, M. Strotkamp, D. Hoffmann, P. Russbueldt, T. Mans, S. Fritzler and S. Schippel

Abstract We report the results of the development of an electron source, based on the emission of electrons from plasma. The plasma electron source delivers emission currents of more than 1 A, based on a laser-induced discharge plasma and grid-controlled electron emission. Circuit times of less than 1 μs and temporal modulation strongly connected to laser burst duration were measured. X-ray imaging verified high brightness and low emittance of the electron beam.

15.1 Introduction

The next generation of computed tomography (CT) is the so-called non-mechanical CT (NM-CT). The advantages of NM-CTs are an increase in temporal and spatial resolution of the 3-D images and simultaneously a decrease in dose rate for patients. Here, detector, X-ray source, power supplies and data-logging electronics, which are now mounted on a rotating gantry, are replaced by a stationary setup [1]. For this conception, where about a thousand separate emitters are on one gantry and which can be individually activated, one of the key components are locally-excited X-ray sources. Each emits X-ray radiation for a duration of a few tens to a few hundreds of microseconds with a switching time less than one microsecond.

J. Wueppen (✉) · M. Strotkamp · D. Hoffmann · P. Russbueldt
Fraunhofer-Institut für Lasertechnik ILT, Steinbachstraße 15, 52074 Aachen, Germany
e-mail: jochen.wueppen@ilt.fraunhofer.de

S. Fritzler
Imaging & IT Division, Components and Vacuum Technology, Siemens AG Healthcare Sector, Guenther-Scharowsky-Str, 21, 91058 Erlangen, Germany

S. Schippel
Layertec GmbH, Ernst-Abbe-Weg, 1, 99441 Mellingen, Germany

T. Mans
Amphos GmbH, 52074 Aachen, Germany

S. Nolte et al. (eds.), *Ultrashort Pulse Laser Technology*,
Springer Series in Optical Sciences 195, DOI 10.1007/978-3-319-17659-8_15

X-ray sources comprise of an electron source, a high voltage to accelerate the electrons and an anode, where the electrons are decelerated and the X-rays are generated. Nowadays used thermionic sources have a high power consumption due to the fact that heating of the cathode is not only necessary during emission but at any time. Therefore an alternative has to be developed, overcoming this problem. Beside the above mentioned temporal modulation capability, the newly developed source has to deliver emission currents up to one Ampere and modulation of the current from ten to hundred percent during microseconds, combined with high brightness, long life-time and low power consumption.

Laser-driven cathodes are of high interest for this application due to the capability of fast switching (temporal and spatial) of lasers. Therefore, two different approaches for electron emission were investigated, the photoelectric effect and a plasma electron source. Additionally, some experiments were carried out to use laser-induced plasma as a high-voltage switch.

15.2 Laser-Induced Electron Emission via Photoelectric Effect

Figure 15.1 shows the setup to measure electron emission by means of the photoelectric effect consisting of a vacuum chamber, a cathode with sample holder, integrated resistor heating and a cylindrical anode made of stainless steel. The anode is operated at high voltage and the cathode signal is detected by means of a linear power resistor with an oscilloscope.

The photoelectric effect was investigated on copper with frequency-tripled femtosecond laser radiation (wavelength $\lambda = 266$ nm, pulse duration $\tau = 200$ fs). The temporal resolution of the electrode was determined at approx. 200 ps, which corresponds to the resolution of the detector (oscilloscopes Tektronix 300 MHz and 1 GHz; see Fig. 15.2).

The laser radiation was focused on the cathode using a lens (f = 150 mm) and the characteristic curves were recorded as a variation of pulse energy and voltage

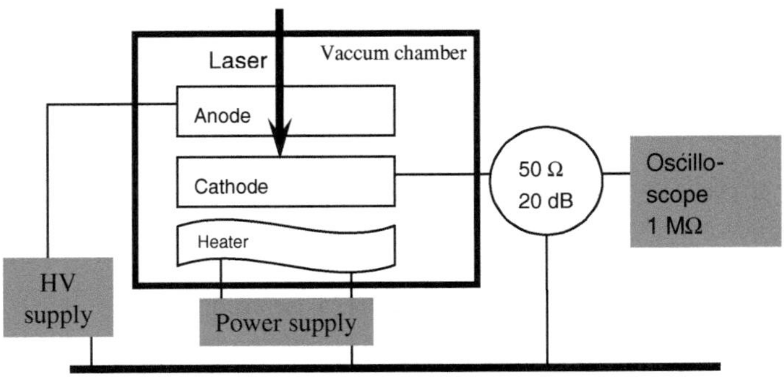

Fig. 15.1 Measuring setup for photoemission due to photoelectric effect

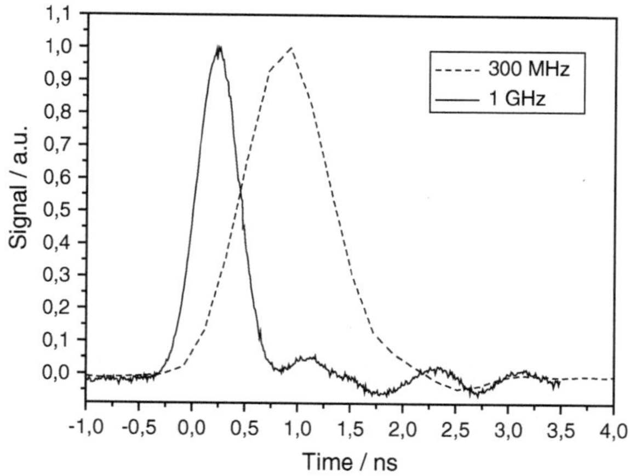

Fig. 15.2 Measured voltage as a function of time after excitation of 1 mm^2 copper with UV femtosecond laser radiation. Measured with 1 GHz bandwidth oscilloscope (*solid line*) and 300 MHz (*dotted line*)

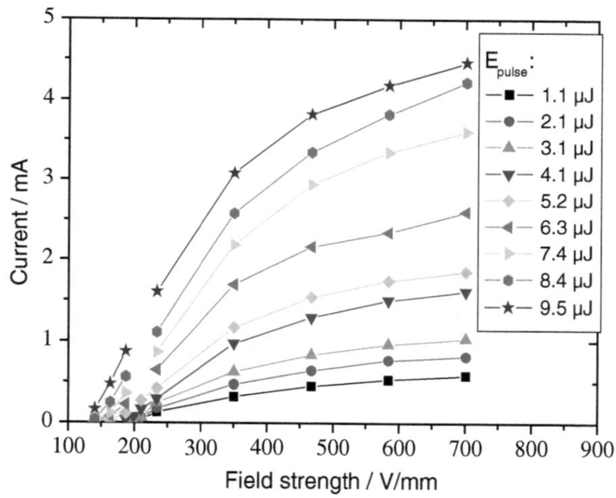

Fig. 15.3 Photoelectric current versus pulse energy and acceleration field after excitation of 1 mm^2 copper with UV femtosecond laser radiation

(Fig. 15.3). At the maximum available field strength of 700 V mm^{-1} an electron current of approx. 5 mA is achieved from 140 μm^2 of a copper cathode at 5.5 MW incident laser power.

Further materials were also examined. It was found, however, that owing to low quantum efficiency, lasers with a very high mean and pulse peak output would be required to emit a mean current of 1 A.

The photoelectric effect is a surface effect, which means that the performance of a photocathode essentially depends on the vacuum conditions. Since electrical outputs of more than 100 kW are envisaged with the target specifications of the X-ray tubes to be developed, these outputs lead to temperatures in excess of 2000 °C on the anode surface. The necessary vacuum condition of 10^{-9} bar is not fulfilled because of the resultant evaporation of the anode. In the test set-ups used, this was observed at electrical outputs of not more than 100 W, which led to a distinct reduction in quantum efficiency as well as to non-reproducible measurement results. Several methods described in the literature for achieving self-cleaning effects (like heating) were tried, but none could be successfully employed.

In view of these results as well as the fact that a 1 A DC current requires mean laser outputs that are technologically not available at present, the investigations into the photoelectric effect as an electron source for an optically induced X-ray tube were not pursued any further.

15.3 Electron Emission from Laser-Induced Plasma

15.3.1 Single-Shot Experiments

For the emission of free electrons from laser-induced plasma, initially single pulse tests were conducted and subsequently tests were performed in burst operation. Here it was found that the currents and current densities generated in this way are more likely to achieve the required specifications than with the other effects.

Figure 15.4 shows the set-up for the cathode and the calculated field distribution used in the single-pulse tests. The laser is focused vertically from the top onto the cathode. In its initial approach the plasma generated by the laser moves vertically to the cathode surface. A ceramic disk made of macor prevents the plasma from spreading directly to the anode. The coaxial arrangement is designed so as to allow rapid measurements and to apply an even electric field in the area of the cathode. The cathode current is measured by shunt resistors.

The emission of electrons takes place perpendicularly to the direction of plasma propagation. This corresponds to the Plasma Edge Cathode concept, as expounded by Zieher et al. [2–4]. The electrons are emitted from the open plasma surface. The emission cannot, however, be maintained for periods longer than a few microseconds, because space charge layers form within the plasma which shield the electrons from the field applied. This arrangement is therefore only suitable for single-pulse investigations.

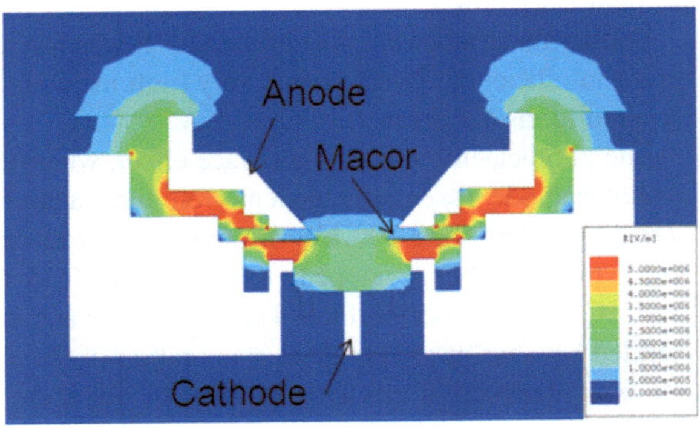

Fig. 15.4 Cross section of the cathode for the single shot experiments and its field distribution. The cathode is axially symmetric

The laser used for the single-pulse tests is a titanium:sapphire laser. The central wavelength is 810 nm. The pulse durations can be set between 100 and 900 fs, with pulse energies in the range of a few μJ. Repetition rates of up to 1 kHz are possible with this laser.

In the single-pulse tests, measurements were taken of the cathode current as a function of the cathode material, the anode voltage, the focus diameter, the pulse length and the laser energy. Exemplarily Fig. 15.5 shows the dependence of the current curve at different focal radii and different laser pulse durations for WTh as the cathode material.

Here the emission duration is in the range of a few 10 ns. The same length of time also applies to other materials. Cu, Al, W, WTh, WLa, WCe, FeCe, Sm, and Fe were tested. It was found that the elements with low ionization energy and small photoelectric work function show advantages concerning the maximum current and the required pulse energy at the ablation threshold.

The dependence of the electron emission and the ablation threshold on the pulse duration proved to be almost negligible for the range from about 100 to 900 fs. As

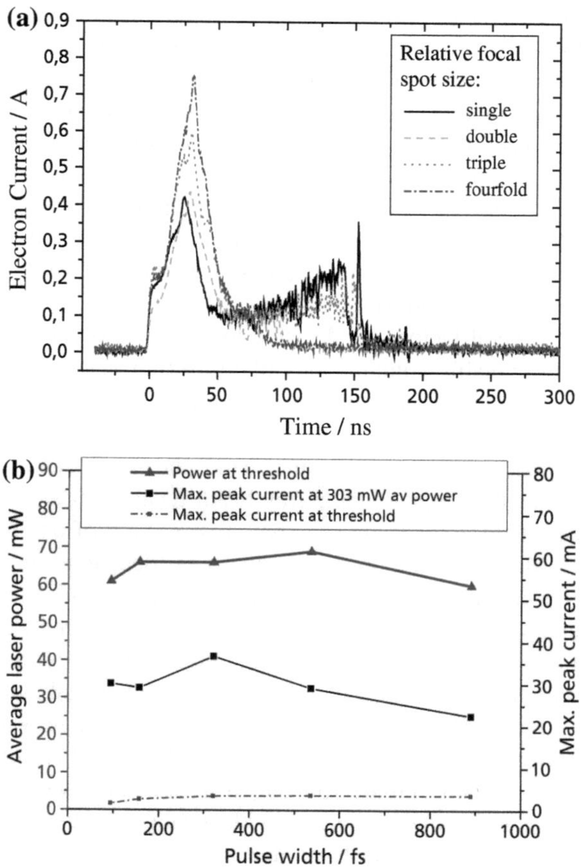

Fig. 15.5 Current curve for different focal radii (**a**). Variation of the laser pulse duration (**b**)

expected, an increase in the anode voltages and the pulse energies corresponds to an increase in the emission current.

15.3.2 Electron Emission in Burst Mode

Due to the electron emission duration, or the plasma life-time, respectively, measured in the single-shot experiments, for continuous emission of electrons an ultra-short pulse laser with a repetition rate of at least 20 MHz has to be applied (in our case 86 MHz). This ensures that there are no interruptions between individual laser pulses during the emission of every single emitter.

As emission durations in the range of 100 μs cannot be achieved with the Plasma Edge Cathode (PEC) [3], grid-controlled emitters were developed [5–7]. Although, the regimes were completely different from those required here: plasma produced by gas discharges, poor vacuum conditions (10^{-3} bar), size and cross section of the electrodes in the range of meters.

The use of an intermediate electrode (grid) prevents a positively charged marginal layer from forming which would shield the acceleration field from the electrons. As a result, DC currents can also be generated in contrast to PEC.

In the set-up the cathode, onto which the laser is focused, is located inside a closed tube. This enables the individual emitters to be separated from each other and also prevents the plasma from spreading in the direction of the anode. There is an opening at only one point. Here the grid is attached in such a way that a discharge takes place between the cathode and the grid when a plasma is present. The emission takes place from this discharge directly at the grid location. Part of the cathode current therefore flows over the grid while the other part is accelerated through the grid to the anode.

Several factors are decisive for the electron emission. To achieve an even current, suitable combinations of the set-up and adjustment parameters have to be found. On the one hand, the size of the electron exit hole or the open emission area significantly influences the current strength. Larger emission holes and areas lead to higher currents. On the other hand, the shape of the electron exit hole also determines the shape of the X-ray beam focal point and therefore cannot be randomly selected (target value is a spot cross section of one by 10 mm).

Moreover, the open emission area (depending on the wire diameter and mesh size of the grid) in part determines the uniformity of the current. Larger relative apertures lead to timely uneven and strong currents and vice versa). Furthermore, the uniformity and the current strength depend on the resistor connected in series with the grid. Here, the high resistor leads to small and even currents and vice versa.

Furthermore, the emission current strength is set via the grid current, and the grid current can be set via the potential of the grid relative to the cathode or the grid series resistor. Figure 15.6 shows as examples two current signals for two different series resistors with otherwise identical parameters. The combination of the settable parameters therefore has a decisive influence.

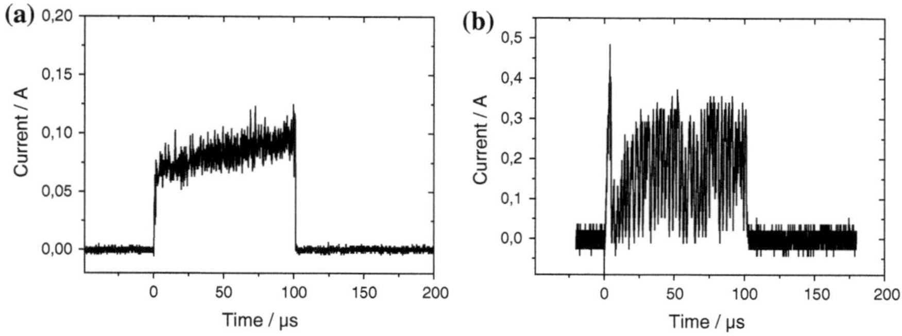

Fig. 15.6 Example of two current signals for high (**a**) and no (**b**) resistance of the grid. All other parameters are identical

The cathode surface is heavily processed by ablation with the laser radiation. As a result, different conditions apply before each pulse burst. Two approaches were therefore taken to solve this problem and to achieve a longer service life with corresponding reproducibility.

A metal (InGa) that is liquid at room temperature was used as cathode material. Here it was found that for long periods the current signal does not disappear, even during continuous irradiation on one place on the cathode. The depressions left by ablation of the material are always filled up again owing to the liquid nature of the cathode. Nevertheless, the ablation pressure causes movements on the surface which prevent an even current signal. Especially in the region of the first 100–300 μs after switching on the laser, there are interruptions in the generation of plasma and thus also in the current.

A second approach is an arrangement which makes it possible to scan the surface of the cathode with the laser. As a result, the surface material is removed layer by layer and drilling of deep holes is avoided.

For the purpose of optimization, the distance between the grid and cathode was varied, work was performed with differently sized pinholes on the inside of the tube, and different tube diameters were tested. The relative easily changeable parameters are the resistor, which is connected in series between the grid and the HV power supply unit of the grid, the mesh size and the wire diameter of the grid as well as the emission area and the voltages applied.

Figure 15.7 shows some of the measured current curves. Graph (a) presents some current curves for different areas of the emission opening with otherwise identical parameters. In graph (b) the anode voltage is varied. At low voltages the current curve is very stable and even; at higher currents the fluctuations increase slightly. The rising and falling slopes at higher current values are not of fundamental importance as they are caused by the transient reactions of the power supply units.

In Fig. 15.8 current/voltage characteristic curves are presented. In graph (a) one can see that with increasing anode voltages the current does not increase as it should in accordance with the Child-Langmuir Law, but reaches a saturation level. The emission is therefore source-limited. In graph (b) a linear dependence of the cathode

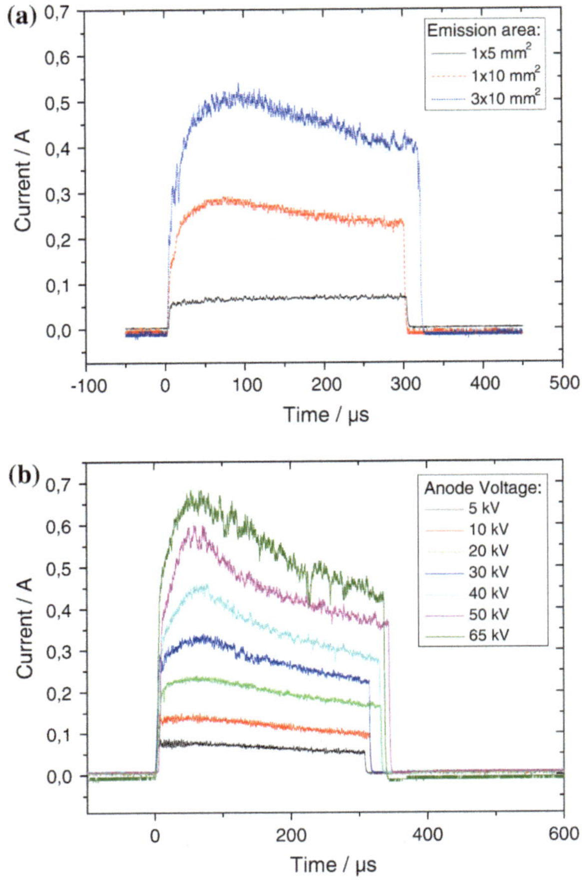

Fig. 15.7 Examples of the current curve for different areas of the emission opening (**a**) and different anode voltages (**b**). The laser burst duration was 300 μs

current on the grid voltage (synonymous with the grid current) can be recognized. The current/voltage characteristic curves were drawn for different combinations of resistors, grids and electron exit openings.

The parameter variations resulted in considerable differences in some cases. Nevertheless the combinations were found which lead to electron currents of more than 1 A (see Fig. 15.8). When larger emission openings areas were used, up to 1.75 A was attained.

15.3.3 X-Ray Radiation

To investigate the emittance, the X-ray radiation generated was studied. By imaging the X-ray beams using a simple pinhole aperture on X-ray films or XCCD cameras it is possible to determine the size of the electron beam on the anode.

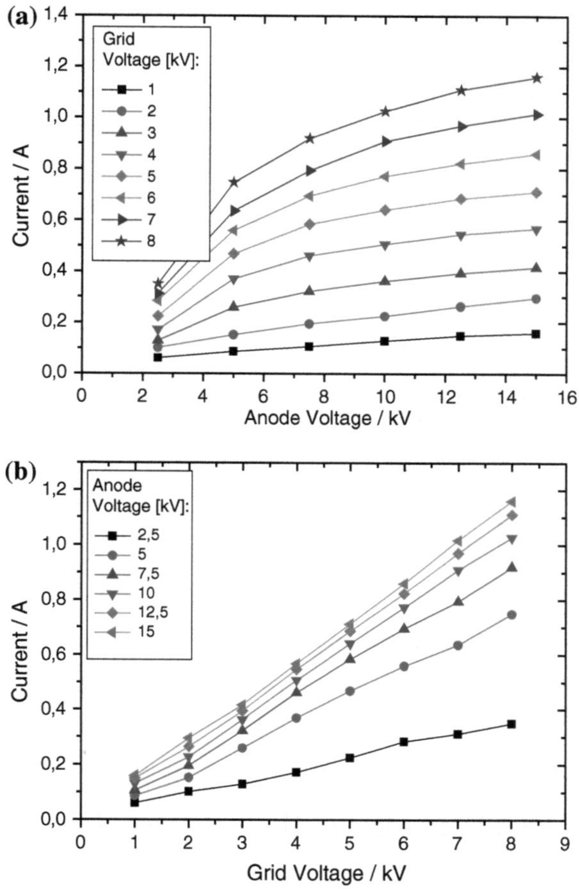

Fig. 15.8 Current/voltage characteristic curves. Emission current versus anode voltage (**a**) and emission current versus grid voltage (**b**)

In Fig. 15.9 the image of an X-ray focal point on the anode is shown. The distribution was captured using an XCCD camera, which detects photons from about 30 keV. In this example the mean current is 600 mA and the anode voltage is 60 kV. The dimensions of the electron beam on the anode amount at 20 % of the maximum to about 10 mm × 1.3 mm. As a result, the focal point is only slightly bigger than the dimensions of the exit opening of (1 × 10) mm^2.

The size of the X-ray focal point is crucial for diagnostic imaging as it determines the resolution in the investigations. To ensure this, the electric field between the electron source and the anode must be designed in such a way that in their trajectories the electrons are focused on a focal point that is as small as possible on the anode. By calculating the field distribution for different designs and thus of the electron trajectories it is ensured that the electron beam is not defocused.

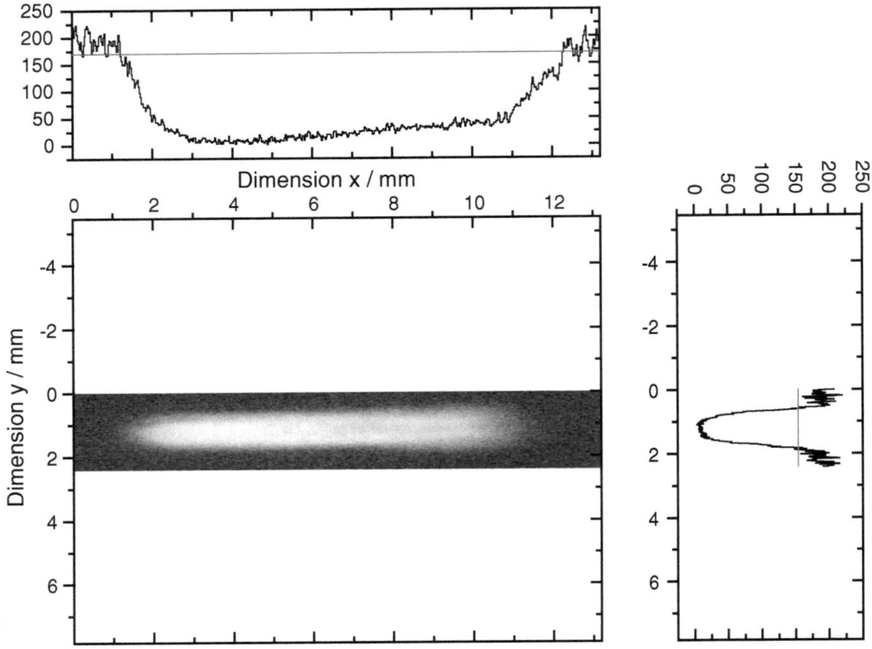

Fig. 15.9 X-ray focal point on the anode. The cross section area of about $1.2 \times 9.8\ mm^2$ is well suited for the application

15.4 Plasma-Based High-Voltage Switch

In contrast to the grid-controlled plasma cathode, where the grid has direct contact to the cathode via the plasma, control grids are used in thermionic or field emitters which primarily influence the field strength on the cathode through appropriate selection of the grid potential. A change in the grid potential may cause considerable difficulties if the changes occur at very short intervals or potentials of several kilovolts have to be switched. A plasma-based solution can be used here as a high-voltage switch.

Some tests were carried out to investigate a laser-controlled plasma switch with the aim of switching voltages of up to 100 kV with low jitter in less than 1 μs. Applying a plasma-based switch the current strength of a conventional cathode is controlled via the grid between the cathode and the anode. Additionally, the x-ray energy can be set directly by controlling the anode voltage.

Such a plasma switch used for control purposes is shown in Fig. 15.10. The laser produces a plasma which creates a conducting contact between two electrodes. A flashover occurs in which the potentials of the electrodes come close to each other via the conducting plasma. For switching off, a second plasma is produced by

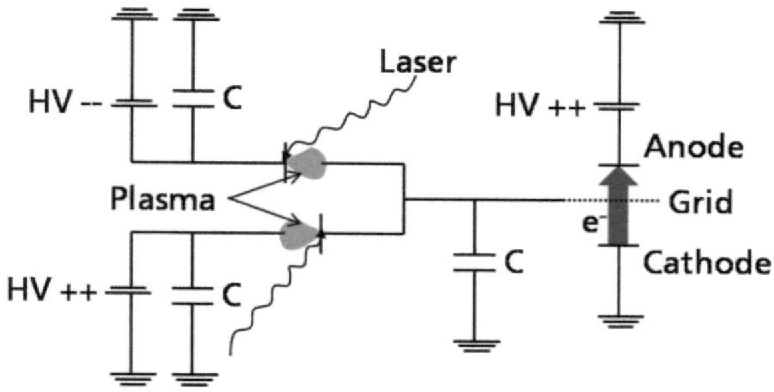

Fig. 15.10 Sketch of the plasma switch. The grid potential is controlled by two laser pulses

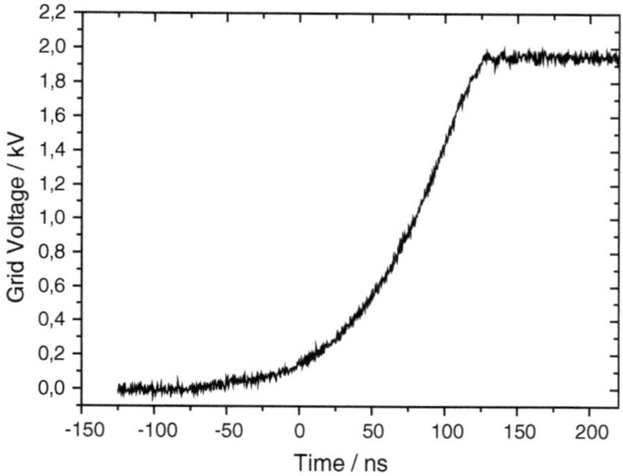

Fig. 15.11 Grid voltage versus time. Within 200 ns the grid has the same potential as the plasma switch electrode

means of a second laser pulse on a second electrode which is oppositely charged to the first electrode.

A fs-laser was used in these experiments with the following parameters: wavelength $\lambda = 800$ nm, pulse duration $\tau \approx 80$ fs, repetition rate $\nu = 1$ kHz, average power $P \approx 1.5$ W. It was found that laser radiation with a relatively low mean output of 30 mW was enough to close the gap between the electrodes. The charging of the grid is shown by way of example in Fig. 15.11. The grid charges up within 200 ns to a potential of 2 kV, which corresponded to the voltage applied on the opposite electrode.

Figure 15.12 shows a typical voltage curve for an interval between the two laser pulses of approx. 280 and 435 ns, respectively. It can be seen that the voltage remains constant over the same period. The asymmetrical curve during charging and discharging can be explained by the greater difference in potential during switching off.

In addition, tests with liquid cathode materials were conducted. The advantage, as already mentioned, is a constant surface from shot to shot which is not knocked out of focus by the ablation. For this purpose a gallium-indium alloy, liquid at room temperature, was used and reproducible results were obtained over several million individual pulses.

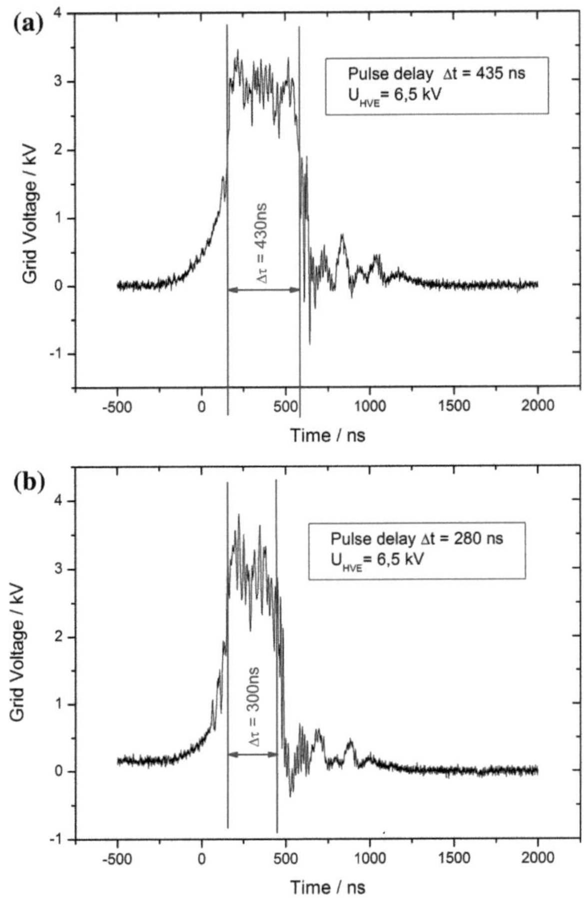

Fig. 15.12 Grid voltage versus time. On- and off-switching times of the grid potential

15.5 Conclusions

In conclusion, an X-ray source was developed based on the emission of electrons from laser-induced plasma. Reproducible pulses of electrons were generated with pulse length in the range of 300 μs. The duration of the electron pulses is strongly connected to the duration of the plasma-generating laser burst. With adapted sets of parameters currents of more than one ampere were accelerated to the anode. The beam cross section was measured by imaging the X-ray beams with a XCCD-camera.

Additionally, some investigations were carried out to use this setup with some changes as a high voltage switch. Here, the laser-induced plasma closes the gap between two electrodes and shifts the potential of one electrode to the level of the other one. At shifting times in the range of 200 ns several kilovolts were switched. In principle the voltage could be increased to higher values.

References

1. Patents: EP0466956A1 und EP0564292A2
2. K.W. Zieher, A plasma edge cathode scheme, in *1986 IEEE International Conference on Plasma Science*, IEEE, Saskatoon, Sask, Canada, New York, NY, USA, 19–21 May 1986. Cat. No. 86CH2317-6, p. 12
3. M.G. Grothaus, K.W. Zieher, Experimental investigation of a plasma edge cathode scheme for pulsed electron beam extraction. J. Appl. Phys. **70**, 7223 (1991)
4. R. Stemprok, K.W. Zieher, Plasma edge cathode for high electron current density, in *9th IEEE International Pulsed Power Conference, Albuquerque, New Mexico, 1993*, *9th IEEE International Pulsed Power Conference,* vol. 2. Digest of Technical Papers. IEEE, New York, NY, USA, 1993, pt. 2, pp. 1002–1005 (Cat.No.93CH3350-6)
5. E. Oks, *Plasma Cathode Electron Sources* (Wiley-VCH, Berlin, 2006). ISBN 9783527406340
6. Y.E. Kreindel, *Plasma Electron Source* (Atomizdat, Moscow, 1977)
7. E.M. Oks, P.M. Schanin, Development of plasma cathode electron guns. Phys. Plasmas **6**(5), 1649–1654 (1999)

Index

S. Nolte et al. (eds.), *Ultrashort Pulse Laser Technology*,
Springer Series in Optical Sciences 195, DOI 10.1007/978-3-319-17659-8

MIX
Papier aus verantwortungsvollen Quellen
Paper from responsible sources
FSC® C105338

If you have any concerns about our products,
you can contact us on
ProductSafety@springernature.com

In case Publisher is established outside the EU,
the EU authorized representative is:
Springer Nature Customer Service Center GmbH
Europaplatz 3, 69115 Heidelberg, Germany

Printed by Libri Plureos GmbH
in Hamburg, Germany